SPIN-STATISTICS CONNECTION AND COMMUTATION RELATIONS

Spin Statistics Related Titles from AIP Conference Proceedings

519 Statistical Physics: 3rd Tohwa University International Conference
Edited by Michio Tokuyama and H. Eugene Stanley, June 2000, 1-56396-940-8

490 Particles and Fields: Eighth Mexican School
Edited by Juan Carlos D'Olivo, Gabriel López Castro, and Myriam Mondragón, November 1999, 1-56396-895-9

488 High Energy Physics at the Millennium: MRST '99
Edited by Pat Kalyniak, Stephen Godfrey, and B. Kamal, October 1999, 1-56396-902-5

484 Trends in Theoretical Physics II
Edited by Horacio Falomir, Ricardo E. Gamboa Saraví, and Fidel A. Schaposnik, July 1999, 1-56396-894-0

464 Latin-American School of Physics XXXI ELAF: New Perspectives on Quantum Mechanics
Edited by Shahen Hacyan, Rocío Jáuregui, and Ramón López-Peña, March 1999, 1-56396-856-8

461 Mysteries, Puzzles, and Paradoxes in Quantum Mechanics
Edited by Rodolfo Bonifacio, March 1999, 1-56396-852-5

457 Trapped Charged Particles and Fundamental Physics
Edited by Daniel H. E. Dubin and Dieter Schneider, January 1999, 1-56396-776-6

453 Particles, Fields, and Gravitation
Edited by Jakub Rembielinski, December 1998, 1-56396-837-1

343 High Energy Spin Physics: Eleventh International Symposium
Edited by Kenneth J. Heller and Sandra L. Smith, 1995, 1-56396-374-4

To learn more about these titles, or the AIP Conference Proceedings Series, please visit the webpage http://www.aip.org/catalog/aboutconf.html

SPIN-STATISTICS CONNECTION AND COMMUTATION RELATIONS

Experimental Tests and Theoretical Implications

Anacapri, Capri Island, Italy   31 May–3 June 2000

EDITORS
Robert C. Hilborn
Amherst College

Guglielmo M. Tino
Università di Napoli "Federico II"

Melville, New York, 2000
AIP CONFERENCE PROCEEDINGS ■ VOLUME 545

Editors:

Robert C. Hilborn
Department of Physics
Amherst College
P. O. Box 5000
Amherst, Massachusetts 01002-5000
USA

E-mail: rchilborn@amherst.edu

Guglielmo M. Tino
Dipartimento di Scienze Fisiche
Università di Napoli
Complesso Universitario di Monte S. Angelo, via Cintia
I-80126 Napoli
ITALY

E-mail: Guglielmo.Tino@na.infn.it

The article on pp. 241–252 was authored by a U. S. Government employee and is not covered by the below mentioned copyright.

Authorization to photocopy items for internal or personal use, beyond the free copying permitted under the 1978 U.S. Copyright Law (see statement below), is granted by the American Institute of Physics for users registered with the Copyright Clearance Center (CCC) Transactional Reporting Service, provided that the base fee of $17.00 per copy is paid directly to CCC, 222 Rosewood Drive, Danvers, MA 01923. For those organizations that have been granted a photocopy license by CCC, a separate system of payment has been arranged. The fee code for users of the Transactional Reporting Service is: 1-56396-974-2/00/$17.00.

© 2000 American Institute of Physics

Individual readers of this volume and nonprofit libraries, acting for them, are permitted to make fair use of the material in it, such as copying an article for use in teaching or research. Permission is granted to quote from this volume in scientific work with the customary acknowledgment of the source. To reprint a figure, table, or other excerpt requires the consent of one of the original authors and notification to AIP. Republication or systematic or multiple reproduction of any material in this volume is permitted only under license from AIP. Address inquiries to Office of Rights and Permissions, Suite 1NO1, 2 Huntington Quadrangle, Melville, N.Y. 11747-4502; phone: 516-576-2268; fax: 516-576-2450; e-mail: rights@aip.org.

L.C. Catalog Card No. 00-109526
ISBN 1-56396-974-2
ISSN 0094-243X
Printed in the United States of America

CONTENTS

Preface ix
Committees x

I. GENERAL THEORY

Quantum Indistinguishability: Spin-Statistics without Relativity
or Field Theory?.. 3
      M. Berry *(invited)* and J. Robbins
The Role of Identity and Entanglement in Quantum Mechanics 16
      G. C. Ghirardi *(invited)*
Quantum Statistics and Dynamical Algebras: Fermions 29
      E. Celeghini and M. Rasetti *(invited)*
Rotational Invariance, the Spin-Statistics Connection,
and the TCP Theorem.. 40
      E. C. G. Sudarshan *(invited)*
Speculations on Spin and Statistics 55
      A. Zee *(invited)*
Bosons and Environment.. 59
      E. Celeghini and M. Rasetti
A Group-Theoretic Approach to Constructions of Nonrelativistic
Spin-Statistics 67
      J. M. Harrison and J. M. Robbins
Pauli Spin-Statistics Theorem and Statistics of Quasiparticles
in a Periodical Lattice 72
      I. G. Kaplan
Permutational Symmetry of Many-Particle Spin-Functions:
the Dirac Identities.. 79
      J. Katriel
Commutation Relations in Mesoscopic Electric Circuits 85
      Y.-Q. Li
Inner Composition Law of Pure-Spin States 92
      V. I. Man'ko, G. Marmo, E. C. G. Sudarshan, and F. Zaccaria
Symmetry Considerations in Quantum Computing 98
      A. Otte and G. Mahler
Symmetrizing the Symmetrization Postulate 104
      M. York

II. THEORIES OF VIOLATIONS OF THE SPIN-STATISTICS CONNECTION

Theories of Violation of Statistics 113
      O. W. Greenberg *(invited)*

Connecting q-Mutator Theory with Experimental Tests
of the Spin-Statistics Connection 128
   R. C. Hilborn (invited)
Quon Theories in Quantum Optics 142
   A. I. Solomon (invited)
Deformed Quantum Statistics of Quons 155
   S. Kirchner and A. Inomata
Generalized Fock Spaces and New Forms of Quantum Statistics............. 162
   A. K. Mishra and G. Rajasekaran
Quantum Field Theory for Orthofermions and Orthobosons 169
   A. K. Mishra and G. Rajasekaran

III. SUPERSYMMETRY

Spin-Statistics Connection and Supersymmetry............................ 179
   F. Iachello (invited)
Boson-Fermion Realization of Lie Algebras and Dynamical
Supersymmetry in Fermion Systems 190
   H. B. Geyer, P. Navrátil, and J. Dobaczewski
Deformed Heisenberg Algebra with Reflection, Anyons,
and Supersymmetry of Parabosons 197
   M. S. Plyushchay

IV. QUANTUM GRAVITY AND RELATED ISSUES

Spin and Statistics in Quantum Gravity 205
   H. F. Dowker (invited) and R. D. Sorkin
Quantum Field Theory on Noncommutative Space-Time
and Its Implication on Spin-Statistics Theorem.......................... 219
   M. Chaichian, A. Demichev, and P. Prešnajder

V. EXPERIMENTAL TESTS OF THE SPIN-STATISTICS CONNECTION AND SYMMETRIZATION POSTULATE

How We Know that Photons are Bosons: Experimental Tests
of Spin-Statistics for Photons 227
   D. DeMille (invited), D. Budker, N. Derr, and E. Deveney
High-Precision Experimental Tests of the Symmetrization
Postulate for Fermions.. 241
   J. D. Gillaspy (invited)
CP Violation and Symmetry of the Neutral Kaon Wave Function............ 253
   I. Mannelli (invited)
Testing the Symmetrization Postulate and the Spin-Statistics
Connection for Nuclei by Molecular Spectroscopy 260
   G. M. Tino (invited)

The Pauli Principle and Ultrahigh-Resolution Spectroscopy of Polyatomic Molecules 274
    C. J. Bordé and C. Chardonnet

Towards an Improved Test of Bose-Einstein Statistics for Photons 281
    D. Brown, D. Budker, and D. P. DeMille

Testing the Pauli Exclusion Principle with Accelerator Mass Spectrometry 288
    D. Javorsek, M. Bourgeois, D. Elmore, E. Fischbach, D. Hillegonds, J. Marder, T. Miller, H. Rohrs, M. Stohler, and S. Vogt

Spectroscopic Tests of the Symmetrization Postulate and of the Statistics for Nuclei in Molecules 295
    G. Modugno, D. Mazzotti, M. Modugno, N. Picqué, G. Giusfredi, P. C. Pastor, P. De Natale, and M. Inguscio

VI. PHILOSOPHICAL ISSUES

Putting a New Spin on Particle Identity 305
    S. French *(invited)*

List of Participants 319
Author Index 321

PREFACE

The connection between the intrinsic spin angular momentum of a particle and the collective behavior of a system of those particles (the so-called spin-statistics connection) lies at the heart of atomic, molecular, condensed matter, nuclear, and elementary particle physics. According to the traditional view, Nature has created two kinds of particles: half-integer spin particles, whose collective behavior is described by Fermi-Dirac statistics and the Pauli Exclusion Principle, and integer-spin particles, whose collective behavior is described by Bose-Einstein statistics. As Richard Feynman noted, this dichotomy, with its profound implications in almost every branch of modern physics, is easy to state but difficult to prove. The infamous Spin-Statistics Theorem, with various versions due to Pauli, Burgoyne, Lüders and others, invokes the machinery of relativistic quantum field theory to establish a result that is important in many situations in which relativity does not seem to play any other role.

Within the past ten years there have been two important developments that challenge the traditional view: (1) the formulation of the q-mutator (quon) theory gives a consistent description of a field theory that violates the spin-statistics theorem, and (2) high-precision experiments have provided significant tests of the spin-statistics connection for electrons, photons, nucleons, and ^{16}O nuclei. Given our own strong interests in the spin-statistics connection and related issues, we believed that it would be timely to host a small conference bringing together theorists, experimentalists, and philosophers whose work bears directly on these issues. This book is a collection of the papers that formed the basis of the conference presentations. We were quite impressed by the active discussions among all of the conference participants and the breadth of physics topics covered by the papers.

Although the issue of finding either a simple proof of the spin-statistics connection (see the papers by Berry, Robbins, Harrison, and Sudarshan, for example), or an experimental demonstration of its violation remains unresolved, the conference proceedings provide a window into this fascinating structure at the foundations of physics.

We wish to thank the International Advisory Committee, the Local Organizing Committee, and the many organizations who supplied financial support for the conference. The Centro Internazionale per la Cultura Scientifica dell'Università di Napoli provided outstanding conference facilities at Anacapri.

Robert C. Hilborn, *Amherst College, USA*
Guglielmo M. Tino, *Università di Napoli, Italy*
Conference Chairs

INTERNATIONAL ADVISORY COMMITTEE

A. P. Balachandran, Syracuse, USA
C. J. Bordé, Paris, France
D. DeMille, New Haven, USA
J. Doyle, Cambridge, USA
S. French, Leeds, UK
J. Gillaspy, Gaithersburg, USA
O. W. Greenberg, College Park, USA
M. Inguscio, Firenze, Italy
G. Marmo, Napoli, Italy
M. Napolitano, Napoli, Italy
M. Rasetti, Torino, Italy
S. Solimeno, Napoli, Italy
R. Sorkin, Syracuse, USA
E. C. G. Sudarshan, Austin, USA
S. Weinberg, Austin, USA

LOCAL ORGANIZING COMMITTEE

G. Gagliardi, Napoli
L. Gianfrani, Napoli
F. Lizzi, Napoli
G. Mangano, Napoli
G. Marmo, Napoli
G. Modugno, Firenze
B. Preziosi, Napoli

IN COLLABORATION WITH

Università di Napoli "Federico II," Dipartimento di Scienze Fisiche
Istituto Italiano per gli Studi Filosofici
Istituto Nazionale di Fisica Nucleare (INFN)
Istituto Nazionale per la Fisica della Materia (INFM)
CNR-Gruppo Nazionale di Struttura della Materia

I. General Theory

Quantum Indistinguishability: Spin-statistics without Relativity or Field Theory?

Michael Berry* and Jonathan Robbins[†]

*H H Wills Physics Laboratory, Tyndall Avenue,
Bristol BS8 1TL, United Kingdom
[†] Basic Research Institute in the Mathematical Sciences,
Hewlett-Packard Laboratories Bristol,
Filton Road, Stoke Gifford, Bristol BS12 6QZ, United Kingdom, and
School of Mathematics, University Walk, Bristol BS8 1TW, United Kingdom

Abstract. We review the formulation of quantum mechanics for identical spinning particles with wavefunctions that are singlevalued when permuted configurations are identified. The identification requires the spins to be smoothly permuted along with position variables, so spin is represented in a position-dependent 'transported basis', rather than the usual fixed basis. The simplest transported basis, constructed in terms of spins represented as pairs of commuting harmonic oscillators, gives the correct connection between spin and statistics. More complicated constructions can give the wrong exchange sign. The theory is generalized to incorporate additional properties such as isospin, colour and strangeness. Some remarks about the relation between this approach and those based on relativitity and/or field theory are given.

1. INTRODUCTION

In the nonrelativistic quantum mechanics of a fixed number of identical particles, the relation between spin and statistics (SS) sits awkwardly on top of the theory, as a separate postulate: it is simply asserted that the wavefunctions of particles with integer-plus-half spin change sign when the variables describing their spin and position are exchanged, whereas wavefunctions for particles whose spin is an integer do not change sign. Here we will investigate the possibility that SS is already contained in the theory, as a hidden consequence of imposing geometrical requirements that follow naturally from indistinguishability. We will give brief nontechnical summaries of, and comment on, the ideas in two recent papers [1] [2].

If SS is an awkward addition to nonrelativistic quantum theory, there is awkwardness too in the attempt to be described here. The explanation of any phenomenon on the basis of a well-established physical theory must begin with assumptions about how the theory is to be applied in the particular case under consideration, followed by deductions that lead unambiguously and precisely to the phenomenon. An explanation is more convincing if it is fruitful, in the sense of successfully predicting phenomena that have not yet been observed.

Here the 'phenomenon' is already known: it is SS, the principle determining how wavefunctions behave under exchange. Of course, this principle has many consequences (e.g. the Pauli exclusion principle) that are fundamental to our description of the world, but it is hard to see how an explanation of SS itself can have any new experimental consequences (though we live in hope that such pessimism will prove wrong). Thus, the degree to which any purported explanation of SS is accepted - once it has been agreed that the deductive part is technically correct - must hinge on the naturalness of the assumptions, and so involves elements of subjectivity and aesthetics. In [1, 2], 'naturalness' was interpreted as: constructing the quantum theory of identical particles using the same principles that are accepted and routinely applied in the quantum physics of nonidentical particles (or individual identical particles).

A further awkwardness is that SS involves not some experimental number whose prediction with increasing accuracy might increase confidence in the theory (like the fine-structure constant), but is simply a sign.

Nevertheless, SS cries out for understanding. There are many derivations based on relativity, beginning with [3], that have been comprehensively reviewed [4], and we will comment a little on these in section 6. And there have been many attempts at derivations that do not involve relativity, based on a variety of different assumptions [5]. We think that previous derivations have lacked a crucial geometrical ingredient, described in section 2, that follows from indistinguishability.

For simplicity, we will concentrate on the understanding of SS for two identical particles with spin S. However, the extension to N particles involved interesting technical challenges that have resulted in beautiful mathematical constructions by Atiyah [6], and we will mention one of these briefly in section 4.

2. SINGLEVALUEDNESS UNDER EXCHANGE

In the position representation, the wavefunction $|\Psi\rangle$ describing the state of two identical particles with spin S depends on the position vectors $\mathbf{r}_1$ and $\mathbf{r}_2$. Only the relative position $\mathbf{r}= \mathbf{r}_2- \mathbf{r}_1$ is relevant here, so we write $|\Psi(\mathbf{r})\rangle$. Exchange of positions corresponds to $\mathbf{r}\to-\mathbf{r}$.

The central assumption in [2] was that the wavefunction for identical particles must be singlevalued under exchange. Thus

$$|\Psi(\mathbf{r})\rangle = |\Psi(-\mathbf{r})\rangle. \qquad (1)$$

At first this seems absurd, but our $|\Psi\rangle$ differs from the more familiar wavefunction in a crucial respect, namely that it has, built into it, the property that exchange of positions $\mathbf{r}\to-\mathbf{r}$ is automatically accompanied by exchange of spin states, so that $\mathbf{r}\to-\mathbf{r}$ corresponds to complete exchange. Only then can indistinguishability be demanded, and the condition (1) imposed. To invoke singlevaluedness for identical particles is not a new idea [7-9]; but the characteristic feature of our approach is the systematic incorporation of spin exchange along with exchange of positions.

To incoporate spin exchange, it is necessary to represent the spin part of the state in a way that is unusual but unavoidable. Following [1], we write the complete state as

$$|\Psi(\mathbf{r})\rangle = \sum_M \psi_M(\mathbf{r})|M(\mathbf{r})\rangle. \qquad (2)$$

Here $M \equiv \{m_1, m_2\}$ labels the spin state of the particles, with m denoting the z component of spin; exchange of spins corresponds to $M \to \overline{M} \equiv \{m_2, m_1\}$. $|M(\mathbf{r})\rangle$ is the *transported spin basis*, that is, a basis for representing spins in a way that depends on the relative position of the particles, with the exchange requirement

$$|M(-\mathbf{r})\rangle = (-1)^K |\overline{M}(\mathbf{r})\rangle, \qquad (3)$$

where K is an integer (see [1, 2]), implying that exchange generates a sign rather than some more general phase factor. The coefficients $\psi_M(\mathbf{r})$ describe the spatial dependence of the state.

The representation (2) is to be contrasted with the familiar expansion in terms of a fixed spin basis $|M\rangle$, namely

$$|\Psi(\mathbf{r})\rangle_{\text{fixed}} = \sum_M \psi_M(\mathbf{r})|M\rangle. \qquad (4)$$

As was shown in [1], the coefficients $\psi_M(\mathbf{r})$ – which are, after all, the physically measurable quantities (up to a single overall phase) - are the same as those in (2) (see also section 3 later). With $|\Psi(\mathbf{r})\rangle_{\text{fixed}}$, there is no spin exchange accompanying position exchange, so indistinguishability is not incorporated and there is no justification for imposing the singlevaluedness requirement (1). With (2), however, the application of (1) implies that any sign change (3) of the transported basis is compensated by a sign change in the coefficients:

$$\psi_{\overline{M}}(-\mathbf{r}) = (-1)^K \psi_M(\mathbf{r}). \qquad (5)$$

This is the usual form in which SS is assumed. Of course, to reproduce *the* SS, rather than a generic form of SS, it is necessary to show that

$$K = 2S. \qquad (6)$$

This requires consideration of the transported basis $|M(\mathbf{r})\rangle$, as will be described in section 3.

It is important to emphasize that with the representation (2), SS, in the form (5), emerges as a quantization condition implied by singlevaluedness. This brings SS into the same framework as other derivations within elementary quantum mechanics. For example the quantization of a component $m\hbar$ of orbital angular momentum using wavefunctions requires singlevaluedness of $\exp(im\phi)$ under $\phi \to \phi + 2\pi$, reflecting the fact that these two angles represent the same point. And in the Aharonov-Bohm effect [10, 11] a similar application of singlevaluedness is required to get a definite (and experimentally confirmed) prediction for quantum scattering by inaccessible magnetic flux. In fact, in every situation that we know in elementary quantum mechanics, wavefunctions representing the same configuration are the same (up to choice of gauge).

In effect, the incorporation of the transported basis as in (2) enables $\mathbf{r}$ and $-\mathbf{r}$ to be regarded as the same point in the configuration space of the two particles. This identification changes the topology of the configuration space, making it nonorientable and non-simply-connected. The space is the direct product of the centre of mass, the separation distance $r=|\mathbf{r}|$, and the projective plane (2-sphere with antipodal points identified) that represents directions $\mathbf{r}/r$. An intrinsic procedure would be to construct quantum mechanics by erecting two-spin bundles on this base space, and this has been systematically carried out [12]. However, we will here continue to use the more elementary approach of regarding **r** as a euclidean vector and then imposing singlevaluedness under **r**→-**r**. (This is analogous to the common and convenient procedure of regarding azimuth angles ϕ as variables with values on the real line $-\infty < \phi < \infty$, and then insisting that functions are periodic, rather than considering functions whose domain is the circle.)

3. TRANSPORTED BASIS

The basis $|M(\mathbf{r})\rangle$ is a set of $(2S+1)^2$ spinors, because the z quantum numbers m_1 and m_2 of the two spins can range from $-S$ to $+S$. Each spinor is required to be a singlevalued and smooth function of **r**. In addition, we impose the parallel-transport requirement

$$\mathbf{A}_{M,M'} \equiv i\langle M'(\mathbf{r})|\nabla M(\mathbf{r})\rangle = 0, \tag{7}$$

to guarantee the vanishing of the curvature of the connection between neighbouring positions **r** ('flat exchange'). (For further discussion, see [12].)

Parallel transport implies that $|M(\mathbf{r})\rangle$ inhabits an ambient space, within which it is smoothly transported, that is larger than the $(2S+1)^2$-dimensional space of the fixed basis $|M\rangle$. Without enlargement, (7) would imply that $|M(\mathbf{r})\rangle$ is independent of **r** and so unable to satisfy the fundamental exchange requirement (3). We find it necessary to emphasize that this enlargement is in no way undesirable or unphysical. Nor is it unfamiliar: in [1] we give the analogy of light in a coiled optical fibre, where transversality implies that in a frame whose z axis is along the the local propagation direction the electric field vector can be described with only two components (x and y) whereas a fixed basis requires all three components. An even simpler analogy is that in a space **r**={x, y, z}, each vector in a field **v**(**r**) possesses only one component when described in a local frame {x_1, x_2, x_3} whose x_3 axis is directed along **v**, but requires three components in the ambient space {x, y, z}.

The transformation between the transported and fixed bases is described by a unitary operator **U**(**r**), such that

$$|\Psi(\mathbf{r})\rangle = \mathbf{U}(\mathbf{r})|\Psi(\mathbf{r})\rangle_{\text{fixed}}, \quad |M(\mathbf{r})\rangle = \mathbf{U}(\mathbf{r})|M\rangle. \tag{8}$$

An immediate consequence is the equality

$$\psi_M(\mathbf{r}) = \langle M(\mathbf{r})|\Psi(\mathbf{r})\rangle = \langle M|\Psi(\mathbf{r})\rangle_{\text{fixed}} \qquad (9)$$

asserted after (4). As described in [1, 2], $\mathbf{U}(\mathbf{r})$ also generates dynamical variables (e.g. momentum and spin) in the transported basis from their more familiar fixed counterparts, thereby guaranteeing that all local physics (for example Schrödinger equations derived from hamiltonians) is the same as in the fixed basis.

The main technical content of [1] was a construction of $\mathbf{U}(\mathbf{r})$. In this, each of the two spins was represented by the formalism of Schwinger [13], by two harmonic oscillators: a_1 and b_1 for one particle, and a_2 and b_2 for the other. Each of these four oscillators had its own creation and annihilation operator, with operators corresponding to different oscillators commuting. The dimension of the ambient space, on which $\mathbf{U}$ acts and within which the $(2S+1)^2$ states $|M(\mathbf{r})\rangle$ are smoothly transported, is the number of ways that $4S$ quanta can be distributed among four oscillators, namely $(4S+1)(4S+2)(4S+3)/6$; for $S=1/2$, there are 10 such states, in contrast to the four fixed-basis and transported-basis states.

Exchange was incorporated using the following insight. For a single spin, interchanging the number of quanta in its a and b oscillators corresponds to replacement of m by $-m$, equivalent to a rotation of the axis of quantization from z to $-z$. Therefore, interchanging the quanta in the 1 and 2 oscillators corresponds to exchanging the spin states of the two particles and can be used to define an 'exchange angular momentum', analogous to spin, which generates 'exchange rotations' $\mathbf{U}(\mathbf{r})$ from z to $\mathbf{r}$, whose effect is precisely to generate a transported basis with the desired exchange property (3). With this construction, in which each spin is decomposed into its 'atomic spin bosons', it was possible to make an explicit calculation of the exchange sign in (3), and the result was the correct SS sign (6). The calculation was extended in I from two to N identical particles.

In [1] we suggested that any construction of the transported basis that was smooth, singlevalued and parallel-transported would lead to the correct exchange sign. This was wrong. In [2] we exhibited two 'perverse' constructions that satisfy these requirements but which, unlike those based on the Schwinger formalism, lead to the wrong sign. The first perverse construction applies to spin $S=0$, and reflects a question frequently asked by people sceptical of the arguments in [1]: can the exchange of two spinless particles be accompanied by a fermionic sign? In this perverse construction, they can. The single transported state is represented as the unit vector

$$|M(\mathbf{r})\rangle = |\{0,0\}(\mathbf{r})\rangle = \mathbf{r}/|\mathbf{r}|. \qquad (10)$$

This is singlevalued, smooth, and parallel-transported, and involves the extended spin space spanned by the three basis states $\mathbf{e}_x$, $\mathbf{e}_y$ and $\mathbf{e}_z$, of which only one (e.g. $\mathbf{e}_z$) corresponds to the fixed-spin state $|\{0,0\}\rangle$. The operator $\mathbf{U}(\mathbf{r})$ is then rotation from $\mathbf{e}_z$ to $\mathbf{r}$. Under $\mathbf{r} \to -\mathbf{r}$, $|M(\mathbf{r})\rangle$ changes sign fermionically, rather than being bosonically invariant.

The second perverse construction in [2] applies to spin $S=1/2$, and consists in replacing all commutators in the Schwinger formalism by anticommutators. For this 'anti-Schwinger' construction, the exchange sign is +1, rather than the fermionic -1.

We have not found a general principle to exclude these perverse constructions, and others that generate the wrong exchange sign. However, all the perverse constructions we have found are defective in one or more ways, described in [2]. For example, the spin-zero fermion construction fails the test of simplicity, because (10) is decomposable into a constant (unity) that satisfies the requirements - and is what Schwinger gives - and a superfluous factor with no intrinsic connection to spin. And the anti-Schwinger construction is special in that it applies only to $S=1/2$ and so fails to describe the statistics of composite objects that can have any spin (there are generalizations of anti-Schwinger for higher spins, but they are cumbersome).

By contrast, the Schwinger construction applies for all spins (and also for all N - see section 4), generates the transported basis without superfluous factors, and also is intrinsically related to spin. In the reformulation of our approach by constructing N-spin bundles on the identified configuration space [12], the Schwinger construction emerges as the simplest implementation of the geometrical requirements.

4. ATIYAH'S CONSTRUCTION for N PARTICLES

In [1] we extended the Schwinger construction to the general case of N particles (with permutations instead of exchanges). The N spins are built from $2N$ oscillators, from whose creation and annihilation operators it is possible to construct 'permutation angular momenta', generating 'permutation rotations' and thence the transported basis states. We showed that any such construction must yield the correct SS sign. But we were unable to exhibit an explicit construction, analogous to the exchange rotation from z to $\mathbf{r}$ for two particles. We reduced the problem to that of finding a unitary $N \times N$ matrix $U_{ij}(\mathbf{R})$, smoothly dependent on the positions $\mathbf{R} = \{\mathbf{r}_1, ... \mathbf{r}_N\}$, with the property that any permutation of the r_i results in the corresponding permutation of its columns, up to an overall phase.

Recently, Atiyah [6] has produced several such constructions. Here we will describe the simplest. The matrix $U_{ij}(\mathbf{R})$ is obtained from the polar decomposition of an $N \times N$ matrix $V_{ij}(\mathbf{R})$, each of whose columns $\mathbf{v}_j(\mathbf{R})$ is associated with the jth particle, so that the required permutation property is assured. It suffices to explain $\mathbf{v}_1(\mathbf{R})$.

Let the directions $(\mathbf{r}_2-\mathbf{r}_1)/|\mathbf{r}_2-\mathbf{r}_1|,...(\mathbf{r}_N-\mathbf{r}_1)/|\mathbf{r}_N-\mathbf{r}_1|$ of the other particles, as seen from 1, be described by their complex stereographic coordinates $\zeta_2,...\zeta_N$ (that is, the real and imaginary parts of ζ_j are the cartesian coordinates of the intersection with the equatorial plane of the line joining the south pole of the unit sphere centred on $\mathbf{r}_1$ to the point where the vector connecting $\mathbf{r}_1$ to $\mathbf{r}_j$ intersects the sphere). Then the components $v_{m,1}$ of $\mathbf{v}_1$ are the coefficients in the expansion

$$P_1(z) = \prod_{n=2}^{N}(z-\zeta_n) = \sum_{m=1}^{N}\frac{z^{m-1}}{\sqrt{(m-1)!(N-m)!}}v_{m,1}. \qquad (11)$$

The orthogonalization leading to $U_{ij}(\mathbf{R})$ requires the columns $\mathbf{v}_j(\mathbf{R})$ to be independent, that is $\det V_{ij}(\mathbf{R}) \neq 0$ for any configuration $\mathbf{R}$ where no two particles coincide. At present this is a plausible conjecture, proved for $N=3$ and some special cases (e.g. N particles in a line), but not generally; this problem remains open. Independence of the columns has however been shown for a more elaborate version of this construction [6].

5. EXTENDED SPIN-STATISTICS RELATIONS FOR PARTICLES WITH ADDITIONAL PROPERTIES

Returning now to two particles, we incorporate into the theory the fact that particles can be characterised not only by position and spin but by one or more further quantum properties, that we denote by P. Examples of P are isospin, strangeness and colour. We denote the values of P by p (assumed discrete), and the pair of values for two particles - and the associated exchanged pair - by

$$P \equiv \{p_1, p_2\}, \quad \overline{P} \equiv \{p_2, p_1\}. \qquad (12)$$

If we regard P as describing different states of identical particles, the argument we employed in [1] to derive the spin-statistics relation can be extended by requiring the state to be singlevalued under full exchange, including $P \to \overline{P}$ as well as $\mathbf{r} \to -\mathbf{r}$.

To implement this idea, we write the state of the two particles as

$$|\Psi_P(\mathbf{r})\rangle = \sum_M \psi_{M,P}(\mathbf{r})|M(\mathbf{r})\rangle, \qquad (13)$$

in which $|M(\mathbf{r})\rangle$ is the same transported spin basis as before, with the exchange sign (3) and (6). Singlevaluedness, that is

$$|\Psi_P(\mathbf{r})\rangle = |\Psi_{\overline{P}}(-\mathbf{r})\rangle, \qquad (14)$$

leads to the extended spin-statistics relation

$$\psi_{\overline{M},\overline{P}}(-\mathbf{r}) = (-1)^{2S} \psi_{M,P}(\mathbf{r}). \qquad (15)$$

This is consistent with the requirement that the original spin-statistics relation must hold when the P state of both particles is the same, that is $P = \overline{P}$.

In the above argument, P has been treated differently from spin, notwithstanding the fact that the operators representing P (e.g. isospin) can have the same mathematical structure as angular momenta. The reason is that such mathematical resemblance conceals a physical difference: it is spin, and not any other property P, that is uniquely related to spatial rotations, because of its connection (section 6) with galilean or Lorentz invariance.

An argument similar to that leading to (15) has been given [14] in the context of Kaluza-Klein theory.

The decision to regard the particles as identical, embodied in (14), needs further discussion. An alternative possibility would be to regard the different values p_1, p_2 as distinguishing the particles. It seems absurd to consider macroscopic objects such as apples and pears as identical particles in different states of quantum fruitiness (P). Nevertheless, it is possible to choose to do this - but the choice is inconsequential, because as is well known it leaves unconstrained the symmetry of the space-spin part of the state - the symmetry of the P part of the state can always be adjusted to satisfy (15). The extended spin-statistics relation has consequences only when superpositions of states with different p are meaningful, and the interactions are such that transitions can occur between them (so that the P physics is coherently entangled with the space-spin physics).

6. SPIN AND RELATIVITY

There is a complicated history of derivations of SS[4] using arguments that rely on relativity, involving successively more refined postulates (causality, absence of negative-energy states, hermitean fields...). This raises the question of the relation between relativistic approaches and our nonrelativistic formulation. On this subject we can make only scattered remarks.

First we should point out that our derivation was nonrelativistic in the sense that it made no use of relativity, and not in the sense of being a low-velocity approximation. Since time never entered our considerations, the exchanges we considered (involving the variables **r** and *M*) can be regarded as taking place at fixed time. But fixed time is not relativistically invariant. Regarded relativistically, our exchanges were spacelike. This makes our arguments appear complementary to the some of the quantum field theoretic ones [15, 16], which involve the creation of pairs of antiparticles, and therefore are based on timelike exchanges.

Second, although we considered only the relation between spin and statistics, and not the origin of spin itself, the widespread belief that spin is unavoidably relativistic has led to doubts about our arguments involving exchange. But the existence of spin is equally a consequence of galilean relativity as of einsteinian relativity. This point has been well made before [17, 18], and it is not necessary to repeat the general arguments. However, we think it worth outlining the galilean spin-1/2 case in the simplest and least technical way, in an argument attributed to Feynman [19]. This is done in the Appendix.

Third, there is the intriguing possibility that the field-theoretic arguments could be made to operate in reverse, in the following sense. Suppose that the nonrelativistic programme outlined here is eventually completed, so that it would become clear that SS is embedded in quantum mechanics in a fundamental way, more primitive than field theory. Then instead of deriving SS by demanding that field theory satisfy certain requirements, such as causality and energy positivity, the knowledge that SS must be true might be invoked to show that field theory already possesses these desirable properties. This would be much more satisfactory, after all, than having to impose them.

Fourth, we are not aware that there exists any relativistic field theory, for particles with spin, that involves the configuration space we use here, where indistinguishability is incorporated geometrically by the identification of permuted configurations.

Fifth, we note that Anandan [14] has presented a relativistic generalization of our construction, in what may be a first step in establishing a bridge to the field-theoretic arguments.

APPENDIX. GALILEAN TRANSFORMATIONS AND SPIN 1/2

For a free particle without spin, with hamiltonian $H=p^2/2m$, Schrödinger's equation

$$i\hbar \frac{\partial}{\partial t}\psi(\mathbf{r},t) = -\frac{\hbar^2}{2m}\nabla_{\mathbf{r}}^2 \psi(\mathbf{r},t) \tag{A1}$$

is Galilean-invariant in the following sense. Under the transformation to

$$t \to t_1 \equiv t-T, \quad \mathbf{r} \to \mathbf{r}_1 \equiv \mathbf{R}\mathbf{r} - \mathbf{v}t - \mathbf{a},$$
$$\psi(\mathbf{r},t) \to \psi_1(\mathbf{r}_1,t_1) \equiv \psi(\mathbf{r},t)\exp\left\{-i\frac{m}{\hbar}\left(\mathbf{v}\cdot\mathbf{r}_1 + \frac{1}{2}v^2 t_1\right)\right\}, \tag{A2}$$

where T is a constant scalar, $\mathbf{a}$ and $\mathbf{v}$ are constant vectors, and $\mathbf{R}$ is a constant rotation matrix, the equation preserves its form:

$$i\hbar \frac{\partial}{\partial t_1}\psi_1(\mathbf{r}_1,t_1) = -\frac{\hbar^2}{2m}\nabla_{\mathbf{r}_1}^2 \psi_1(\mathbf{r}_1,t_1). \tag{A3}$$

For a particle with spin 1/2, this invariance is obviously shared by the two-spinor Schrödinger equation generated by the free 2x2 matrix Hamiltonian

$$\mathbf{H} = \frac{1}{2m}(\mathbf{S}\cdot\mathbf{p})^2 = \frac{1}{2m}p^2 \mathbf{1}, \tag{A4}$$

where $\mathbf{S}$ is the vector of Pauli matrices.

In both cases, external fields with potentials $\mathbf{A}(\mathbf{r},t)$, $V(\mathbf{r},t)$ can then be introduced by minimal coupling to the particle's charge q through

$$\mathbf{p} \to \mathbf{p} - q\mathbf{A}(\mathbf{r},t). \tag{A5}$$

and addition of $qV(\mathbf{r},t)$ to $\mathbf{H}$. In the spin 1/2 case, coupling to the first equation in (A4) leads to

$$\mathbf{H} = \frac{1}{2m}(\mathbf{S}\cdot(\mathbf{p}-q\mathbf{A}))^2 + qV(\mathbf{r},t)$$
$$= \frac{1}{2m}[\mathbf{p}-q\mathbf{A}(\mathbf{r},t)]^2 \mathbf{1} - \frac{q\hbar}{2m}\mathbf{S}\cdot\mathbf{B}(\mathbf{r},t) + qV(\mathbf{r},t).$$
(A6)

where $\mathbf{B}=\nabla\times\mathbf{A}$. This is the Pauli equation, with the spin operator $\mathbf{s}=h\mathbf{S}/4\pi$ coupled to the magnetic field with a magnetic moment $\mathbf{m}$, that is

$$\frac{q\hbar}{2m}\mathbf{S}\cdot\mathbf{B} = \mathbf{m}\cdot\mathbf{B}, \quad \text{where} \quad \mathbf{m} = \frac{q\hbar}{2m}\mathbf{S} = \frac{q\mathbf{s}}{m}.$$
(A7)

This $\mathbf{m}$, originating in a free equation that is invariant under galilean transformations, is the same – that is, it has the same gyromagnetic ratio - as that in the corresponding Dirac equation, which is Lorentz-invariant.

REFERENCES

1. Berry, M. V., and Robbins, J. M., Indistinguishability for quantum particles: spin, statistics and the geometric phase. *Proc. Roy. Soc. Lond.* **A453**, 1771-1790 (1997).
2. Berry, M. V., and Robbins, J. M., Quantum indistinguishability: alternative constructions of the transported basis. *J. Phys. A (Letters)* **33**, L207-L214.
3. Pauli, W., Connection between spin and statistics. *Phys. Rev.* **58**, 716-722 (1940).
4. Duck, I., and Sudarshan, E. C. G., *Pauli and the spin-statistics theorem*, World Scientific, Singapore, 1997.
5. Duck, I., and Sudarshan, E. C. G., Toward an understanding of the spin-statistics theorem. *Am. J. Phys.* **66**, 284-303 (1998).
6. Atiyah, M., Geometry of classical particles. *Asian Journal of Mathematics, in press* (2000).
7. Laidlaw, M. G. G., and DeWitt, C. M., Feynman functional integrals for systems of indistinguishable particles. *Phys. Rev. D.* **3**, 1375-1378 (1971).
8. Leinaas, J. M., and Myrheim, J., On the theory of identical particles. *Nuovo Cim.* **37B**, 1-23 (1977).
9. Sorkin, R., Particle statistics in three dimensions. *Phys. Rev. D.* **27**, 1787-1792 (1983).
10. Aharonov, Y., and Bohm, D., Significance of electromagnetic potentials in the quantum theory. *Phys. Rev.* **115**, 485-491 (1959).
11. Olariu, S., Popescu, I., and I.I., I., The quantum effects of electromagnetic fluxes. *Revs. Mod. Phys.* **57**, 339-436 (1985).
12. Robbins, J. M., *in preparation* (2000).
13. Schwinger, J., "On angular momentum", in *Quantum theory of angular momentum* (L. C. Biedenharn, and H. Van Dam, Eds.), Academic Press, New York, 1965, pp. 229-279.
14. Anandan, J., Spin-statistics connection and relativistic Kaluza-Klein space-time. Physics Letters A **248**, 124-130 (1998).
15. Balachandran, A. P., Daughton, A., Gu, Z.-C., Sorkin, R. D., Marmo, G., and Srivastava, A. M., Spin-statistics theorems without relativity or field theory. *Int. J. Mod. Phys.* **A8**, 2993-3044 (1993).

16. Feynman, R. P., "The reason for antiparticles", in *The 1986 Dirac memorial lectures* (R. P. Feynman, and S. Weinberg, Eds.), Cambridge University Press, New York, 1987.
17. Levy-Leblond, J.-M., Nonrelativistic particles and wave equations. *Commun. Math. Phys* **6**, 286-311 (1967).
18. Levy-Leblond, J.-M., The pedagogical role and epistemological significance of group theory in quantum mechanics. *Riv. Nuovo. Cim* **4**, 99-143 (1974).
19. Mackintosh, A. R., The Stern-Gerlach experiment, electron spin and intermediate quantum mechanics. *Eur. J. Phys.* **4**, 97-106 (1983).

The Role of Identity and Entanglement in Quantum Mechanics

GianCarlo Ghirardi

Department of Theoretical Physics, University of Trieste, Italy

Abstract. We review the main aspects of entanglement and we prove some general theorems concerning entangled states. In order to satisfy the symmetry requirements for systems containing identical constituents one is unavoidably led to consider entangled states. One could then naively conclude that the identity gives rise, by itself, to peculiar and problematic aspects of physical systems. This is by no means the case: the symmetrization or antisymmetrization of factorized states is an absolutely benign procedure and does not pose any specific conceptual problem apart from the one of the impossibility of attributing their elemetary constituents to one or another of two far-away systems. We briefly reconsider the same problem within the context of recently proposed models of dynamical reduction based on nonlinear and stochastic modifications of the standard evolution.

INTRODUCTION

Erwin Schrödinger [1] has appropriately identified entanglement as *the characteristic trait of Quantum Mechanics, the one that enforces its entire departure from classical lines of thought*. In fact it plays an absolutely primary (and problematic) role in connection with: the EPR situation, quantum nonlocality, the possibility of attributing properties to individual physical systems, the measurement problem and the conception of the universe. It also plays an extremely relevant (and promising) role for quantum criptography, quantum teleportation and, hopefully, for quantum computation. Entanglement emerges in connection with composite systems.

In this paper we present first of all a general analysis of the formal aspects of entanglement and of its most relevant implications, with particular refrence to the problem of attributing objective properties to individual physical systems. We then pass to analyze the case in which the considered composite systems contain identical constituents and we investigate the entanglement induced by the symmetry requirements which have to be imposed to the statevector in such a case. Finally we comment on some recent proposals to overcome the so called macro-objectification problem and we outline their nice features concerning entangled systems with identical constituents.

ENTANGLEMENT AND INDIVIDUAL PROPERTIES

Let us begin by considering a composite system $S = S_1 + S_2$ and the associated Hilbert space $\mathcal{H} = \mathcal{H}^{(1)} \otimes \mathcal{H}^{(2)}$. In such a case one has to face two radically different situations concerning the states of the system. It can be described by a factorized state:

$$|\Psi(1,2)>=|\Phi(1)>\otimes|\Xi(2)> \qquad (1)$$

or by an entangled one:

$$|\Psi(1,2)>=\sum_{i,j}c_{ij}|\Phi_i(1)>\otimes|\Xi_j(2)>, \quad c_{ij}\neq a_i b_j. \qquad (2)$$

Let us also recall some basic aspects of the problem of property attribution to individual physical systems within quantum mechanics (QM). If one assumes that any self-adjoint operator corresponds to an observable and one accepts that when the theory attributes probability 1 to an outcome then "there is an objective element of physical reality - an individual property" corresponding to the considered observable and outcome - one could state that Q.M. has taught us that one cannot pretend to attach too many properties to a system (due to the noncommutativity, in general, of operators associated to different observables). However, in the case of a factorized state of a composite system $S = S_1 + S_2$, its constituents have the properties corresponding to the set of all operators which have each of the factors of the statevector as a common eigenvector. On the contrary, in the case of entangled states the constituents may also have no property at all and, in general, they cannot have "as many properties" as those allowed by the theory for an isolated system.

Trivial examples are represented by a spin 1/2 particle in an arbitrary spin state

$$\begin{pmatrix} a \\ b \end{pmatrix}, \quad |a|^2+|b|^2=1 \qquad (3)$$

and by a pair of such particles in the singlet state. In the first case the particle has the property of having, objectively - i.e. independently of any measurement - its spin *up* in the direction $\mathbf{n}=(2\,\mathrm{Re}\,ab^*,-2\,\mathrm{Im}\,ab^*,1-2|b|^2)$. On the contrary, in the second case, if consideration is given to the physical properties of particle S_1, since they are determined by the reduced statistical operator obtained by taking the partial trace of the statistical operator for the composite system on the Hilbert space $\mathcal{H}^{(2)}$, which is given by 1/2 the identity operator, there follows that there is a nonepistemic probability of getting one of the two outcomes ± 1 in a spin measurement along any arbitrary direction. The particles do not have any spin property at all!

DISTINGUISHABLE PARTICLES: THE GENERAL CASE

In this Section we consider a system $S = S_1 + S_2 + ... + S_N$ containing N distinguishable particles in a pure state $|\Psi(1,2,...,N)>$ of the appropriate Hilbert space $\mathcal{H} = \mathcal{H}^{(1)} \otimes \mathcal{H}^{(2)} \otimes ... \otimes \mathcal{H}^{(N)}$, and we divide them in two groups $S_A = S_1 + S_2 + ... + S_M$ and $S_B = S_{M+1} + S_{M+2} + ... + S_N$. Let us denote as $\Pi(\Omega = \omega|\Psi)$ the probability of getting the eigenvalue ω in a measurement of the observable Ω when the state is $|\Psi>$, and let us adopt the following:

Definition: The set of particles $\{1,2,...,M\}$ is not entangled with the set of the remaining ones iff there exist a projection operator $E^{(1,M)}$ associated to a one-dimensional manifold of the Hilbert space $\mathcal{H}^{(1,M)} = \mathcal{H}^{(1)} \otimes \mathcal{H}^{(2)} \otimes ... \otimes \mathcal{H}^{(M)}$ such that: $\Pi(E^{(1,M)} = 1|\Psi(1,2,...,N)) = 1$.

One then easily proves the following:

Theorem: The set of particles $\{1,2,...,M\}$ is not entangled with the set of the remaining ones iff $|\Psi(1,2,...,N)>$ is factorized.

Proof: if $|\Psi(1,2,...,N)> = |\Phi(1,2,...,M)> \otimes |\Xi(M+1,M+2,...,N)>$ the reduced statistical operator $\tilde{\rho}^{(1,M)} \equiv Tr^{(M+1,...,N)} \rho(1,2,...,N)$ is a projection operator on a one dimensional manifold, $\tilde{\rho}^{(1,M)} = |\Phi(1,2,...,M)><\Phi(1,2,...,M)|$, so that $\Pi(\tilde{\rho}^{(1,M)} = 1|\Psi(1,2,...,N)) = 1$.

Suppose now that there exist a one dimensional projection operator $E^{(1,M)}$ of $\mathcal{H}^{(1,M)}$ such that $\Pi(E^{(1,M)} = 1|\Psi(1,2,...,N)) = 1$. We have:

$$\Pi(E^{(1,M)} = 1|\Psi(1,2,...,N)) = Tr^{(1,...,N)}\{E^{(1,M)}|\Psi(1,2,...,N)><\Psi(1,2,...,N)|\} = $$
$$Tr^{(1,...,M)}\{E^{(1,M)}Tr^{(M+1,...,N)}[|\Psi(1,2,...,N)><\Psi(1,2,...,N)|]\} = Tr^{(1,...,M)}\{E^{(1,M)}\tilde{\rho}^{(1,M)}\} \quad (4)$$

Writing $E^{(1,M)} = |\varphi><\varphi|$, $|\varphi> \in \mathcal{H}^{(1,M)}$ we have, from the last expression:

$$\Pi(E^{(1,M)} = 1|\Psi(1,2,...,N)) = Tr^{(1,...,M)}\{E^{(1,M)}\tilde{\rho}^{(1,M)}\} = <\varphi|\tilde{\rho}^{(1,M)}|\varphi>. \quad (5)$$

Since $\tilde{\rho}^{(1,M)}$ is a statistical operator, the condition $\Pi(E^{(1,M)} = 1|\Psi(1,2,...,N)) = 1$ implies then $\tilde{\rho}^{(1,M)} = |\varphi><\varphi|$.

We consider now the state $|\Psi(1,2,...,N)>$ and we express it by resorting to the von Neumann biorthonormal decomposition in terms of states of $\mathcal{H}^{(1,M)}$ and $\mathcal{H}^{(M+1,N)}$ (if there is any accidental degeneracy we can dispose of it as we want):

$$|\Psi(1,2,...,N)> = \sum_k \pi_k |\phi_k(1,2,...,M)> \otimes |\xi_k(M+1,M+2,...,N)> \quad (6)$$

with real and positive π_k. Equation (6) implies:

$$Tr^{(M+1,...,N)}|\Psi(1,2,...,N)><\Psi(1,2,...,N)| = \sum_k \pi_k^2 |\phi_k(1,2,...,M)><\phi_k(1,2,...,M)| \quad (7)$$

which, due to the orthogonality of the states $|\phi_k(1,2,...,M)>$ can coincide with $\tilde{\rho}^{(1,M)} = |\varphi><\varphi|$ iff the sum in (6) contains only one term, the corresponding π_k taking the value 1. Thus $|\Psi(1,2,...,N)>$ is factorized.

Q.E.D.

The above analysis has shown that one could also have adopted the following:
Definition: The set of particles $\{1,2,...,M\}$ is not entangled with the set of the remaining ones iff the reduced statistical operator $\tilde{\rho}^{(1,M)}$ is a projection operator onto a one dimensional manifold of $\mathcal{H}^{(1,M)}$.

We conclude this Section by recalling that, when the state is factorized, its subsystems S_A and S_B have as many properties as the theory allows them to have.

DEGREES OF ENTANGLEMENT

Let us reconsider an individual composite system in the pure state $|\Psi(1,2,...,N)>$ and the reduced statistical operator $\tilde{\rho}^{(1,M)}$ referring to the first M particles. As we know this is a trace-class, trace-one, semi-positive definite operator. We are interested in its Range $R(\tilde{\rho}^{(1,M)})$. Three cases are possible:

1. $R(\tilde{\rho}^{(1,M)})$ is a one-dimensional manifold of $\mathcal{H}^{(1,M)}$,

2. $R(\tilde{\rho}^{(1,M)})$ is a proper submanifold of $\mathcal{H}^{(1,M)}$ of dimension ≥ 2,

3. $R(\tilde{\rho}^{(1,M)})$ coincides with $\mathcal{H}^{(1,M)}$.

Let us analyze them.

1. The first case has already been discussed in all details. In particular:

   1a. $|\Psi(1,2,...,N)>$ is factorized,
   1b. $\tilde{\rho}^{(1,M)}$ is a projection operator (on a one-dimensional manifold),
   1c. The subsystem S_A of the first M particles has as many objective properties as the theory allows any system to have.

2. In the second case let us denote as $E^{(R)}$ the projection operator onto $R(\tilde{\rho}^{(1,M)})$. Then:

$$\Pi(E^{(R)} = 1|\Psi(1,2,...,N)) = Tr^{(1,...,M)}(E^{(R)}\tilde{\rho}^{(1,M)}) = Tr^{(1,...,M)}(\tilde{\rho}^{(1,M)}) = 1. \quad (8)$$

Accordingly, given any self-adjoint operator $\Omega^{(1,M)}$ of $\mathcal{H}^{(1,M)}$ which commutes with $E^{(R)}$ and considering the subset B (a Borel set) of its spectrum such that the corresponding "eigenvectors" are eigenstates of $E^{(R)}$, we can state that the subsystem S_A of the first M particles has the objective property $\Omega^{(1,M)} \in B$. As a particular case one can consider an operator $\Sigma^{(1,M)}$ such that, for it, $R(\tilde{\rho}^{(1,M)})$ is a degenerate eigenmanifold corresponding to the eigenvalue s. Then: $\Pi(\Sigma^{(1,M)} = s|\Psi(1,2,...,N)) = 1$ and the subsystem has an objective property.

3. In the third case, if consideration is given to an arbitrary projection operator E of $\mathcal{H}^{(1,M)}$ and one imposes that $\Pi(E = 1|\Psi(1,2,...,N)) = 1$, one has, by considering the spectral representation of $\tilde{\rho}^{(1,M)}$ ($\tilde{\rho}^{(1,M)} = \sum_i p_i |\varphi_i><\varphi_i|$):

$$\Pi(E = 1|\Psi(1,2,...,N)) = Tr^{(1,...,M)}\left(E\tilde{\rho}^{(1,M)}\right) = \sum_i p_i <\varphi_i|E|\varphi_i> = \sum_i p_i \||E|\varphi_i>\|^2 = 1. \quad (9)$$

Since $p_i > 0$, $\sum_i p_i = 1$, $\||E|\varphi_i>\|^2 \leq 1$, the above relation can be satisfied iff $E|\varphi_i> = |\varphi_i>$, $\forall i$. This condition, taking into account that the set $\{|\varphi_i>\}$ is a complete orthonormal set, implies that the projection E is the identity operator of $\mathcal{H}^{(1,M)}$.

Concluding: in the third case, there is no projection operator exception made for the identity, of which one can predict with certainty the outcome: the subsystem S_A of the first M particles has no property whatsoever; Entanglement can be such that the constituents of a composite system have lost any individual property (obviously the system as a whole has still some property). This leads to the conclusion that since in the long run everything interacts with everything and, in general, interactions induce entanglement, only the universe as a whole has properties, its parts do not have individual properties. Thus: the *Undivided Universe* of D. Bohm or the *Unbroken Whole* of B. d'Espagnat.

CORRELATED SUBSYSTEMS

Note that in cases 2 and 3 of the previous Section, if one considers once more the von Neumann biorthonormal decomposition:

$$|\Psi(1,2,...,N)> = \sum_k \pi_k |\phi_k(1,2,...,M)> \otimes |\xi_k(M+1,M+2,...,N)> \quad (10)$$

and one takes into account self-adjoint operators $\Gamma^{(1,M)}$ and $\Lambda^{(M+1,N)}$ having the othonormal states $|\phi_k(1,2,...,M)>$ and $|\xi_k(M+1,M+2,...,N)>$ as nondegenerate eigenstates, then, in spite of the fact that the corresponding outcomes γ_k and λ_k have nonepistemic probabilities π_k^2 of being obtained in a measurement, the outcomes themselves are perfectly correlated.

Accordingly, a measurement of $\Lambda^{(M+1,N)}$ and the registration of the outcome allows us to predict with probability 1 the outcome of a prospective measurement of $\Gamma^{(1,M)}$. An objective property emerges istantaneously at-a-distance. The nonlocal features associated with entanglement and the violation of Bell's inequality make their appearence.

SOME ELEMENTARY EXAMPLES

Obviously the symmetry requirements that quantum theory imposes to systems containing identical constituents imply that such systems are (practically always) associated to entangled states. Does this fact, by itself, give rise to puzzling situations similar to those of composite systems in entangled states? This is an interesting question and some authors have given a positive answer to it. In what follows I will show that, while, in general, one has to be worried by the occurrence of entangled states when trying to give a reasonable quantum foundation to classical physics (and I will briefly point out one of the possible ways out), one can safely ignore the implications of the entanglement arising exclusively from the symmetry requirements for identical constituents.

To begin with, let us consider a factorized state of two (distinguishable) fermions of spin 1/2:

$$|\Psi_{FD}(1,2)\rangle = \alpha(1)\phi_k(1)\beta(2)\xi_j(2). \tag{11}$$

Here we have denoted, as usual, as $\alpha(i)$ and $\beta(i)$ ($i=1,2$), the eigenstates of $\sigma_z^{(i)}$ corresponding to the eigenvalues +1 and -1, respectively, and we assume that $\phi_k(1)$ and $\xi_j(2)$ are eigenvectors of two self-adjoint operators $\Lambda^{(1)}$ and $\Gamma^{(2)}$ belonging to appropriate eigenvalues λ_k and γ_j, respectively. For such a state it is undoubtely legitimate (in the spirit of Einstein's realism) to claim that there is a fermion of type 1 which has its spin *up* (along z) and possesses the objective property $\Lambda^{(1)} = \lambda_k$ and a fermion of type 2 which has its spin *down* and possesses the objective property $\Gamma^{(2)} = \gamma_j$, quite independently of our performing, or not, the corresponding tests.

In what follows, in place of precise eigenstates of two observables, we will consider, for simplicity and to have the opportunity of dealing with typical EPR-Bohm like situations, states corresponding to a particle "being in a precise region", i.e. in a state whose configuration space wavefunction has compact support.

Accordingly, let $\psi(\mathbf{r})$ and $\phi(\mathbf{r})$ have as support two bounded and non overlapping compact sets $D[\psi]$, $D[\phi]$ with $D[\psi] \cap D[\phi] = \emptyset$, and let us consider, for the system of a proton and a neutron, the analogue of state (11):

$$|\Psi_{FD}(p,n)\rangle = \alpha(p)\psi(\mathbf{r}_p)\beta(n)\phi(\mathbf{r}_n). \tag{12}$$

For such a state we can claim that "There is a proton in $D[\psi]$ with spin *up* and a neutron in $D[\phi]$ with spin *down*."

Let us compare now the previous state with the following entangled one:

$$|\tilde{\Psi}_{ED}(p,n)\rangle = \frac{1}{\sqrt{2}}\left[\alpha(p)\psi(\mathbf{r}_p)\beta(n)\phi(\mathbf{r}_n) - \beta(p)\phi(\mathbf{r}_p)\alpha(n)\psi(\mathbf{r}_n)\right] \tag{13}$$

For such a state "there is an epistemic probability equal to 1/2 that the particle in $D[\psi]$ is a proton or a neutron. In spite of that, such a particle has certainly its spin *up*".

We could have also considered the state

$$|\hat{\Psi}_{ED}(p,n)\rangle = \frac{1}{\sqrt{2}}\left[\alpha(p)\beta(n) - \beta(p)\alpha(n)\right]\psi(\mathbf{r}_p)\phi(\mathbf{r}_n) \tag{14}$$

for which: "There is a proton in $D[\psi]$ and a neutron in $D[\phi]$" but no certain claim can be made about their spin states, and the state:

$$|\Psi_{ED}^*(p,n)> = \frac{1}{\sqrt{2}}[\alpha(p)\beta(n) - \beta(p)\alpha(n)] \otimes \qquad (15)$$
$$[\psi(\mathbf{r}_p)\phi(\mathbf{r}_n) + \phi(\mathbf{r}_p)\psi(\mathbf{r}_n)]$$

for which no claim about locations and spin properties of both particles can be made.

IDENTITY AND ENTANGLEMENT

In the case of identical particles (e.g., two electrons) the analogous of our original factorized state (11), i.e.,

$$|\Psi_{FI}(e,e)> = \alpha(1)\psi(\mathbf{r}_1)\beta(2)\phi(\mathbf{r}_2), \qquad (16)$$

is unacceptable for symmetry reasons. However, let us antisymmetrize it:

$$|\tilde{\Psi}_{EI}(e,e)> = \frac{1}{\sqrt{2}}[\alpha(1)\psi(\mathbf{r}_1)\beta(2)\phi(\mathbf{r}_2) - \beta(1)\phi(\mathbf{r}_1)\alpha(2)\psi(\mathbf{r}_2)] \qquad (17)$$

For such a state it is still perfectly legitimate to state that: "In $D[\psi]$ there is an electron with spin *up* and in $D[\phi]$ there is an electron with spin *down*". What is no more legitimate is to claim that the electron in one region is the one we have conventionally called "electron 1" or "2". This is a trivial fact that reflects simply the indistinguishability of identical systems within a quantum context.

On the contrary, the state analogous to (15) for distinguishable particles:

$$|\Psi_{EI}^*(e,e)> = \frac{1}{\sqrt{2}}[\alpha(1)\beta(2) - \beta(1)\alpha(2)] \otimes [\psi(\mathbf{r}_1)\phi(\mathbf{r}_2) + \phi(\mathbf{r}_1)\psi(\mathbf{r}_2)] \qquad (18)$$

does not consent to make any statement about the spin properties of either of the electrons.

IDENTICAL FERMIONS: A MORE FORMAL PROOF

Within the non-relativistic second quantized formalism, let $a^\dagger(\mathbf{r},\gamma)$, $a(\mathbf{r},\gamma)$ be the creation and annihilation operators for a fermion of spin component γ (taking the values $\{\alpha,\beta\}$) at position $\mathbf{r}$, satisfying the usual anticommutation relations. Within such a formalism, state (17) reads:

$$|\tilde{\Psi}_{EI}> = \int_{All\,space} d\tilde{\mathbf{r}}_1 \int_{All\,space} d\tilde{\mathbf{r}}_2 \psi(\tilde{\mathbf{r}}_1)\phi(\tilde{\mathbf{r}}_2) a^\dagger(\tilde{\mathbf{r}}_1,\alpha) a^\dagger(\tilde{\mathbf{r}}_2,\beta)|0> \qquad (19)$$

Let us consider the number operator for fermions with spin α in the volume V

$$N_{V,\alpha} = \int_V d\mathbf{r}\, a^\dagger(\mathbf{r},\alpha) a(\mathbf{r},\alpha) \qquad (20)$$

where V is such that $D[\Psi] \subseteq V$, $D[\Phi] \cap V = \emptyset$, and let us apply it to the state (19). We have, as easily checked:

$$N_{V,\alpha}|\tilde{\Psi}_{EI}\rangle = |\tilde{\Psi}_{EI}\rangle. \tag{21}$$

In a completely similar way, if one considers a volume $\tilde{V}$ such that $D[\Phi] \subseteq \tilde{V}$, $D[\Psi] \cap \tilde{V} = \emptyset$ and the number operator $N_{\tilde{V},\beta}$ counting the particles with spin down in $\tilde{V}$, one has:

$$N_{\tilde{V},\beta}|\tilde{\Psi}_{EI}\rangle = |\tilde{\Psi}_{EI}\rangle. \tag{22}$$

Equations (20) and (22) show that $|\tilde{\Psi}_{EI}\rangle$ is an eigenstate of the two considered number operators belonging to the eigenvalue 1. Accordingly, it is perfectly legitimate to state that, for an individual composite system of two electrons in such a state: there is an electron in V with spin *up* and one in $\tilde{V}$ with spin *down*.

Summarizing: the entanglement arising from the pure symmetry requests on a system with identical constituents does not give rise to any embarrassing situation concerning properties when the state is obtained by antisymmetrizing (symmetrizing) a factorized state. Actually this statement has a much more general validity, as shown by the theorem of the next Section, which is due to A. Bassi, G.C. Ghirardi, L. Marinatto and T. Weber [2].

INDISTINGUSHABLE PARTICLES: A GENERAL THEOREM

For simplicity we will confine our considerations to the case of two identical particles. The proof holds both for boson and fermion systems. Let

$$P^{(i)}, \quad i = \{1,2\}, \qquad P^{(i)} = |\Phi^{(i)}\rangle\langle\Phi^{(i)}| \tag{23}$$

be a projection operator associated to a one dimensional manifold of the Hilbert space of the *i-th* particle and let us consider the projection operator of the whole space:

$$\mathbf{P}_{AtL1} = P^{(1)} \otimes 1^{(2)} + 1^{(1)} \otimes P^{(2)} - P^{(1)} \otimes P^{(2)}. \tag{24}$$

Note that its expectation value on a state $|\Psi(1,2)\rangle$ gives *the probability that there is At Least One particle possessing the properties identified by the state* $|\Phi\rangle$.

We can now state our:

Theorem: Be $S = S_1 + S_2$ a system composed of two identical particles, associated to the pure state $|\Psi(1,2)\rangle$. The condition:

$$Tr^{(1+2)}[\mathbf{P}_{AtL1}|\Psi(1,2)\rangle\langle\Psi(1,2)|] \equiv \langle\Psi(1,2)|\mathbf{P}_{AtL1}|\Psi(1,2)\rangle = 1 \tag{25}$$

implies and is implied by the fact that the state $|\Psi(1,2)>$ is obtained by symmetrizing or antisymmetrizing a factorized state of the two particles.

Note that the above theorem makes rigorous the fact that the symmetrization or antisymmetrization procedures by themselves do not forbid to consider one of the particles (obviously one cannot say which one) as objectively possessing the maximum set of properties that quantum mechanics allows any system to have.

Proof: If $|\Psi(1,2)>$ is obtained by symmetrizing or antisymmetrizing a factorized state of two identical particles:

$$|\Psi(1,2)>= N\{|\varphi^{(1)}\chi^{(2)}>\pm|\chi^{(1)}\varphi^{(2)}>\} \qquad (26)$$

expressing the state $|\chi^{(i)}>$ as follows

$$\chi^{(i)}>= \alpha|\varphi^{(1)}>+\beta|\varphi_\perp^{(1)}>, \qquad (27)$$

one gets immediately

$$<\Psi(1,2)|\mathbf{p}_{AtL1}|\Psi(1,2)>= \frac{2(1\pm|\alpha|^2)}{2(1\pm|\alpha|^2)}=1 \qquad (28)$$

Alternatively, since $\mathbf{p}_{AtL1}$ is a projection operator

$$\left[<\Psi(1,2)|\mathbf{p}_{AtL1}|\Psi(1,2)>=1\right] \supset \left[\|\mathbf{p}_{AtL1}|\Psi(1,2)>\|=1\right]$$
$$\supset \left[\mathbf{p}_{AtL1}|\Psi(1,2)>=|\Psi(1,2)>\right] \qquad (29)$$

If one chooses a c.o.n. set of one particle states whose first element $|\Phi_0>=|\Phi>$, one has:

$$\mathbf{p}_{AtL1}|\Psi(1,2)>= \mathbf{p}_{AtL1}\sum_j c_{ij}|\Phi_i^{(1)}>\otimes|\Phi_j^{(2)}>=$$
$$|\Phi_0^{(1)}>\otimes\left[\sum_j c_{0j}|\Phi_j^{(2)}>\right]+\left[\sum_j c_{j0}|\Phi_j^{(1)}>\right]\otimes|\Phi_0^{(2)}>-c_{00}|\Phi_0^{(1)}>\otimes|\Phi_0^{(2)}> \qquad (30)$$

so that, calling $|\Xi^{(i)}>=\left[\sum_j c_{j0}|\Phi_j^{(i)}>\right]$ one has:

$$|\Psi(1,2)>\propto |\Phi^{(1)}>|\Xi^{(2)}>-|\Xi^{(1)}>|\Phi^{(2)}> \qquad (31)$$

for fermions, and

$$|\Psi(1,2)>\propto \left\{|\Phi^{(1)}>\left[|\Xi^{(2)}>-\frac{c_{00}}{2}|\Phi^{(2)}>\right]+\left[|\Xi^{(1)}>-\frac{c_{00}}{2}|\Phi^{(1)}>\right]|\Phi^{(2)}>\right\} \qquad (32)$$

for bosons.   Q.E.D.

ENTANGLEMENT INVOLVING MACROSCOPIC SYSTEMS

As it is well known, the most embarassing aspects of entanglement, arise in connection with macroscopic systems. Typically the final state of a measurement process involves entangled states of the measured microsystem and the measuring apparatus, in particular situations in which different microstates are entagled with macroscopically different states of the measuring instruments. In such a situation, as discussed previously, one cannot legitimately attribute definite macroscopic properties to the apparatus itself.

Without reviewing in details the so called measurement or macro-objectification problem which is well known to everybody, let us consider a macroscopic system which is made of two macroscopic subsystems, having different shapes and different space locations. We start by considering a quite reasonable situation: we have two bulks of matter, one having the shape of a sphere and the second the one of a cube, which are in precisely definite positions:

$$|\Psi> = |Sphere \quad here> \otimes |Cube \quad there> \qquad (33)$$

here and *there* denoting two disticnt and disjoint locations of the centres of mass of the two pieces constituting the body. For such a state one can legitimately claim that there is a macroscopic bulk of matter having, e.g., normal density and the shape of a sphere near to me and another bulk with the same density and the shape of a cube located in a far away position.

Up to this moment we have not taken into account that the two macrosystems are, in reality, made up of an extremely large number of identical elementary particles (protons, neutrons and electrons). State (33) has then to be modified to take into account the quantum mechanical requests concerning identical particles. Accordingly it has to be replaced by the state:

$$|\Psi> = A\left(|Sphere \quad here> \otimes |Cube \quad there>\right) \qquad (34)$$

where A is the operator which antisymmetrizes the state to which it is applied on all identical fermions appearing in it.

As it has been discussed in great detail in the previous Sections, the state (34) is a state for which it is perfectly legitimate to continue to assert that there is a macroscopic bulk of matter having the shape of a sphere near to me and another bulk located in a far away position. The only thing one cannot claim any more is which ones of the electrons, protons, neutrons, ..., belong to the sphere and which ones belong to the cube.

On the other hand, as we all know, due to the linear nature of the theory, states much more embarrassing than the one just considered are allowed within a quantum context. In particular, if consideration is given to the state:

$$|\tilde{\Psi}> = N(|Sphere \quad here> \otimes |Cube \quad there> + \\ |Cube \quad here> \otimes |Sphere \quad there>) \qquad (35)$$

N a normalization factor, as well as to the properly antisymmetrized version of it:

$$|\tilde{\Psi}>= A \ \{N(|Sphere \quad here> \otimes |Cube \quad there> + \\ |Cube \quad here> \otimes |Sphere \quad there>)\} \quad (36)$$

for both of them there is no matter of fact concerning the "objective" location of the cube and the sphere.

ENTANGLEMENT, IDENTITY AND DYNAMICAL REDUCTION

Recently various proposals aimed to overcome problems of the kind we have just mentioned, i.e. the macro-objectification problem, have been put forward. Among them, one of the most widely discussed is the so called dynamical reduction program. The idea is quite simple. One accepts that the quantum description of individual physical systems is complete, i.e. that knowledge of the statevector represent the most accurate information one can get about a given physical system, but one entertains the idea that [3] *Schrödinger's equation is not always right*. In particular, one considers nonlinear and stochastic modifications of it which, without contradicting any known fact about microsystems lead, on the basis of the unique dynamics which governes all physical processes, to the dynamical suppression of superpositions of macroscopically distinguishable states.

The first model of this type [4], QMSL (Quantum Mechanics with Spontaneous Localizations) did not respect the symmetry requirements for a system of identical particles. Subsequently a variant of it, usually referred to as the CSL (Continuous Spontaneous Localization) model has been worked out [5]. What matters for our argument is the fact that CSL is a theory which respects the symmetry requirements for identical constituents and implies the emergence of definite properties for any macrosystem. If the symmetry requirements would, by themselves, forbid the consideration of isolated systems and the assignement of (almost) perfectly definite properties to them even in the macroscopic case, the above program would turn out to be unviable. Therefore, for our purposes it is interesting to look at the CSL model in one of its recent versions and to call attention to its implications for macroscopic systems.

The CSL model is based on a linear stochastic evolution equation which does not preserve the norm of the statevector, but only its average value. The equation replacing the standard one is the Stratonivich stochastic equation:

$$\frac{d|\Psi_w(t)>}{dt} = \left[-\frac{i}{\hbar}\hat{H} + \frac{1}{m_0}\int d\mathbf{r}\hat{M}(\mathbf{r})w(\mathbf{r},t) - \frac{\gamma}{m_0^2}\int d\mathbf{r}\hat{M}^2(\mathbf{r}) \right]|\Psi_w(t)> \quad (37)$$

where $\hat{H}$ is the quantum hamiltonian of the system and m_0 the nucleon mass. The operator $\hat{M}(\mathbf{r})$ has the form:

$$\hat{M}(\mathbf{r}) = \sum_k m^{(k)}\hat{N}^{(k)}(\mathbf{r}), \quad (38)$$

$m^{(k)}$ being the mass of the particles of type k (k=proton, neutron, etc.) and $N^{(k)}(\mathbf{r})$ gives the average density of particles of this type in an appropriate volume (of about $10^{-15} cm^3$ - see below) around the point $\mathbf{r}$:

$$\hat{N}^{(k)}(\mathbf{r}) = \left[\frac{\alpha}{2\pi}\right]^{\frac{3}{2}} \sum_s \int d\mathbf{q}\, e^{-\frac{\alpha}{2}(\mathbf{q}-\mathbf{r})^2} a^\dagger_{(k)}(\mathbf{q},s) a_{(k)}(\mathbf{q},s). \tag{39}$$

In this equation $a^\dagger_{(k)}(\mathbf{q},s)$, $a_{(k)}(\mathbf{q},s)$ are the creation and annihilation operators of a particle of type k at point $\mathbf{q}$ with spin component s, satisfying the canonical commutaion or anticommutation relations. The parameter α is assumed to take the value (characterizing also the QMSL model) $10^{10}\, cm^{-2}$ and the parameter γ is given by $\gamma = \lambda(4\pi/\alpha)^{3/2}$, where $\lambda = 10^{-16}\, sec^{-1}$, is the quantity representing (in the spirit of the original model) the mean frequency at which the stochastic "localizations", i.e. the random processes striving to make definite the mass distribution, take spontaneously place. The c-number functions $w(\mathbf{r},t)$ describe a continuous family of stochastic processes satisfying:

$$<<w(\mathbf{r},t)>> = 0; \quad <<w(\mathbf{r},t)w(\mathbf{r}',t')>> = \gamma\delta(\mathbf{r}-\mathbf{r}')\delta(t-t'). \tag{40}$$

The physical meaning of the model is made precise by the following prescription: if a homogeneous ensemble (a pure case) is associated at the initial time $t=0$ to the initial statevector $|\Psi,0>$, then the ensemble at time t is the union of homogeneous ensembles associated to the normalized statevectors $|\Psi_w,t>/\||\Psi_w,t>\|$ where $|\Psi_w,t>$ is the solution of Eq. (37) with the assigned initial conditions and for the specific stochastic process $w(t)$ which has occurred in the time interval $(0,t)$. The probability density for such subensembles is not the "raw" one $P_{Raw}[w]$ associated to the Gaussian process $w(t)$ (in the interval $(0,t)$), but the "cooked" one $P_{Cooked}[w]$:

$$P_{Cooked}[w] = P_{Raw}[w]\||\Psi_w,t>\|^2. \tag{41}$$

It is easily proven that, with the above prescriptions, the non-hamiltonian terms in the dynamical equation drive (with the appropriate probabilities) the statevector of each individual member of the ensemble into one of the "common eigenmanifolds" of the commuting operators $\hat{M}(\mathbf{r})$, i.e., the nonhamiltonian dynamical terms strive to make definite the average (over volume elements of the order of $(\alpha)^{-3/2}$) mass density distributions of the whole universe.

With reference to state (36), the CSL theory forbids the persistence of the superposition of the macroscopically different situations associated to its terms. In a split second such a state is then reduced either to

$$|\Psi, Sh> = A\left(|Sphere\ here> \otimes |Cube\ there>\right) \quad (42)$$

or to

$$|\Psi, Ch> = A\left(|Cube\ here> \otimes |Sphere\ there>\right) \quad (43)$$

The picture should now be clear. A state like |Cube here> involves a macroscopic difference of the mass density from the one associated to the state |Sphere here> in appropriate space regions (those in which the sphere and the cube do not overlap). The fact that CSL forbids the persistence of the superposition implies that what is here being a cube or a sphere is completely independent from our performing an observation and, in general, from our perceptions.

CONCLUSIONS

In this paper we have analized the implications of the symmetry requirements imposed to the wavefunction by the identity of the constituents of composite systems and we have shown that no new puzzling situation emerges with respect to the case of distinguishable particles. The embarrassing aspects of the entanglement derive from the occurrence of states which would be entangled even if no symmetry requirement would be taken into account, while the entanglement brought in by the conditions required by the identity of the constituents does not give rise to any further conceptual problem. This fact is at the very basis of the possibility of solving the macro-objectification problem by considering dynamical mechanisms which tend to render definite the macroscopic aspects of the world around us, in accordance with our definite perceptions about it.

ACKNOWLEDGMENTS

I acknowledge illuminating discussions with A. Bassi, L. Marinatto and T. Weber.

REFERENCES

[1] Schrödinger, E, *Proc. Cambridge Philos Soc.*, **31**, 555-563 (1935).
[2] Bassi, A., Ghirardi, G.C., Marinatto L. and Weber T., *in preparation*.
[3] Bell, J.S., in: *Schrödinger, Centenary Celebration of a Polymath*, C.W. Kilminster ed., Cambridge University Press, Cambridge, 1987, pp.40-52.
[4] Ghirardi, G.C., Rimini, A. and Weber, T., *Phys. Rev.* D **34**, 470- 491 (1986).
[5] Pearle, P., *Phys. Rev.*, A **39**, 2277-2289 (1989), Ghirardi, G.C., Pearle P. and Rimini, A., *Phys. Rev.*, A **42**, 78-89 (1990), Ghirardi, G.C., Grassi R. and Benatti, F., *Found. Phys.*, **25**, 5-38 (1995).

Quantum Statistics and Dynamical Algebras: Fermions

Enrico Celeghini* and Mario Rasetti[†]

*Dipartimento di Fisica and Sezione INFN
Università di Firenze, I50125 Firenze, Italy
† Dipartimento di Fisica and Unità INFM
Politecnico di Torino, I10129 Torino, Italy*

Abstract. In the frame of a general scheme aimed to relate dynamical algebras with quantum statistics, the possibility is discussed that interacting fermions may be described by the dynamical algebra $osp(2|2)$ instead of $h(1|1)$. This is exemplified by the application to superconductivity, for which the analogue in the new scheme of the BCS theory is constructed. The results are quite similar, showing that the proposed description of fermions is consistent also with such a complex phenomenology.

Outline of the strategy and summary of the results

Main goal of this contribution is to propose a novel algebraic scheme to describe coupled fermions, in the frame of an effort that the authors have been pursuing in the past few years aiming to clarifying the deep relation connecting quantum statistics to the algebra of physical observables [1], [2].

In such new scheme interacting electrons will be dealt with in a Fock picture as belonging to the odd sector of the superalgebra $osp(2|2)$ ($osp(1|2)$ when parity invariance holds).

The proposed structure is innovative mainly in that Cooper pairs (living in the even sector of $osp(1|2)$, isomorphic to $su(1,1)$) are true bosons *i.e.* particles with unlimited occupation numbers (see also [3]). In view of this latter feature it appears suggestive to rephrase in this new scheme the theory of low T_c superconductivity and investigate possible differences with the conventional theory. We shall show that the results are, rather unexpectedly, identical with those of BCS theory, confirming that the theoretical frame proposed is consistent with the complex phenomenology associated with the superconductive phase transition.

In **BCS** theory electrons are described, as in the vacuum, by the superalgebra $h(1|1)$. The attractive interaction due to phonons is considered as a perturbation that does not change the space of states. Cooper pairs are thus hard-core bosons: they have integer spin but the square of their creation operator equals zero (a feature that should imply that the onset of superconductivity cannot in fact be ascribed, as it is common wisdom, due to Bose-Einstein condensation [4]). Indeed, if Cooper pairs were true bosons, they should be consistently described by an infinite dimensional representation of the Weyl-Heisenberg algebra $h(1)$ or by a discrete series bounded below of $su(1,1)$ [3]. The algebra of pairs closes instead a two-dimensional (singlet) representation of $su(2)$.

We shall argue that the BCS interaction should modify the space of states in a way consistent with the requirement that Cooper bound states are true particles (**cooperons**) endowed with *bosons* statistics, which can live together with single electrons, which behave instead as *fermions*. Because this is incompatible with $h(1|1)$, the dynamical algebra must be changed: we propose it becomes $osp(2|2)$ ($osp(1|2)$ imposing parity invariance).

In other words, we shall claim that $osp(1|2)$ is the *true algebra of superconductivity* (when no parity violation occurs, e.g. no external field is present).

The superconductive phase transition is related in this picture with the spontaneous breaking (contraction) of $osp(2|2)$ into $h(1|1) \oplus h(1|1)$, that describes the normal phase in which no bound states exist.

The general scheme

That the quantum statistics including both electrons and cooperons on equal footing should be connected with the $\mathbb{Z}_2$-graded algebra $osp(1|2)$ derives just from the basic physics on which the BCS theory stands:

⋆ Physics requires the existence of creation and annihilation operators for electrons with both spin up [$a_\uparrow^+$ and $a_\uparrow^-$] and spin down [$a_\downarrow^+$ and $a_\downarrow^-$]. In view of the spin-statistics theorem these operators must belong to the odd sector of a $\mathbb{Z}_2$-graded algebra.

⋆ In the absence of pairing these operators must define normal electrons, hence they should generate two orthogonal $h(1|1)$'s: for this one needs two further operators, h_σ, $\sigma \in \{\uparrow,\downarrow\}$, such that:

$$\{a_\sigma^+, a_\sigma^-\} = h_{-\sigma} \, , \, (a_\sigma^\pm)^2 = 0 \, , \, [h_\sigma, a_{-\sigma}^\pm] = 0 \, , \tag{1}$$

while

$$\{a_\uparrow^\pm, a_\downarrow^\pm\} = 0 \, , \tag{2}$$

and all other relations are zero.

The two required $h(1|1)$'s are generated by $\{a_\uparrow^+, a_\uparrow^-, h_\downarrow\}$ and $\{a_\downarrow^+, a_\downarrow^-, h_\uparrow\}$, respectively, and the full structure is just the $h(1|1) \oplus h(1|1)$, adopted in BCS.

⋆ Because creation and annihilation operators for different spins are assumed to anticommute, BCS theory is forced to use products of operators to describe Cooper pairs. However products live outside the algebra, and an algebrically coherent scheme should then further include in the algebra creation and annihilation operators $c^\pm$ for the particles (of integer spin) generated by binding pairs of electrons.

Such operators should of course be bilinear in the electron operators and because of the statistics requirement (pairs of fermions must give rise to bosons) they should belong to the even sector of the superalgebra.

The only possibility to fullfill all the above requisites is to replace the anticommutators;

$$\{a_\uparrow{}^\pm, a_\downarrow{}^\pm\} = 0 \implies \{a_\uparrow{}^\pm, a_\downarrow{}^\pm\} = c^\pm . \tag{3}$$

These relations show that the algebra still contains the two $h(1|1)$'s but not their direct sum $h(1|1) \oplus h(1|1)$ as subalgebras.

The new anticommutation relations express the physical requirement that, due to interaction, electrons belonging to a pair should loose their fermionic nature, and behave as true bosons. As such, cooperons should have unlimited occupation numbers.

In conclusion, physics appears to ask for a superalgebra with 8 generators:

4 **odd** $\{a_\uparrow{}^+, a_\downarrow{}^+, a_\uparrow{}^-, a_\downarrow{}^-\}$;
4 **even** $\{c^+, c^-, h_\uparrow, h_\downarrow\}$.

An inquiry among simple superalgebras of small dimension indicates at dimension 8 only one superalgebra: $osp(2|2)$ (classified also as $sl(1|2)$) [5].

The other Jordan products are:
• *odd-odd*

$$\{a_\uparrow{}^\pm, a_\downarrow{}^\mp\} = 0 , \tag{4}$$

• *even-odd*

$$[h_\sigma, a_\sigma{}^\pm] = \pm a_\sigma{}^\pm , \quad [c^\pm, a_\sigma{}^\pm] = 0 , \quad [c^\pm, a_\sigma{}^\mp] = \mp a_{-\sigma}{}^\pm , \tag{5}$$

• *even-even*: (generate the subalgebra $su(1,1) \oplus u(1)$)

$$[h_\uparrow, h_\downarrow] = 0 , \quad [c^-, c^+] = 4(h_\uparrow + h_\downarrow) , \quad [h_\sigma, c^\pm] = \pm c^\pm . \tag{6}$$

Cooperons are represented in the even sub-algebra $su(1,1)$, generated by $\{c^+, c^-, h \equiv h_\uparrow + h_\downarrow\}$ (that describes the "correct" statistics of bosons according to [3]).

The subalgebra of $osp(2|2)$ physically most relevant is $osp(1|2)$, generated by two odd, $\{a^\pm \equiv a_\uparrow^\pm + a_\downarrow^\pm\}$, and three even, $\{c^\pm, h\}$, operators. $osp(1|2)$ contains cooperons along with electrons and commutes with spin parity, while $osp(2|2)$ does not. $osp(1|2)$ can thus be considered a natural candidate for the (super-) algebra of superconductivity in the absence of magnetic field.

The independent relations of $osp(1|2)$ can be assumed as those generated by $\{a^+, a^-; h\}$ with $c^\pm \equiv \left(a^\pm\right)^2$:

$$[h, a^\pm] = \pm a^\pm \quad , \quad \{a^+, a^-\} = 2h \ . \tag{7}$$

Let us address now representations. In view of the physical application, we impose first the hermiticity condition $\left(a^+\right)^\dagger = a^-$, requesting as well that the spectrum of h is bounded below and positive.

We select $h = n + \frac{1}{2}$, n being the occupation number of electrons, so that the numerical value of the matrix elements of $a^\pm$ are identical with those in $h(1)$. This representation of $osp(1|2)$ reduces indeed under the subalgebra $su(1,1)$ into the two representations belonging to the discrete series $\mathcal{D}_{\frac{1}{4}}^+$ and $\mathcal{D}_{\frac{3}{4}}^+$.

Completion of the structure is straightforwardly obtained by recalling that in the multi-mode case, where an additional label related to the wave-vector quantum number **k** should be introduced, all the Jordan products for different labels should be zero, *i.e.* even-even and even-odd operators should **commute**, whereas single electron creation and annihilation (odd-odd) operators should **anticommute**, as customary.

A few comments

In the picture proposed the cooperon acquires the nature of true particle as far as statistics is concerned. At this level, however, the theory does not distinguish otherwise between a cooperon and a pair of non-interacting electrons. Moreover, the "reduced" hamiltonian $\mathcal{H}_r$ below will include – as in BCS theory – only cooperons of total momentum zero.

The scheme could be properly extended by resorting to a (possibly infinite-dimensional) $\mathbb{Z}_2$-graded Kac-Moody algebra endowed with cooperon operators of momentum different from zero. As we are interested here only in the equilibrium thermodynamics and not in the transport properties we disregard for the moment this latter possibility.

Application to Superconductivity

The main idea of the BCS theory of superconductivity (see *e.g.* ref. [6], whose conceptual scheme we shall closely follow, though within a different algebraic structure, in the sequel) stays in the feature that after the role of phonons has been properly taken into account, and the Cooper hypothesis on the existence of electron bound states is implemented, the system is described by the *reduced* Hamiltonian of interacting electrons:

$$H_r = \sum_{\mathbf{k},\sigma}(\varepsilon_\mathbf{k} - \mu)\, n_{\mathbf{k},\sigma} + \sum_{\mathbf{p},\mathbf{q}} W_{\mathbf{p},\mathbf{q}}\, a^+_{\mathbf{p},\uparrow} a^+_{-\mathbf{p},\downarrow} a^-_{\mathbf{q},\uparrow} a^-_{-\mathbf{q},\downarrow}\,. \tag{8}$$

Here $a^\pm_{\mathbf{k},\sigma}$, $n_{\mathbf{k},\sigma}$ are creation, annihilation and number operators for electrons of quantum numbers $(\mathbf{k},\uparrow)$, and $(-\mathbf{k},\downarrow)$ and the sum over $\mathbf{p},\mathbf{q}$ is actually over $\varepsilon_\mathbf{p}, \varepsilon_\mathbf{q}$ restricted to lie in half the energy shell of thickness $\sim \hbar\bar{\omega} \approx k_B T$ about the Fermi surface. μ is the chemical potential, $\hbar\bar{\omega} \propto k_B T$.

In the BCS mean-field approach, H_r is transformed into $H_{m.f.} = \sum_\mathbf{k} H_\mathbf{k}$, with

$$H_\mathbf{k} = \epsilon_\mathbf{k}(n_{\mathbf{k},\uparrow} + n_{-\mathbf{k},\downarrow}) + \Delta_\mathbf{k}\left(a^+_{\mathbf{k},\uparrow} a^+_{-\mathbf{k},\downarrow} + \text{h.c.}\right) - \Delta_\mathbf{k}^2\,, \tag{9}$$

where $\epsilon_\mathbf{k} \equiv \varepsilon_\mathbf{k} - \mu$, and $\Delta_\mathbf{k} \doteq \langle a^+_{\mathbf{k},\uparrow} a^+_{-\mathbf{k},\downarrow}\rangle \sum_{\mathbf{k}'} W_{\mathbf{k},\mathbf{k}'}$. Here $\langle \mathcal{O}\rangle = \text{Tr}\{\mathcal{O}\,e^{-\beta H}\}/\text{Tr}\{e^{-\beta H}\}$ denotes the thermodynamic expectation value of operator $\mathcal{O}$.

Each mean-field hamiltonian $H_\mathbf{k}$ belongs to the (compact) dynamical algebra $su(2)_\mathbf{k}$ generated by $\{n_{\mathbf{k},\uparrow} + n_{-\mathbf{k},\downarrow}, a^+_{\mathbf{k},\uparrow} a^+_{-\mathbf{k},\downarrow}, a^-_{\mathbf{k},\uparrow} a^-_{-\mathbf{k},\downarrow}\}$.

As indicated above, we assume, as it is done in the BCS theory, that Cooper pairs are made of electrons with opposite momenta and spins: wave-vector $\mathbf{k}$ is attached to $\uparrow$ and $-\mathbf{k}$ to $\downarrow$, and we denote respectively by k and $-k$ the corresponding multi-index. Then $c_k^\pm = \{a^\pm_{\mathbf{k},\uparrow}, a^\pm_{-\mathbf{k},\downarrow}\}$.

$osp(1|2)_k$ is generated by $\{c_k^\pm, a_k^\pm = a^\pm_{\mathbf{k},\uparrow} + a^\pm_{-\mathbf{k},\downarrow}, h_k = h_{\mathbf{k},\uparrow} + h_{-\mathbf{k},\downarrow}\}$. $k \equiv (\mathbf{k},\uparrow)$ is now to be interpreted simply as a label characterizing the different copies of $osp(1|2)$, in the representation with $h_k \equiv n_k + \frac{1}{2}$.

Hamiltonian H_r is to be consistently replaced by:

$$\mathcal{H}_r = \sum_k \epsilon_k n_k + \sum_{p,q} W_{p,q}\, c_p^+ c_q^-\,. \tag{10}$$

The total number of electrons is $N_e \equiv \sum_k n_k$, where n_k counts the electrons, possibly bound into cooperons, in the state labelled by k.

Cooper pairs are true bosons, as they can occupy each state in a number limited only by $\frac{1}{2}N_e$ (∞ in the thermodynamic limit). Manifestly $\mathcal{H}_r \in su(1,1)$, allowing us to deal with multi-cooperon states without need of resorting to any mean-field reduction.

We study first ground state $|\psi_{g.s.}\rangle$,

$$|\psi_{g.s.}\rangle = \prod_k |\psi_{g.s.}\rangle_k\,. \tag{11}$$

$|\psi_{g.s.}\rangle_k$ is a superposition of all possible states coupled in pairs (the total number of electron assumed to be even) of total momentum zero.

Contrary to BCS theory, where only two states (with zero or one cooperon) can be mixed, we can realize our mixing of many $[\leq \frac{1}{2}N_e]$ states resorting to the coherent state representation induced by the unitary operator

$$\mathcal{U} \equiv \prod_k \mathcal{U}_k \quad , \quad \mathcal{U}_k = e^{-i\varphi_k \left(c_k^+ + c_k^-\right)} , \qquad (12)$$

and $|\psi_{g.s.}\rangle$ can be obtained variationally by introducing first the trial state vectors

$$|\psi'_{g.s.}\rangle_k = \mathcal{U}_k(\varphi_k)|0\rangle_k , \qquad (13)$$

and finding then the set of parameters $\{\varphi_k\}$ which minimizes the energy functional $\langle \psi'_{g.s.}|\mathcal{H}_r|\psi'_{g.s.}\rangle$.
Together with this, the additional condition $\langle \psi'_{g.s.}|N_e|\psi'_{g.s.}\rangle = \nu N$ must be implemented, in order to obtain the value of μ necessary to impose the required filling ν (N being the number of lattice sites).
Solution to the resulting system of $N+1$ equations, N, one, variational, for each k and an extra one to implement the filling requirement, provides us the values of the set $\{\varphi_k\}$ and of μ.
Introducing the gap $\Delta_k \equiv -\sum_p W_{k,p} \sinh(4\varphi_p)$, the variational equations lead to $\tanh(4\varphi_k) = \Delta_k/\epsilon_k$, and hence to the gap equation

$$\Delta_k = -\sum_p W_{k,p} \frac{\Delta_p}{\mathcal{E}_p} \quad , \quad \mathcal{E}_p \equiv \sqrt{\epsilon_p^2 - \Delta_p^2} . \qquad (14)$$

Notice the minus sign within the square root.
Implemented upon keeping into account filling constraint

$$\sum_k \sinh^2(2\varphi_k) = \nu N , \qquad (15)$$

the gap equations provide a self-consistent value of Δ_k and μ.
The ground state $|\psi_{g.s.}\rangle$ is constructed by inserting the solutions thus obtained $\{\varphi_k\}$ into the trial state:

$$|\psi_{g.s.}\rangle_k = \frac{1}{\sqrt{\cosh(2\varphi_k)}} \sum_n (-i\alpha_k)^n \sqrt{\binom{2n}{n}} |2n\rangle_k , \qquad (16)$$

where $\alpha_k = \frac{1}{2}\tanh(2\varphi_k)$.
Excited states over this ground state can be constructed, resorting the same transformation $\mathcal{U}$, as

$$|\psi_k(n)\rangle = \mathcal{U}|n\rangle_k . \qquad (17)$$

The spectrum of $\mathcal{H}_r$ can be more easily written rotating operators

$$\tilde{c}_k^\pm \doteq \mathcal{U}^{-1} c_k^\pm \mathcal{U} \quad , \quad \tilde{a}_k^\pm \doteq \mathcal{U}^{-1} a_k^\pm \mathcal{U} \quad , \quad \tilde{h}_k \doteq \mathcal{U}^{-1} h_k \mathcal{U} \, , \tag{18}$$

instead of states $\{|n\rangle_k\}$: indeed, for any operator $\mathcal{O}$, $_k\langle \psi_k(m)|\mathcal{O}|\psi_k(n)\rangle_k = {}_k\langle m|\tilde{\mathcal{O}}|n\rangle_k$.

Energy eigenstates are the Fock states $|\{n_k\}\rangle = \prod_k |n_k\rangle$ characterized by the collection of electron numbers $\{n_k\}$; energy eigenvalues are $E_{\{n_k\}} \equiv \langle\{n_k\}|\tilde{\mathcal{H}}_r|\{n_k\}\rangle$. While the ground state energy is

$$E_{g.s.} \equiv E_{\{0\}} = \sum_k \left(\epsilon_k \sinh^2(2\varphi_k) - \frac{1}{4} \Delta_k \sinh(4\varphi_k) \right) , \tag{19}$$

the energy of the generic excited state is

$$E_{\{n_k\}} = E_{g.s.} + \sum_i \mathcal{E}_{k_i} n_{k_i} + 2 \sum_{i,j} W_{k_i,k_j} \frac{\Delta_{k_i} \Delta_{k_j}}{\mathcal{E}_{k_i} \mathcal{E}_{k_j}} n_{k_i} n_{k_j} , \tag{20}$$

analogous to that of BCS, but holding for any set of occupation numbers $\{n_k\}$. It is worth noticing that the form of $E_{\{n_k\}}$ implies the existence of a gap, whose amplitude is of the order of $-\sum_{\mathbf{k}'} W_{\mathbf{k},\mathbf{k}'}$ (one should keep in mind that the $W_{\mathbf{k},\mathbf{k}'}$'s are negative), namely of Δ_k.

It is interesting, as well as surprising, that the *osp*(1|2) scheme gives all results of BCS theory. In order to check this, we have to move to finite temperature. We do so by introducing first probability $P_k(n_k)$ that the state labelled by k be $|\psi_k(n_k)\rangle$. The probability of state $|\psi_{\{n_k\}}\rangle$ is then given by $\prod_k P_k(n_k)$, with $\sum_{n_k} P_k(n_k) = 1$.

The $P_k(n_k)$'s in turn must be expressed in terms of the (*temperature-dependent*) quantities $\{f_k\}$, each f_k representing the probability that the single state k is occupied. As $n_k \leq N_e$, $P_k(n_k)$ can be easily recovered in the thermodynamic limit $N_e \to \infty$: indeed, the probability for state $|\psi_{\{n_k\}}\rangle$ is proportional to $f^{n_k}(1-f)^{N_e-n_k}$, and one finds in the limit (provided $f_k < \frac{1}{2}$)

$$P_k(n_k) = \frac{1-2f_k}{1-f_k} \left(\frac{f_k}{1-f_k} \right)^{n_k} . \tag{21}$$

The internal energy is then given by

$$U = \sum_{\{n_k\}} \prod_k P_k(n_k) \, E_{\{n_k\}} \tag{22}$$

$$= \frac{1}{2} \sum_k \frac{\epsilon_k}{1-2f_k} \left[\cosh(4\varphi_k) - 1 + 2f_k \right] \tag{23}$$

$$+ \sum_{p,q} W_{p,q} \frac{\sinh(4\varphi_p)}{1-2f_p} \frac{\sinh(4\varphi_q)}{1-2f_q} , \tag{24}$$

while the **entropy** is

$$S = -\kappa \sum_k \sum_{n_k} P_k(n_k) \ln P_k(n_k) \qquad (25)$$

$$= -\kappa \sum_k \left[\frac{f_k}{1-2f_k} \ln \frac{f_k}{1-f_k} + \ln \frac{1-2f_k}{1-f_k} \right] . \qquad (26)$$

The new set of variational parameters $\{\varphi_k\}$ and the set of single-state occupation probabilities $\{f_k\}$, all temperature dependent, can now be obtained minimizing the **free energy** $F = U - TS$.
With $\Delta_k(T) \equiv -\sum_p W_{p,k} \sinh(4\varphi_p)/(1-2f_p)$, one finds the Fermi-like expression for f_k

$$f_k = \frac{1}{\exp{(\mathcal{E}_k/\kappa T)} + 1} , \qquad (27)$$

where now $\mathcal{E}_k = \mathcal{E}_k(T) \equiv \sqrt{\epsilon_k^2 - \Delta_k(T)^2}$, while for the $\{\varphi_k\}$'s the formula $\tanh{(4\varphi_k)} = \Delta_k(T)/\epsilon_k$ holds (analogous to that derived for $T = 0$).
The consistency equation for $\Delta_k(T)$ is now itself temperature dependent:

$$\Delta_k = -\sum_p W_{k,p} \frac{\Delta_p}{\mathcal{E}_p} \coth \frac{\mathcal{E}_p}{2\kappa T} , \qquad (28)$$

(notice the term coth), while the filling equation reads

$$\sum_k \left[\cosh{(4\varphi_k)} \frac{f_k}{1-2f_k} + \sinh^2{(2\varphi_k)} \right] = \nu N . \qquad (29)$$

At the formal level, one immedialtely notices two main differences with respect to the **BCS** solution:

- the *minus* sign within the square root in the definition of $\mathcal{E}_k$, instead of the *plus* of the customary BCS solution [6];
- the *hyperbolic cotangent* – due to the non-compactness of $su(1,1)$ (vs. the the *circular tangent* of the (compact) $su(2)$ case).

These two features together, however, lead to the same results as BCS theory. This is essentially due to the fact that hamiltonian $\mathcal{H}_r$ describes now a set of true bosons, and the chemical potential μ correctly turns out, when the filling condition is implemented, to be *negative*, so that Δ_k can consistently remain $\leq (\varepsilon_\mathbf{k} - \mu)$.

Conclusions

Numerical results show that, in the customary assumption that $\Delta_k(T)$ is independent on k, $\Delta = \Delta(T)$ has the correct BCS behaviour:

- monotonically decreasing in T, with the characteristic features of an order parameter:

  i) $\Delta \to \Delta(0)$ exponentially for $T \to 0$;

  ii) $\Delta \to 0$ for $T \to T_c-$ with critical exponent $\frac{1}{2}$ (notice that here no mean-field assumption was made: the "mean-field" value found for the critical exponent is to be ascribed to the "classical" features of the coherent states adopted for the variational scheme).

- $\Delta(T)$ identically zero for $T > T_c$.

$\Delta(0)$ (the spectrum gap, under the present assumptions) and T_c are solutions to the **gap equation** and filling equation, rewritten in integral form under the current hypotheses $W_{k,p} \sim$ constant $= -\mathcal{W}$ ($\mathcal{W} > 0$) in a shell of thickness $2\hbar\bar{\omega} \ll \varepsilon_F$ around the Fermi surface and zero elsewhere ($g(\varepsilon)$ denotes the density of states) for $T = 0$ and $T = T_c$:

$$2\hbar\bar{\omega}\,\mathcal{W}\,g(\varepsilon_F) = \sqrt{(\varepsilon_F - \mu_0)^2 - \Delta(0)^2}\,, \tag{30}$$

$$2\hbar\bar{\omega}\,\mathcal{W}\,g(\varepsilon_F) = (\varepsilon_F - \mu_c)\tanh\left(\frac{\varepsilon_F - \mu_c}{2\kappa T_c}\right)\,, \tag{31}$$

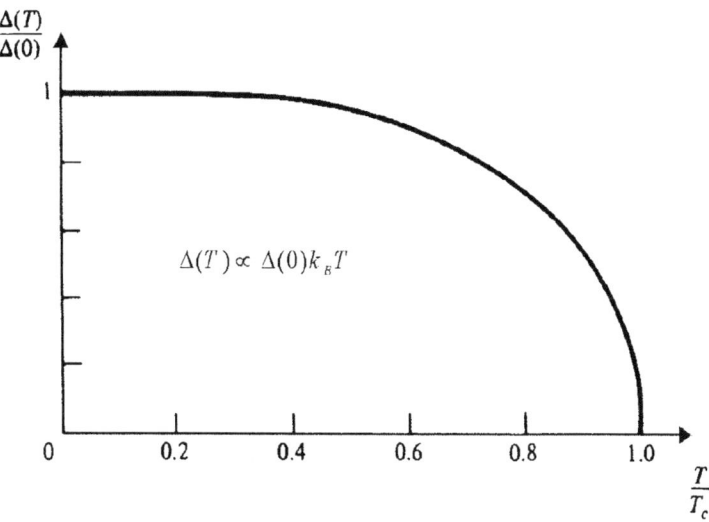

FIGURE 1. Order parameter Δ vs. T in dimensionless units.

$$\int_0^{\varepsilon_F+\hbar\bar\omega} \left(\frac{\varepsilon-\mu_0}{\sqrt{(\varepsilon-\mu_0)^2-\Delta(0)^2}}-1\right) g(\varepsilon)d\varepsilon = 2\nu, \tag{32}$$

$$\int_0^{\varepsilon_F+\hbar\bar\omega} \frac{1}{\exp\left(\frac{\varepsilon_F-\mu_c}{\kappa T_c}\right)-1} g(\varepsilon)d\varepsilon = \nu, \tag{33}$$

from which also the chemical potential at $T=0$ and $T=T_c$ (μ_0 and μ_c) can be derived.

Solutions of the equation for the *order parameter* $\Delta(T)$ and the chemical potential $\mu(T)$ allow for the explicit calculation of the internal energy as a function of temperature, whence the specific heat $C_V \doteq \frac{\partial U}{\partial T}$ can be derived. Figures 1. and 2. show Δ and C_V vs. T. Notice the behaviour of the order parameter, typical of a second order phase transition, and of the specific heath (linear, as for *normal* electrons for $T > T_c$, and of the form $\propto T^{\frac{3}{2}}$ for $T < T_c$, characteristic of Bose condensate).

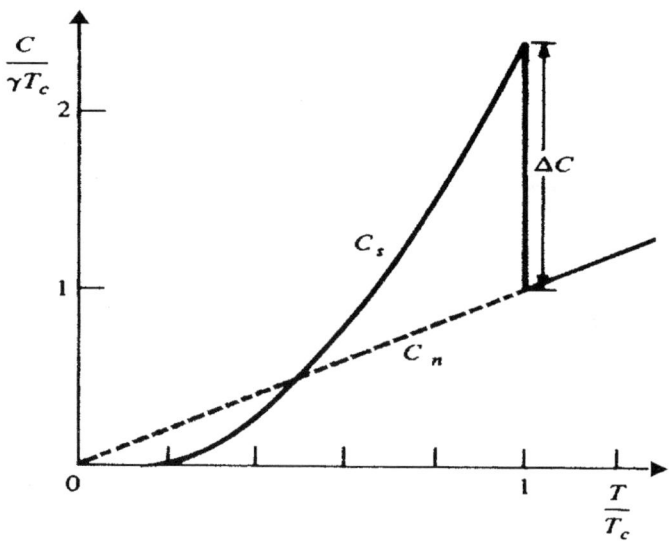

FIGURE 2. Specific heat C vs. T in dimensionless units. γ is the normal electrons (n) specific heat slope. s denotes the superconductive regime.

REFERENCES

1. Celeghini E., Rasetti M., and Vitiello G., *J. Phys. A: Math. Gen.* **28**, L239-L244 (1995).
2. Celeghini E., and Rasetti M., *Int. J. Mod. Phys.* **B10**, 1625-1636 (1996).
3. Celeghini E., and Rasetti M., *Phys. Rev. Lett.* **80**, 3424-3427 (1998).
4. Randeria M., *Crossover from BCS Theory to Bose-Einstein Condensation*, in *Bose Einstein Condensation*, Griffin A., Snoke D., and Stringari S., eds.; Cambridge University Press, Cambridge, 1994.
5. Abe E., *Hopf Algebras*, Cambridge Tracts in Math. 74, Cambridge University Press, Cambridge, 1980.
6. Feynman R.P., *Lectures in Statistical Mechanics*, A. Benjamin, Reading, MA, 1972.

Rotational Invariance, the Spin-Statistics Connection and the TCP Theorem

E.C.G. Sudarshan

University of Texas, Austin, TX, USA

Abstract. Quantum Field Theory formulated in terms of hermitian fields automatically leads to a spin-statistics connection when invariance under rotations is required. In three (or more) dimensions of space this implies Bose statistics for integer spin fields and Fermi statistics for half-integer spin fields. One should recall that spin-½ fields in three dimensions have two nonhermitian or four hermitian components. This automatic doubling of the number of components enables one to define a pseudoscalar matrix, and this in turn allows one to prove the TCP theorem for rotationally invariant field theories. In two space dimensions one obtains anyon statistics independent of the "spin". For the quantum mechanics of identical particles we obtain only the possibility of either statistics for either spin as long as the spatial dimension is three (or higher). For two space dimensions we get anyon statistics. This difference is due to the contractibility of closed loops in three or more dimensions. The relation to the arguments of Broyles, of Bacry and of Berry and Robbins is discussed.

1. INTRODUCTION

In the description of dynamics of identical particles in either quantum theory or classical theory the Hamiltonian is symmetric under any permutation of the particles; therefore, if the initial conditions have any definite symmetry, it would be preserved by the dynamics. It therefore follows that the problem of symmetries of the dynamics is purely a kinematics problem. Since in classical physics the state is usually described by a phase space density, which evolves linearly in time, the only choice, is to use symmetric kinematics. This may also be seen as the vestige of Bose statistics for integer spin particles. If we transcribe quantum dynamics in terms of Moyal brackets in phase space and keep only the lowest order terms in Planck's constant we will have a classical Poisson bracket formulation for classical fields which have integer spin. If we use more sophisticated classical dynamics and kinematics in terms of spinor variables, like in the Proca description of ideal fluid dynamics, we could legitimately raise the possibility of anticommuting spinor 'urfields'. Ideal fluid dynamics per se can be given a canonical formulation in terms of Clebsch potentials. In classical particle mechanics we always deal with integer spin particles and phase space densities which must, of necessity, be totally symmetric. If there are substructures, which yield the particles as bilinear composites, we could entertain more general possibilities. In passing we recognize that the question of the kinematic symmetries for field theories is distinct from the corresponding question for particles.

It is natural to ask if there is a continuous path connecting a configuration of particles with another configuration of particles with some of the particles permuted, would this not settle the problem of statistics? Several attempts have been made to date. The simplest is by H. Bacry [1] who took a configuration of two spin-½ particles polarized, perpendicular to their separation, and in opposite directions. He then showed that if we carry out a rotation through π of the system of two particles about the center of their separation, we get a configuration identical with the original configuration. So if we conclude that the wave function returns to its value at the end of this rotation, then we can deduce Fermi statistics for electrons. But there is a hidden assumption. If we chose commuting Bose operators on the vacuum to create the two electrons, such a path will change the wave function by a factor of -1. In the space of configurations the beginning and ending configurations are identical; so we have a closed loop. But not all closed loops automatically preserve the wave function. It will be preserved if the "loop" can be continually contracted to the identity. But the loop considered by Bacry is not so contractible. In general, if we have a closed loop in the space of configurations there is no guarantee that the wave function is preserved unless the configuration space is simply connected [2].

This question has been studied in its full generality by T.D. Imbo and collaborators [3]. If we consider a base manifold M, which may be R^3 or it could itself be multiply connected, and we have N identical particles, the configurations in which two particles are exchanged are not different. So the possible manifold is

$$\frac{M \otimes M \otimes \ldots \otimes M}{S_N} = \frac{M^N}{S_N}, \tag{1}$$

where S_N is the permutation group on N variables. But this is not a differentiable manifold unless we remove Δ_N, the set of points where two or more coordinates coincide. We do so. The appropriate manifold is then

$$S = \frac{M^N - \Delta_N}{S_N}. \tag{2}$$

This is a differentiable manifold but it is multiply connected. The connectivity is summarized in its loop group, the first homology group $\pi_1(S)$, which may be nontrivial. If M is simple connected, we have

$$\pi_1(S) = \pi_1\left(\frac{M^N - \Delta_N}{S_N}\right) = S_N, \tag{3}$$

provided we work in three or more dimensions. The generic wave functions are multi-valued, formalizing a representation of S_N. Apart from the completely symmetric and completely antisymmetric representations, there are no further one-dimensional representations. Since we wish to have one-dimensional representations, only the Bose and Fermi statistics obtain. Of course, all the parastatistics are also allowed if we relax the requirement of only one-dimensional representations. For dimension one, the space S is not even connected. For dimension two, the homotopy group is the braid group

$$B_N(M) = \pi_1(S) = \pi_1\left(\frac{M^N - \Delta_N}{S_N}\right) \qquad (4)$$

There is a continuous family of one-dimensional representations of the braid group labeled by a phase $0 \leq \theta \leq 2\pi$. These particles are called anyons.

Note that there is no correlation between the representations of the connectivity group $\pi_1(S)$ and the spin of the particles.

The argument of Bacry is equivalent to disregarding all except the trivial identity representation of the homotopy group. The existence of a continuous path following a loop gives no guarantee of the representation being trivial. If one asserts that the wavefunction remains unchanged, it is an additional explicit *assumption* [1,2].

2. REQUIREMENTS FOR A QUANTUM FIELD THEORY

In the quantum mechanics of particles, we have a description in terms of wave functions (or state vectors), which enter bilinearly in the computation of expectation values. Therefore a completely antisymmetric as well as completely symmetric wave function gives us completely symmetric expectation values. Even if the permutation of identical particles gives a phase change, not necessarily ±1, it will cancel out in the rule for taking expectation values. Since wave functions allow for superposition of any two states to form a new state, all the wavefunctions of a collection of identical particles must have the same behavior. We may then deal with the symmetry properties of a collection of identical particles without special reference to any state or collection of states.

For particles that can be separated from each other, we need to have some restrictions on the symmetry due to the requirement of "cluster decomposition" [4]. If a set of $M+N$ particles has a wave function with a definite symmetry property, it should be such that when the cluster of M particles is arbitrarily far from the cluster of N particles each cluster has the "same" symmetry property. To require this is to assume that the clusters can be separated arbitrarily far: this property does not obtain for "particles" such as quarks which are "confined".

In quantum theory the permutations of the particles form a group of operations which will have a set of possible representations. In this manner, finding the appropriate "statistics" for a particle species can be reduced to finding the representations of the groups that interchange configurations [3].

Within the framework of relativistic quantum field theory, W.Pauli [5] demonstrated a spin-statistics relation. He used Lagrangian field theory with scalar or spinor wavefields. This well known and celebrated result asserts that scalar (or pseudoscalar) fields cannot be consistently quantized in accordance with Fermi statistics; similarly, spinor fields cannot be consistently quantized in accordance with Bose statistics. Thus the theorem is of wide applicability provided we deal only with Bose or Fermi quantizations. Since there are other possibilities, we need to reexamine this question. Pauli's method of proof suffers from the defect that it applies directly for free fields only. When interactions are included, it is a hoped that these same conclusions obtain. Since the statistics is a kinematics question (since dynamics preserves the symmetry of the initial conditions), a proper formulation can eliminate this objection.

But the main reasons why we should search for a formulation not invoking relativistic invariance are two-fold. For one, all the places where the spin-statistics relation is most important like atomic and molecular structure, blackbody radiation, and condensed matter physics, the velocities of the electrons are sufficiently small so that relativistic effects can be neglected. For the second, we have quantized sound waves with phonons, the quantum electrodynamics of wave-guides, and the excitation spectrum of an ideal fluid. So it would be desirable to find a proof of the spin-statistics relation without directly invoking relativistic invariance or particle statistics with renormalized fields. It is nevertheless taken for granted that the interactions do not affect the statistics. However, in 2+1 dimensions the special theory with a Chern-Simons kinematic term [5] in interaction with a charge-current transmutes its statistics, with the low frequency components obeying one statistics and the high frequency components obeying another statistics [6]. The 2+1-dimensional theory is special in that in addition to Bose and Fermi statistics, a one-parameter family of anyon statistics may obtain. But it does caution us in asserting that the free-field commutation relations imply the interacting-particle statistics.

3. FIELD STATISTICS WITHOUT PARTICLES

A direct study of the field commutation relations without specific reference to the interaction is provided by axiomatic quantum field theory. Such a formulation was provided by A.S. Wightman. The Wightman scheme defines the quantized fields as operator-valued distributions acting on a Hilbert space. The fields admit a family of automorphisms corresponding to different Poincaré frames, and these are unitarily implemented. The energy is always nonnegative and there is a unique vacuum state [8], which is left invariant by the family of Poincaré automorphisms. No specific form of interactions is postulated, or required. Within this scheme, N. Burgoyne [9] has given a proof of the connection between the spin of the field (tensor or spinor) and its commutation properties. The proof goes as follows. Consider the vacuum expectation values of the product of two field operators at two space-time points that are separated by a space like interval. From Poincaré invariance we deduce

$$\langle 0|\phi(x)\phi(y)|0\rangle = F(x-y) \tag{5}$$

where F is a Lorentz-invariant function of the separation. They must then be functions of the Lorentz-invariant quantities that can be constructed from the interval. For time-like intervals both $(x-y)^2 = (\vec{x}-\vec{y})^2 - c^2(x^o-y^o)^2$ and $\theta(x^o-y^o)^2$ are invariants, the first symmetric and the second antisymmetric in x and y. But for space-like intervals, only the symmetric invariant $(x-y)^2$ obtains. It follows that

$$\langle 0|\phi(x)\phi(y)|0\rangle = \langle 0|\phi(y)\phi(x)|0\rangle, \quad x-y \text{ spacelike}, \tag{6}$$

and hence the commutator has zero vacuum expectation value. If we now demand that the field is quantized with anticommutators,

$$\phi(x)\phi(y) + \phi(y)\phi(x) = 0, \quad c^2(x^o-y^o)^2 < (\vec{x}-\vec{y})^2, \tag{7}$$

it will follow that

$$\langle 0|\phi(x)\phi(y)|0\rangle = 0. \tag{8}$$

This implies that

$$\phi(x)|0\rangle = 0 \tag{9}$$

and that the field is trivially zero. For spinor fields a similar analysis leads to the conclusion that

$$\langle 0|\psi_r(x)\psi_s(y)|0\rangle = -\langle 0|\psi_s(x)\psi_r(y)|0\rangle \tag{10}$$

for $x-y$ space-like. Consequently, imposition of the requirement that the commutators of two fields with space-like separation vanish gives

$$\langle 0|\psi_r(x)\psi_s(y)|0\rangle = 0 \quad \psi_r(y)|0\rangle = 0, \tag{11}$$

so the field is again trivial. We note that the Burgoyne proof is also a negative result: integer spin fields cannot obey Fermi statistics and half-integer spin fields cannot obey Bose statistics. But the field has no immediate particle interpretation; if there were, then the negative proof may be applied to the corresponding particles.

Similar results are also obtained within the framework of the C* algebra formulation [10] of D.Kastler and R.Haag. Again there is no immediate particle interpretation, and the theorems are negative results.

We can take the lessons of both Pauli's proof and Burgoyne's and ask if we can deduce these negative results from weaker assumptions, especially releasing relativistic invariance. We can further ask if we can deduce a positive result that the tensor fields <u>must</u> satisfy Bose statistics and that the spinor fields <u>must</u> satisfy Fermi statistics. This can be accomplished in a suitably restricted Lagrangian field theory invariant under the Newtonian group N(3,1) comprising space and time translations and space rotations. There must be some restrictions, since we could consider non-relativistic quantum field theories with either statistics for both tensor and spinor fields.

Before giving the requirements on the N(3,1) invariant fields, we need to answer a question raised by internal symmetries. What if we do have internal symmetry multiplets of tensor (spinor) fields which overcome the limitations imposed by

Burgoyne, by making the vacuum expectation values antisymmetric in the internal symmetry variables? Does this reverse the conclusions? A careful analysis eliminates this possibility. Any internal symmetry dependence can be diagonalized by suitable global linear combinations. An antisymmetric matrix would have pure imaginary eigenvalues of either sign. The action density is then pure imaginary and of either sign. Such a field theory is not acceptable. So we cannot evade the Burgoyne constraints by invoking antisymmetry in internal symmetry labels.

4. EXOTIC STATISTICS

It is interesting and useful to ask what can be asserted about particle statistics without relativity and without fields [9]. In this case we specify the configuration space of N particles as

$$S = \frac{R^3 \times R^3 \times ... \times R^3 - \Delta}{S_N}, \quad (12)$$

where Δ is the set of points in which two or more sets of coordinates coincide and S_N is the symmetric group in N variables permuting the different R^3 spaces. This space (with Δ removed) is a manifold on which the wavefunctions can be described.

We can define a set of wavefunctions that transform among themselves when we consider a non-contractible loop in S. It is possible for these wave functions to furnish a representation of S_N. But except for the completely symmetric and completely antisymmetric representations, all the others are nontrivial "parastatistics" [10].

This set of possible statistics emerges from the work of E.P. Wigner and of H.S. Green [11]. They observed that the equations of motion of an assembly of particles realized by the Fock states of a collection of creation and annihilation operators a_j, $a_j^\dagger$ are those of harmonic oscillators

$$i\dot{a}_j = [a_j, H], \quad (13)$$

$$H = \tfrac{1}{2}\sum \omega_k (a_k^\dagger a_k \pm a_k a_k^\dagger), \quad (14)$$

so that

$$i\dot{a}_j = \omega_j a_j \qquad i\dot{a}_j^\dagger = -\omega_j a_j^\dagger. \quad (15)$$

But these do not require the canonical relation

$$a_j a_k^\dagger \pm a_k^\dagger a_j = \delta_{jk}. \quad (16)$$

It suffices to have

$$\tfrac{1}{2}\left[\sum \omega_k (a_k^\dagger a_k \pm a_k a_k^\dagger), a_l\right] = i\dot{a}_l. \quad (17)$$

Since this must be true independent of the way the orthonormal modes j, k,... are chosen, it follows that we need the generalized commutation relation

$$[a_j a_k^\dagger \pm a_k^\dagger a_j, a_l] = 2\delta_{kl} a_j$$
$$[a_j a_k^\dagger \pm a_k^\dagger a_j, a_l^\dagger] = 2\delta_{kl} a_j^\dagger \quad . \tag{18}$$

We may supplement these with

$$[a_j a_k \pm a_k a_j, a_l] = 0$$
$$[a_j a_k \pm a_k a_j, a_l^\dagger] = 2(\delta_{kl} a_j \pm \delta_{jl} a_k) \tag{19}$$

and its adjoint. These were the generalized statistics that H.S. Green studied. Wigner had raised the question for one degree of freedom for the harmonic oscillator. H.S. Green gave a generic construction of operators which obey these commutation relations. We shall call this the "Green ansatz" and discuss it below.

But before discussing the Green ansatz let us study the generalized Green commutation relations. First let us take para-Fermi statistics with the commutation relations

$$[a_j^\dagger a_k - a_k a_j^\dagger, a_l] = -2\delta_{jl} a_k \tag{20}$$
$$[a_j^\dagger a_k - a_k a_j^\dagger, a_l^\dagger] = 2\delta_{kl} a_k^\dagger \quad . \tag{21}$$

We can use them in this form to find all representations of these commutation relations. But it will be more transparent to label the various degrees of freedom from 1 to N and introduce

$$J_{1,2N+1} = \tfrac{1}{2}[a_1 + a_1^\dagger] \quad J_{2,2N+1} = i\tfrac{1}{2}[a_1 - a_1^\dagger]$$
$$\cdots \tag{22}$$
$$J_{2N+1,2N+1} = \tfrac{1}{2}[a_N + a_N^\dagger] \quad J_{2N+2,2N+1} = i\tfrac{1}{2}[a_N - a_N^\dagger]$$

$$J_{1,3} = \tfrac{1}{4}[a_1 + a_1^\dagger, a_2 + a_2^\dagger] \quad J_{1,2} = i\tfrac{1}{4}[a_1 + a_1^\dagger, a_2 - a_2^\dagger]$$
$$\cdots \tag{23}$$
$$J_{2N-1,2N} = i\tfrac{1}{4}[a_N + a_N^\dagger, a_N - a_N^\dagger] \quad J_{2N-2,2N} = \tfrac{1}{4}[a_{N-1} + a_{N-1}^\dagger, a_N - a_N^\dagger].$$

Then $J_{\mu\nu}$, for $1 \leq \mu, \nu \leq 2N+1$, obey the commutation relation of the orthogonal group $B_N \equiv O(2N+1)$ in odd dimensions [12]. Therefore the representations we need are all the representation of B_N, reducible or irreducible. This is a generalization of a result of H.J. Bhabha [13]. All the representation of B_N are known; the irreducible ones are listed by F.D. Murnaghan. In particular the standard Fermi realization of this algebra is the fundamental irreducible 2^N dimensional spinor.

5. THE GREEN ANSATZ

The Green ansatz for this case is to take R families of mutually commuting Fermi operators

$$\left[b_j^{(r)}, b_k^{\dagger(s)}\right] = 0; \quad \left[b_j^{(r)}, b_k^{(s)}\right] = \left[b_j^{\dagger(r)}, b_k^{\dagger(s)}\right] = 0, \quad j \neq k \quad r \neq s. \quad (24)$$

Then choose

$$a_j = \sum b_j^{(r)} \quad a_j^\dagger = \sum b_j^{\dagger(r)}. \quad (25)$$

This ansatz solves Green's commutation relations. It simply corresponds to the vector additions of the spins of R independent spinors. All irreducible representations of B_N can obtained by the reduction of this R-fold product of spinors of B_N. Except in the trivial case of $R=1$, the Green ansatz is a reducible realization. The labels (r,s) are inert and are the color labels, and the unitary group in R variables is the color symmetry. We may restrict the representation of color symmetry arbitrarily to the symmetric identity representation. Then our physics would be colorless.

For the para-Bose case the identification is somewhat more complicated. Consider the commutation relation of the bilinear forms $a_j a_k + a_k a_j$, $a_j a_k^\dagger + a_k^\dagger a_j$, and $a_j^\dagger a_k^\dagger + a_k^\dagger a_j^\dagger$. They satisfy the commutation relations of $C_N = S_p(2N)$, the symmetric group in N pairs of variables [14]. The operators a_j, $a_j^\dagger$ may be considered as grading the C_N. And every representation of this graded Lie algebra consists of a pair of representations of C_N. These representations are also given be Murnaghan [15].

The Green ansatz for para-Bose commutation relations is to take R mutually anticommuting Bose algebras

$$\left[b_j^{(r)}, b_k^{(r)}\right] = 0; \quad \left[b_j^{(r)}, b_k^{\dagger(r)}\right] = \delta_{jk} \quad (26)$$

$$b_j^{(r)} b_k^{(s)} = -b_k^{(s)} b_j^{(r)}$$
$$b_j^{(r)} b_k^{\dagger(s)} = -b_k^{\dagger(s)} b_j^{(r)} \quad r \neq s, j \neq k \quad (27)$$

and construct

$$a_j = \sum b_j^{(r)} \quad a_j^\dagger = \sum b_j^{\dagger(r)}. \quad (28)$$

Again we get realizations of these para-Bose commutation relations. All these realizations are reducible with the exception of the trivial $R=1$ case. The internal color labels r,s are inert. We may insist only on the singlet representation of the unitary color group.

It turns out, however, that for finite N, the Green ansatz does not give all the realizations of C_N. In addition to the discrete family of representations obtained by the Green ansatz, there is a class of continuous representations which cannot be so obtained. This is most clearly seen for $N=1$ where the Lie algebra is that of $O(2,1)$. The representations of $O(2,1)$ contain the discrete series of representations and a continuous set of representations. However, if $N \to \infty$ and we still have a Fock realization of the oscillators, Messiah and Greenberg have shown that only the Green ansatz obtains.

Let us now raise the question of cluster decomposition. The only irreducible representation of O(2N+1) that remains irreducible when it is broken up into clusters is the fundamental spinor representation apart from the trivial representation. So any generalization of the Fermi anticommutation relations are reducible. The Green ansatz furnishes such a realization. Unfortunately it is not the most general one with such a cluster property. This problem was studied by J.L. Richards and E.C.G. Sudarshan, who demanded that entropy be an extensive quantity [16].

Another problem with parastatistics is that in a field theory the vacuum may not be cyclic. There are states not connected with the vacuum by a finite number of applications of the field. It is not known if the requirement of color singlet states cures this problem.

In concluding this section we restate the result, the irreducible realizations of the Green commutation relations are *not equivalent* to color. Conversely the Green ansatz which realizes color is *not irreducible*.

If the modes corresponding to a_j, $a_j^\dagger$ are for a spinning particle, then rotation and translation invariance in space puts further restrictions. By invariant integration with respect to the indices j, k in the commutator

$$\left[a_j a_k + a_k a_j, a_l^\dagger\right] = 2\left(\delta_{jl} a_k + \delta_{kl} a_j\right), \tag{29}$$

we obtain

$$B_{jk}\left[a_j a_k + a_k a_j, a_l^\dagger\right] = B_{jk} a_k + B_{jl} a_j. \tag{30}$$

If the invariant 'scalar' matrix B_{jk} is symmetric, there is no problem, but for B_{jk} antisymmetric the left side vanishes and the right side does not vanish leading to a contradictions. So para-Bose commutation relations are not consistent with half-integer spin. Similarly, one can show that para–Fermi commutation relations are consistent for tensor fields.

6. SPIN-STATISTICS IN LAGRANGIAN FIELD THEORY

Introduce a Lagrangian density for a multi-component field φ_r linear in the time derivative:

$$L = \dot{\phi}_r A_{rs} \phi_s + m\phi_r B_{rs} \phi_s + L_{int}, \tag{31}$$

where the repeated indices are summed over. Since the Lagrangian density is invariant with respect to rotations, we can conclude that the matrices A_{rs} and B_{rs} have definite symmetry properties [17]. B_{rs} is symmetric for tensor fields and antisymmetric for spinor fields. Due to the presence of the time derivative, A_{rs} has the opposite symmetry; it is antisymmetric for tensor fields and symmetric for spinor fields.

For this Lagrangian theory with the action functional
$$A = \int dt \int d^3x \, L(\vec{x}, t) ,\qquad(32)$$
the Weiss-Schwinger action principle then asserts
$$i\delta\phi_r(\vec{x},t) = [\phi_r(\vec{x},t), \delta A] \qquad(33)$$
$$\delta A = \dot{\phi}_r A_{rs} \delta\phi_s . \qquad(34)$$
Depending on whether the variation of the field $\delta\phi_s$ commutes or anticommutes with the field, we get the commutation relations
$$[\varphi_r(\vec{x},t), \varphi_s(\vec{y},t) A_{sr}] = \delta_{rs} \delta(\vec{x}-\vec{y}) \qquad(35)$$
or
$$[\varphi_r(\vec{x},t), \varphi_r(\vec{y},t)]_\pm = \left(A^{-1}\right)_{rs} \delta(\vec{x}-\vec{y}) . \qquad(36)$$
For commutation relations, A and hence A^{-1} must be antisymmetric, which is true for tensor fields and only for them. For anticommutation relation, A and hence A^{-1} must be symmetric, which is true for spinor fields and only for spinor fields.

The Action Principle and Newtonian group invariance of the action lead to the assertion that Bose quantization using commutators is only for tensor fields while Fermi quantization using anticommutators is only for spinor fields. To the extent that we have a particle interpretation, integer spin particles are bosons and half-integer spin particles are fermions. No use has been made of Lorentz invariance. The conclusion is therefore valid for ideal fluid hydrodynamics, which is Galilean invariant, and for phonons.

The red herring of antisymmetry in the internal symmetry labels providing the wrong spin-statistics connection is inadmissible for reasons given earlier in connection with Poincaré invariant theories.

For the spinor fields, the doubling of the number of components to deal with hermitian fields is seen as necessary also for providing the required number of canonical field components.

But if we try to quantize a scalar field canonically using anticommutators, we get an inconsistent theory. If the anticommutators satisfy
$$\begin{aligned}\{\phi(\vec{x},t), \phi(\vec{y},t)\}_+ &= 0 \\ \{\phi(\vec{x},t), \pi(\vec{y},t)\}_+ &= i\delta(\vec{x}-\vec{y}) \\ \{\pi(\vec{x},t), \pi(\vec{y},t)\}_+ &= 0 \end{aligned} \qquad(37)$$
the theory is inconsistent; the vector space on which $\phi(\vec{x},t)$ operates cannot have a positive-definite metric.

This is not avoided by making $\phi(\vec{x},t)$ nonhermitian or a spinor field. For a nonhermitian scalar field with the anticommutators

$$\{\phi(\vec{x},t),\phi(\vec{y},t)\}_+ = \{\phi(\vec{x},t),\phi^\dagger(\vec{y},t)\}_+ = 0$$
$$\{\pi(\vec{x},t),\pi(\vec{y},t)\}_+ = \{\pi(\vec{x},t),\pi^\dagger(\vec{y},t)\}_+ = 0 \tag{38}$$
$$\{\phi(\vec{x},t),\pi^\dagger(\vec{y},t)\}_+ = i\delta(\vec{x}-\vec{y}) = \{\phi^\dagger(\vec{x},t),\pi(\vec{y},t)\}_+ ,$$

the same inconsistency obtains. So, for a nonhermitian spinor field obeying such anticommutation relations, the theory is also inconsistent.

Pauli's proof, as mentioned above applies to free fields with a vacuum, for which the states of the fields are spanned by suitable symmetrized states of particles. In such a field theory the proof of the commutation relations is applicable immediately to particles. Pauli's proof may thus be thought of as applying to particles or to fields.

However, Lagrangian quantum field theory leads to infinities in direct calculation. While in many of these cases we can circumvent the infinities by renormalization, the renormalized field does not obey finite commutation relations. So for interacting fields we have either field commutation relations with unrenormalized fields or for noninteracting fields. In quantum theory a quantized field is equivalent to assemblies of identical particles. The statistics of the particles (that is the symmetry or antisymmetry or any other behavior) can be coded into the commutation properties of the field. If $\phi(x)$ is a field defined at time t, the Bose statistics of the quanta of the field depend on whether the canonical commutation relations are obeyed by the field and its conjugate. If $\phi(\vec{x},t)$ contains both creation and destruction operators in the form

$$\phi(\vec{x},t) = \sum \frac{1}{\sqrt{2\omega}}\left(a_{\vec{k}}e^{i\vec{k}\cdot\vec{x}-i\omega t} + a_{\vec{k}}^\dagger e^{i\vec{k}\cdot\vec{x}+i\omega t}\right), \tag{39}$$

the canonical conjugate is its time derivative

$$\pi(\vec{x},t) = \dot{\phi}(\vec{x},t) = \sum \frac{1}{\sqrt{2\omega}}\left(-i\omega a_{\vec{k}}e^{i\vec{k}\cdot\vec{x}-i\omega t} + i\omega a_{\vec{k}}^\dagger e^{i\vec{k}\cdot\vec{x}+i\omega t}\right), \tag{40}$$

and they obey the canonical relation

$$[\phi(\vec{x},t),\pi(\vec{x},t)] = \sum_{k,j}[a_{\vec{k}},a_j^\dagger]e^{i\vec{k}\cdot(\vec{x}-\vec{y})} = \sum e^{i\vec{k}\cdot(\vec{x}-\vec{y})} = \delta(\vec{x}-\vec{y}) . \tag{41}$$

But what if we had two fields

$$\psi(\vec{x},t) = \sum a_{\vec{k}}e^{i\vec{k}\cdot\vec{x}-i\omega t}$$
$$\psi(\vec{x},t) = \sum a_{\vec{k}}^\dagger e^{i\vec{k}\cdot\vec{x}+i\omega t} \quad ? \tag{42}$$

Then we may choose either commutators or anticommutators to have the canonical forms

$$\psi(\vec{x},t)\psi^\dagger(\vec{x},t) \pm \psi^\dagger(\vec{x},t)\psi(\vec{x},t) = \delta(\vec{x}-\vec{y}) . \tag{43}$$

For example if $\psi,\psi^\dagger$ are both scalar fields, or $\psi,\psi^\dagger$ are both spinor fields, either commutators or anticommutators can be used to quantize the field.

7. TCP THEOREM IN QUANTUM FIELD THEORY WITHOUT LORENTZ INVARIANCE

Returning to the three-dimensional space, not only can we provide the spin statistics theorem but also the TCP theorem [18]. We outline the arguments for spin-0 and spin-½ cases.

For spin-0, we have a scalar field ϕ and its conjugate π. We may define the transformations of ϕ and π under parity P, charge conjugation C, and weak time reversal T as follows

$$\phi(\vec{x},t) \to_P \eta_P \phi(-\vec{x},t)$$
$$\to_C \eta_C \phi(-\vec{x},t) \quad (44)$$
$$\to_T \eta_P \eta_C \phi(\vec{x},-t)$$
$$\pi(\vec{x},t) \to_P \eta_P \pi(-\vec{x},t)$$
$$\to_C \eta_C \pi(\vec{x},t) \quad (45)$$
$$\to_T \eta_P \eta_C \pi(\vec{x},-t) \quad.$$

Hence under TCP, we have

$$\phi(\vec{x},t) \to \phi(-\vec{x},-t)$$
$$\pi(\vec{x},t) \to \pi(-\vec{x},-t) \quad. \quad (46)$$

Then the kinematic term in the Lagrangian is unaltered. The (nonderivative) interaction terms also retain their form and the action is unchanged.

For spin-½ fields the kinematic terms are such that they yield the real equation of motion

$$\dot{\psi}_r = \left\{ \left(\vec{\alpha} \cdot \vec{\nabla} \right)_{rs} + B_{rs} \right\} \psi_s \quad. \quad (47)$$

This results from the Lagrangian

$$L_{min} = \psi_r \delta_{rs} \dot{\psi}_s + \vec{\nabla} \cdot \psi_r \vec{\alpha}_{rs} \psi_s + \psi_r B_{rs} \psi_s \quad. \quad (48)$$

The discrete transformations are

$$\psi_r(\vec{x},t) \to_P \eta_P (i\beta)_{rs} \psi_s(-\vec{x},t)$$
$$\to_C \eta_C (i\beta\gamma_5)_{rs} \psi_s(\vec{x},t) \quad (49)$$
$$\to_T \eta_C \eta_P (i\beta\gamma_5)_{rs} \psi_s(\vec{x},-t) \quad.$$

The case of two space dimensions needs to be addressed separately. Except for spin-0, all other fields with spin have two hermitian components. This is most clearly seen for two-dimensional vectors $\vec{A}$ and $\vec{B}$. Then both $\vec{A} \cdot \vec{B}$ and $\vec{A} \times \vec{B}$ are invariants, the first being symmetric and the second antisymmetric in $\vec{A}$ and $\vec{B}$. The generic kinematic Lagrangian density contains a suitable linear combination of the two invariants

$$\dot{\phi}_r A_{rs} \phi_s \sim \cos\theta (\dot{\phi}_r \phi_s) + \sin\theta (\phi_r^* \dot{\phi}_s)$$
$$A_{rs} = \cos\theta \, \delta_{rs} + \sin\theta \, i\varepsilon_{rs} \quad. \quad (50)$$

The factor of i is necessary to assure the hermiticity of A. This angle θ is related to the anyon angle θ which defines the commutation property

$$\phi_r(x)\phi_s(y) = e^{i\theta}\phi_s(y)\phi_r(x) \ . \tag{51}$$

This is true for other spins including spin-½. So every possibility in the one-dimensional representations of the (abelianized) braid group $B_N(R^2)$ may be realized in field theory.

Consider a Newtonian quantum field, Poincaré covariant fields as well as Galilean covariant fields. But its scope is wider. We want to incorporate the fundamental Kirchhoff's law that *emissivity* and *absorptivity* are proportional. We implement this fundamental requirement by insisting that both kinematics and dynamics are expressed in terms of fields that contain creation and annihilation operators together as the positive and negative frequency parts of a single field. One easy way to bring this about is to use only hermitian fields in the formalism. A nonhermitian field can be so transcribed using two hermitian fields.

The equations of motion of tensor fields can be chosen real with the rotation group acting on the field components by real matrices. For spinor fields we need to double the components since the irreducible spinor in three dimensions is intrinsically complex. The usual Pauli matrices σ_1, σ_2, σ_3, have only σ_2 pure imaginary and the corresponding rotations purely real. When we double the components we may define an independent set τ_1, τ_2, τ_3 of Pauli matrices and choose:

$$\begin{aligned}\sigma_1 &\to \sigma_1 \otimes \tau_2 \\ \sigma_2 &\to \sigma_2 \otimes 1 \\ \sigma_3 &\to \sigma_3 \otimes \tau_2 \ .\end{aligned} \tag{52}$$

Along with this we can define a set of vector matrices

$$\begin{aligned}\alpha_1 &\to \sigma_3 \otimes \tau_3 \\ \alpha_2 &\to 1 \otimes \tau_1 \\ \alpha_3 &\to \sigma_1 \otimes \tau_2\end{aligned} \tag{53}$$

and a scalar matrix

$$i\beta \to i1 \otimes \tau_2 \ , \tag{54}$$

so that we may consider

$$\psi(\vec{x},t) \to i\gamma_5 \psi(-\vec{x},-t) \ . \tag{55}$$

To sum up,

$$\begin{aligned}\alpha_1 &= \sigma_3 \otimes \tau_3 \\ \alpha_2 &= 1 \otimes \tau_1 \\ \alpha_3 &= \sigma_1 \otimes \tau_2 \\ \beta &= 1 \otimes \tau_2 \\ \gamma_5 &= \sigma_2 \otimes \tau_1\end{aligned} \tag{56}$$

These transformations are real and preserve the kinematic term. Since all dynamical terms must involve the spinor fields bilinearly, the action is invariant under this combined transformation. The action is thus TCP invariant; and hence the dynamics. Again we stress that relativistic invariance has *not* be used in the derivation.

8. SUMMARY

In summary we note that the usual Pauli derivation and the Burgoyne derivation are negative results, and it is not clear to what extent they apply to interacting particles. The same applies to the C* algebraic results. This situation is not satisfactory and needs be improved. The study of the homotopy of the configuration spaces of identical particle showed what possible statistics may apply. But no deduction can be made by showing a continuous loop in the configuration space. The wave function may return to its original assigned values or to any other according to the representation of the homotopy group that applies. If we assume that only the trivial realization of the homotopy group is realized, we can "deduce" the spin-statistics connection. This would be a natural assumption but nevertheless an *assumption*.

To go beyond these arguments and to derive a positive result asserting the connection between spin and statistics, we recognize that the symmetry of the fundamental bilinear forms of two tensors or two spinors are respectively symmetric and antisymmetric. This basic property, together with the use of hermitian fields exclusively, allows us to *deduce* the spin-statistics connection. When the number of space dimensions is reduced to two, we have a generic bilinear form neither symmetric nor antisymmetric and parameterized by an angle [19].

The statement about the symmetry of the fundamental invariant bilinear forms does not hold in space of arbitrarily high dimension. The tensor invariant continues to be a bilinear symmetric form. But for spinors, the form may also be symmetric. In such a case even spinors must be quantized with Bose statistics [20]. Existence of internal symmetries cannot alter the spin-statistics relation.

We have also shown that in a local Newtonian quantum field theory the TCP theorem is *automatically valid*.

REFERENCES

1. Bacry, H., *Am. J. Phys.* **63**, 297–298 (1995).
2. Berry, M.V., and Robbins, J.M, *Proc. Roy. Soc.* **A453**, 1771–1790 (1997).
3. Imbo, T.D., and Sudarshan, E.C.G., *Phys. Rev. Lett.* **60**, 481–483 (1988); Imbo, T.D., Imbo, C.S., and Sudarshan, E.C.G., *Ann. Inst. Henri. Poincaré* **49**, 387–396 (1988).

4. Streater, G.F., and Wightman, A.S., *PCT, Spin, Statistics and All That*, W. A. Benjamin, New York, 1964.
5. Pauli, W., *Progr. Theor. Phys.* **5**, 526–543 (1950).
6. Polyakov, M.A., *Lett. Mod. Phys.* A **3**, 325 (1998).
7. Wightman, A.S., *Phys. Rev.* **101**, 860–865 (1956).
8. Sudarshan, E. C. G., *J. Math. Phys.* **4**, 1029–1036 (1963).
9. Burgoyne, N., *Nuovo Cimento* **8**, 607–609 (1958).
10. Haag, R., and Kastler, D., *J. Math. Phys.* **5**, 848–861 (1964).
11. Wigner, E.P., *Phys. Rev.* **77**, 711 (1950).
12. Ryan, C., and Sudarshan, E.C.G., *Nucl. Phys.* **47**, 207–211 (1963).
13. Bhabha, H.J., *Rev. Mod. Phys.* **21**, 451 (1949).
14. Kamefuchi, S., and Takahashi, Y., *Nucl. Phys.* **36**, 177–206 (1962).
15. Murnaghan, F.D, *Theory of Group Representations*, Johns Hopkins Press, Baltimore, 1938.
16. Richard, J.L., and Sudarshan, E.C.G, *J. Math. Phys.* **14**, 1170-1175 (1973).
17. Duck, I.M., and Sudarshan, E.C.G., *Am. J. Phys.* **66**, 284–303 (1998).
18. Luders, G., *Z. f. Physik* **133**, 325–330 (1954).
19. Wilczek, F, *Int. J. Mod. Phys. A* **3**, 2827–2853 (1998).
20. Boya, L.J., and Sudarshan, E.C.G., unpublished.

Speculations on Spin and Statistics

A. Zee

Institute for Theoretical Physics
University of California
Santa Barbara, CA 93106 USA

Abstract. In this paper I first survey some of the peculiar features of quantum statistics in quantum field theory. I then discuss the possibility of building known particles from bosons bound to monopoles.

Surely one of the most amazing features of quantum physics is the appearance of two types of particles: bosons and fermions. While bosons are gregarious, fermions are loners. The wave function acquires a factor of +1 upon the interchange of two bosons, and a factor of −1 upon the interchange of two fermions.

The nature of quantum statistics [1] lies at the crux of our difficulties in making further progress, since the glory days of the 1970's, in understanding fundamental physics. The problem goes back to Weisskopf's discovery in 1934 that while the self-energy of a fermion diverges only logarithmically, the self-energy of a boson diverges quadratically. (In fact, Weisskopf had first obtained the erroneous result, noted by Furry, that the fermion mass diverges linearly.) In classical electrodynamics, the self-energy of a charged particle is just the Coulomb energy contained in the surrounding electric field and this diverges like the inverse of the size of the particle. In other words, the self-energy diverges linearly. Thus, fermi statistics suppresses the self-energy divergence by one power, while bose statistics make the divergence worse.

As Weisskopf explained in a 1939 paper [2], the difference in the nature of the divergence can be understood physically in terms of quantum statistics. The "bad" behavior of bosons has to do with their gregariousness. A fermion would push away the fermions contained in the Dirac sea, thus creating a cavity in the vacuum charge distribution surrounding it. Hence its self-energy is less singular than would be the case were quantum statistics not taken into account. A boson does the opposite.

With the advent of grand unified theory, it was realized that the Weisskopf phenomenon means big trouble. In the modern interpretation of renormalization, the self-energy divergence is understood as the difference in the mass (or mass squared) of a particle at one energy scale versus another energy scale. (Another mysterious consequence of statistics is that fermions obey linear equations of motion while bosons obey quadratic equations. Renormalization comes in the mass for fermions, and in the mass squared for bosons.) The mass of a fermion varies logarithmically with energy scale, while the mass squared of a boson varies quadratically. This is a disaster. If we fix the mass of the Higgs field, a bosonic field, to have some generic value at the grand-unification scale, typically 10^{15} GeV, then it would have a value order 10^{15} GeV at the low-energy weak-interaction scale. We would have to fine tune the mass of the Higgs field at the unification scale in order for it to have at the weak-interaction scale a value typical of the weak-interaction scale, rather than a value typical of the grand-unification scale.

Incidentally, Weisskopf's 1939 paper [2] contains the following prescient statement. "This may indicate that a theory of particles obeying Bose statistics must involve new features at this critical length, or at energies corresponding to this length; whereas a theory of particles obeying the exclusion principle is probably consistent down to much smaller lengths or up to much higher energies." In other words, theory with an elementary Higgs field cannot be pushed up to grand unification energies. (Notice also that the terms bosons and fermions were unknown in the late thirties.)

It was partly to solve this so-called hierarchy problem that some particle theorists started traveling down the long road to supersymmetry, and thence to superstrings. Supersymmetry was originally motivated by a desire to connect fermions and bosons. Unfortunately, it manages to connect the known fermions to unknown bosons and the known bosons to unknown fermions. In superstring theories, for which supersymmetry is an essential ingredient, fermionic and bosonic degrees of freedom are necessarily treated on the same footing.

Ever since the work of Fermi and Yang, who tried to construct the pion as a nucleon-antinucleon composite, there has been a general prejudice that fermions are somehow more fundamental than bosons, for the obvious reason that if you put two fermions together you get a boson while if you put two bosons together you still get a boson. For example, in the now discredited technicolor theory, the bose Higgs field is regarded as a composite of techniquarks. Some people also feel that since fermions are described by Grassmanian variables and since Grassmanian variables are weirder than c-numbers, fermions ought to be more fundamental.

I feel that we should also look at the other possibility: perhaps bosons are the fundamental constructs. At first sight, this seems absurd: how can we construct an object with half-integral spin out of objects with integral spin?

In fact, it has been known since the 1930s to the likes of Tamm and Fierz that the eigenstates of a charged boson in the presence of a magnetic monopole have half-integral angular momentum. Physically, this can be seen by evaluating the angular momentum contained in the crossed electic and magnetic fields.

We now know that monopoles exist in non-abelian gauge theories. A composite made of a monopole and a boson in a suitable representation acts as a fermion. Similarly, a composite made of a monopole and a fermion in a suitable representation acts as a boson. Roughly speaking, what happens is that when you interchange two such composites, you pick up an extra phase from the non-abelian gauge potentials. The physics is not entirely dissimilar from the physics responsible for fractional statistics.

Also, Witten has shown that the skyrmions in a non-linear sigma model constructed out of boson fields can be quantized as fermions. The effective action contains a so-called Wess-Zumino term. Upon interchange of two such lumps of bose fields, the Wess-Zumino term produces an additional phase. This is again not entirely dissimilar from what happens in the fractional statistics case, in which a so-called Hopf term can produce a phase upon the interchange of two lumps of bose fields. So it is certainly not excluded that the world is constructed out of bose fields.

In studying grand unified theories based on groups such as SO(10) and SO(18), Wilczek and I [3] have noticed another intriguing possibility. In SO(10) theory, each family of quarks and leptons is assigned to a 16-dimensional representation. It is a matter of mathematics that each of the 16 states can be labeled by five + or - signs, with the constraint that the product of the five signs has to be equal to +1. For example, the red down quark corresponds to (+-++-) and the electron to (+----). It almost brings to mind the binary coding used in computer science.

A fermion in the presence of a monopole can go into a bound state of exactly zero energy. There are thus two degenerate states: the monopole with or without the zero-energy fermion added. Charge conjugation symmetry, together with the fact that the two states differ by one unit of fermion number, implies that the two degenerate states have fermion number +1/2 and -1/2. Suppose we have five different types of fermions, with a zero-energy bound state for each type, so that we can fill these states independently. Then we have produced the mathematical classification, if not the physical structure, of the known quarks and leptons. The red down quark would consist of three fermions, of type 1, 3, and 4, bound to a monopole.

As Wilczek and I noted [3], there are formidable obstacles to be overcome before this sort of scenario can be called a model, let alone a theory. In fact, since we have to bind an odd number of fermions to the monopole, we end up with a boson. As long as we are speculating, we can of course throw in an extra "inert" boson to make the composite fermionic. A more serious drawback is that monopoles have a long-ranged interaction between them, which is somehow not seen between quarks and leptons. (Here we are not speaking of the "usual" long-ranged interactions associated with the photon and the gluons. These are included in the standard way by gauging the SU(5) that transforms the five types of fermions into each other.)

If we are enamored of the idea that the fundamental entities of the world are bosonic, then we can imagine that each of the five types of fermions is itself a bound state of a boson and another kind of monopole. The picture becomes more baroque: now quarks and leptons are bound states of bosons with two types of monopoles.

Alternatively, we can bind an odd number of bosons (of which there are five types), instead of fermions, to the monopole. The spin and statistics would come out correctly, as fermionic. But then we do not have any understanding on why the occupation number of each of the five zero energy states is restricted to 0 and 1, rather than ranging over all integers 0, 1, 2, ... Perhaps there are excited states of quarks and leptons, a generic feature of such composite models, that experimentalists have yet to discover.

These remarks are obviously speculative but they may serve to encourage further work towards a deeper understanding of the nature of quantum statistics.

ACKNOWLEDGEMENTS

I like to thank R. C. Hilborn, G. Gagliardi, E. C. G. Sudarshan, G. M. Tino, and S. Weinberg for inviting me to a most enjoyable meeting, and D. DeMille and O. W. Greenberg for illuminating discussions.

REFERENCES

1. This paper is based in part on. Zee, A., "Musings on Quantum Statistics," in *Quantum Coherence, Proceedings of the International Conference on Fundamental Aspects of Quantum Theory to Celebrate 30 Years of the Aharonov-Bohm Effect*, edited by J. S. Anandan, World Scientific, Singapore, 1990.
2. Weisskopf, V. F., *Phys. Rev.* **56**, 72 (1939).
3. Wilczek, F. and Zee, A., *Phys. Rev. D* **25**, 553–565 (1982).

Bosons and Environment

Enrico Celeghini* and Mario Rasetti[†]

*Dipartimento di Fisica and Sezione INFN
Università di Firenze, I50125 Firenze, Italy
[†]Dipartimento di Fisica and Unità INFM
Politecnico di Torino, I10129 Torino, Italy

Abstract. The role of the background in bosonic quantum statistics is discussed in the frame of a new approach in terms of coherent states. Bose gas is not the only physical situation where bosons are detected. In particle physics bosons are dealt with in a context where the number of observed particles is finite: here the relevant feature are the canonical commutation relations, which we shall show to be related to a Boltzmann-like distribution. A further case is Bose condensate in harmonic traps, where discrete spectrum leads us to predict a new critical temperature. All these cases are analysed in the unified approach proposed, showing that all differences can be ascribed to the role of the background.

Introduction

Bosons are discussed in different areas of physics with the implicit assumption that none of their properties is related to the characteristics of the Fock space where their states are defined (continuum or discrete) or to the number of particles (finite or, essentially, infinite). We discuss here different physical paradigms, all in Fock space and with unlimited occupation numbers, showing that such assumption is incorrect: different environments are not equivalent and unexpected relations with the background can be found.

We introduce first the algebraic approach to quantum statistics recently developed, showing that an equivalent but more powerful definition of bosons can be stated in terms of coherent states of the non compact algebra $su(1,1)$ [1]. In the necessary definition of consistent limit procedures for each physical problem, the analyticity properties imposed by group theory will be shown to play an essential role. The applications considered will deal with particle physics (PP), Bose gas, and Bose-Einstein condensation (BEC). In PP the interest is focused on canonical commutation relations (CCR), the energy spectrum is continuum and the number of particles is limited. The Bose gas is well exemplified by the blackbody radiation – once more with continuum spectrum but centered on Gibbs and Bose postulates and described in terms of finite density (and infinite number) of particles – and we consider it only to compare it with BEC where, together with the postulates, one has discrete spectrum and finite number of particles.

It will be shown that in PP bosons satisfy a Boltzmann-like statistics, and just because of that they are related to $h(1)$. There follows that they obey CCR, which are thus the correct commutation relations in this physical situation (even though, as we shall see, not in others). We shall then argue that the standard approach working for the blackbody, cannot be extended to BEC: in the latter for $T < T_c$, finite density approximation fails as the spectrum discreteness becomes relevant and a new physics emerges. In particular equipartition does not hold any longer and a new critical temperature is found, where the specific heat exhibits a spike.

Algebras and Coherent States

A relation between each statistics and a well defined algebra, has been found in 1998 [1] in terms of coherent states [2]. This has allowed us to select specific raising and lowering operators among those that one can arbitrarily introduce into Fock space. Bose-Einstein definition of bosons is indeed given by requiring the density matrix ρ to be a multiple of the identity $\mathbf{I}$ on the Fock space $\mathcal{F}$, with no need of referring to any specific operator. One is then free to consider the operators $a^\pm$ relating spaces with different number of particles – $\mathcal{F}_n$ and $\mathcal{F}_{n+1}$ – such that

$$a^+|n\rangle = f(n+1)|n+1\rangle, \tag{1}$$

where $f(n)$ is an arbitrary function. We can assume $f(n) \equiv \sqrt{n}$ but also e.g. n or e^n. The choice of $a^\pm$ usually done – that derives from the Weyl-Heisenberg algebra $h(1)$ of position and momentum operators and implies CCR – is arbitrary or, more precisely, appears to be an independent postulate. In what follows such postulate will be shown, following [1], to be related to a specific quantum statistics even though, at first sight, an unsatisfactory one: only a careful discussion of the limit procedure (taking into account correctly the structure of the Fock space) clarifies that for bosons in the physical background of PP (and only in it) consistency holds.

We describe now briefly the procedure that leads from each algebra to its characteristic number of states of the system $W\{\bar{n}\}$; more details can be found in ref. [1]. Let us start from $h(1)$ (often considered as "the algebra" of bosons) and its coherent states. If we follow Condon-Shortley conventions, we have

$$a^+|n\rangle = \sqrt{n+1}\,|n+1\rangle, \qquad e^{a^+}|0\rangle = \sum \frac{1}{\sqrt{n!}}|n\rangle.$$

If we disregard such conventions and do not fix any gauge, an arbitrary phase (possibly a function also of an external parameter t, that we may assume to describe time) must be considered in front of each addendum of the coherent state, and the coefficients in the sum are exactly those needed to describe the 1-mode Boltzmann statistics (see *e.g.* [3]). Analogously, 2-modes are realized operating on the 2-mode vacuum $|0,0\rangle$ with $a_1^+ + a_2^+$:

$$e^{a_1^+ + a_2^+} |0,0\rangle = \sum \frac{e^{i\phi(n_1,n_2,t)}}{\sqrt{n_1! n_2!}} |n_1, n_2\rangle .$$

In order to generalize to M-modes we have now only to consider the M-mode vacuum and the operator $\sum_{i=1}^{M} a_i^+$ (in a more formal language, we operate on the highest weight of the direct product representation with the iterated coalgebra $\Delta^M(a^+) = \sum a_i^+$ [4]):

$$\exp\left[\Delta^M\left(a^+\right)\right] |0,0,\ldots,0\rangle = \sum_{\{\bar{n}\}} \frac{e^{i\phi(\bar{n},t)}}{\sqrt{n_1! n_2! \ldots n_M!}} |n_1, n_2, \ldots, n_M\rangle . \quad (2)$$

This is equivalent to

$$W\{\bar{n}\} \propto \prod_i \frac{1}{n_i!} , \quad (3)$$

i.e. to the definition of Boltzmann statistics [3], that we thus obtain from the algebra $h(1)$ – which contains the canonical commutation relations $[\psi(x), \psi^\dagger(y)] = \delta(x-y)$ as the relations $[a_j^-, a_\ell^+] = \delta_{j\ell}$ written in configuration space. Vice versa eq.(3) *i.e.* Boltzmann statistics can be rewritten as eq.(2) and thus requires $f(n) = \sqrt{n}$ that implies $h(1)$ and its CCR.

We look now for an algebra, if any, related to Bose-Einstein statistics. It turns out to be sufficient selecting $f(n) = n$ to obtain

$$\exp\left[\Delta^M\left(a^+\right)\right] |0,0,\ldots,0\rangle = \sum_{\{\bar{n}\}} e^{i\phi(\bar{n},t)} |n_1, n_2, \ldots, n_M\rangle . \quad (4)$$

Manifestly in this case all states in Fock space have equal probability of being occupied. Bose statistics appears thus to be related to $f(n) = n$. With this choice eq. (1) leads us to identifying a^+ with the raising operator of the representation $\mathcal{D}_{1/2}^+$ of $su(1,1)$. Eq.(4) describes thus the M-mode coherent states of the $(\mathcal{D}_{1/2}^+)^{\otimes M}$ representation of $su(1,1)$ (see [1]).

Canonical Commutation Relations

The results discussed in the previous section – surprising as they may sound – relate on the one hand CCR to the Boltzmann statistics and on the other bosons to $su(1,1)$. This apparently uncertain description signals indeed a more subtle and complex situation: bosons behave differently in different physical contexts. To demonstrate such thesis, we start from eq.(4) and we follow [3], collecting energy levels into cells, each containing g levels. If cell k, of energy ϵ_k, contains N_k particles, then the number of states corresponding to the collection of occupation numbers $\{\bar{N}\}$ is given by

$$W\{\bar{N}\} \propto \prod_k \frac{\Gamma(N_k + g)}{\Gamma(g)\Gamma(N_k + 1)}. \tag{5}$$

Eq.(5), that can be assumed as starting point of any bosons description in quantum statistics, is nothing but, as discussed in [1], the weight of the coherent state corresponding to the representation $\mathcal{D}^+_{g/2}$ of $su(1,1)$. By construction, such weight provides the statistics of a finite number of bosons in Fock space with discrete spectrum. However, physical situations encountered in experiments (except for magnetic traps where BEC is currently investigated) are typically different. Appropriate limit procedures must thus be considered for eq.(5); these procedures, peculiar of each problem, must be included in the consistent definition of what bosons are in each physical situation.

In PP the spectrum is continuum (*e.g.* in momentum space) while the number of particles remains finite: the correct limit procedure on eq.(5) is thus given by the limit for $g \to \infty$ at N_k fixed [5]:

$$W\{\bar{N}\} \propto \lim_{g\to\infty} \prod_k \frac{\Gamma(N_k + g)}{\Gamma(g)\Gamma(N_k + 1)} = \prod_k \frac{g^{N_k}}{N_k!}. \tag{6}$$

Eq.(6) is exactly equivalent to eq.(3) (and to eq.(2)) from which it can be directly obtained with the same procedure of collecting elementary levels in cells that allows to obtain eq.(5) from eq.(4) [3]. This means that a finite number of bosons, in a continuum Fock space, satisfies Boltzmann's statistics and, thus, CCR. This result, which appears to contradict common wisdom, is actually consistent with all properties of bosons: all statistics – Bose, Boltzmann and Fermi – give the same distribution in these conditions as, in the physics of the continuum, only states with occupation numbers limited to zero and one have probabilities different from zero. The only statistical property that survives the limit is indeed the symmetry of the wave functions and the related Bose-Einstein correlation [6] [7].

Bose-Einstein Condensation

For the blackbody the limit is different: once more $g \to \infty$ (because the configuration space is continuum) but $N_k \to \infty$ as well (as blackbody is a theory of gases). Now the limit prescription is that the density remain finite: $0 < N_k/g \equiv \delta_k < \infty$. In this way one recovers the usual formulas [1].

Let us now compare it with Bose-Einstein condensation in harmonic traps. A detailed technical discussion can be found in ref. [8]; we consider here only the

---

[1] One should not forget that the usual formalism is affected by a small but conceptually relevant difficulty: as for large energy $\delta \approx 0$, the limit prescriptions are not satisfied and a not zero probability is predicted for photons of energy greater than that of the entire blackbody.

aspects more closely related with statistics. As a finite fraction of particles migrate into the fundamental level, $\delta_0 \approx \infty$ and the blackbody prescriptions are badly violated. In the standard approach, an *ad hoc* delta function is introduced in the density of states to allow for a macroscopic occupation of the ground state: a procedure which is not adequate in that it does not determine the ground state occupation. The theory of BEC for $T < T_c$ should be constructed on a different basis, namely not in terms of density but of N – the total number of particles – (finite) in a discrete spectrum. In particular the continuum approximation not only does not permit a rigorous treatment of the macroscopic single state occupation characteristic of BEC, but is far from the experimental situation, where typically $h\nu/k_B \approx 10\,nK$ whereas the estimated temperature $T_c \approx 10^2\,nK$.

The solution to the problem can be found coming back to eq.(5), without any limit. We can thus easily look for the most probable distribution, maximizing the logarithm of $W\{\bar{N}\}$ (constrained by the two conditions $\sum_i N_i = N$ and $\sum_i \epsilon_i N_i = E$) with respect to all N_i's. One obtains in this way the exact equation for N_i and g finite:

$$\psi(N_i + g) - \psi(N_i + 1) = \alpha + \beta\epsilon_i \equiv h_i, \tag{7}$$

where $\alpha = \alpha(g)$ and $\beta = \beta(g)$ are the two Lagrange multipliers, clearly depending on the value of g, while ψ is the digamma function [5].

For large argument ψ coincides with the logarithm, thus for $N_i \gg 1$ and $g \gg 1$ it is easy to obtain the Bose-Einstein distribution we are well accustomed to:

$$\delta = \frac{1}{e^h - 1}, \tag{8}$$

but these limit assumptions do not hold either in PP (where the N_i's are finite) nor in harmonic traps (where both g and the N_i's are finite). As the digamma function has poles for negative integers only, eq.(7) is defined almost everywhere in the complex plane and one could conjecture that it provides the solution for BEC. Unfortunately it is not so: in the presence of a physical 3-dimensional harmonic confining potential, levels are not degenerate (as different frequencies cannot be exactly equal) *i.e.* g should be set equal to 1, and this leads for eq.(7) to a perfectly acceptable but void equation (left hand side identically zero, implying $\alpha(1) = \beta(1) = 0$). The way out is found in the analyticity of group theory representations: while in the combinatorial approach g is a multiplicity and thus an integer, in theory of representations it is simply a parameter that can assume any strictly positive real value. We can thus find by analytical continuation for $g \to 1$ the equation for not degenerate levels we need. N_i results to be the solution of the implicit equation

$$\psi'(N_i + 1) = \alpha' + \beta'\epsilon_i \equiv h'_i, \tag{9}$$

where $\alpha' \equiv \lim_{g \to 1} \alpha(g)/(g-1)$ and $\beta' \equiv \lim_{g \to 1} \beta(g)/(g-1)$ are the new Lagrange multipliers to be determined by the constraints of fixed N and E, and ψ' is the trigamma function [5], derivative of ψ. Eq.(9) is one-to-one and can be inverted, giving us the correct formula for discrete spectrum, to be used instead of eq.(8):

$$N_i = [\psi']^{-1}(h'_i) - 1 \quad \text{for} \quad h'_i < \pi^2/6 \quad \text{and} \quad N_i = 0 \quad \text{for} \quad h'_i \geq \pi^2/6, \quad (10)$$

$[\psi']^{-1}$ denoting the inverse function of ψ' and $\pi^2/6 \equiv \psi'(1)$. The discrete description can now be implemented as follows: one assigns first the values of N and E; from these the two Lagrange multipliers $\alpha' = \alpha'(N, E)$, $\beta' = \beta'(N, E)$ are then obtained imposing the constraints with N_i given by the defining equation (10); finally, inserting $\alpha'(N, E)$ and $\beta'(N, E)$ in (10) itself, one gets $N_i \equiv N_i(\alpha', \beta') = N_i(N, E)$.

By inspection, the two formulas (8) and (10) exhibit a similar structure, because they have a pole with residue 1 for $h = 0$ and they go to zero for large values of the argument. As condensation derives from such a behavior, combined with the feature that the number of states increases quadratically with energy, both descriptions are correct in predicting the collapse of the atoms in the fundamental level, also if the continuous approach is actually unable to describe what happens for $T < T_c$. The subtlety here is that the very concept of temperature, as defined in the theory of gases, cannot be straightforwardly extended to the condensate.

For E/N large ($T > T_c$, classical region) continuum quantum statistics should predict indeed for the second Lagrange multiplier $\beta' \propto (E/N)^{-1}$, while we find $\beta' \propto (E/N)^{-4}$. Also for T large the energy is therefore not proportional to the inverse Lagrange multiplier β'^{-1} but to the more complex expression $(\pi^2/6 - \alpha')\beta'^{-1}$. Temperature can then no longer be obtained, as for a gas, by $T = (k_B \beta')^{-1}$: the whole scheme (including the very definition of temperature and its relation with energy, i.e. equipartition) must be reconsidered.

In order to obtain the temperature we resort to the more basic notion of entropy, adopting its Shannon information theoretical definition:

$$S = -k_B \sum p_i \ln p_i,$$

where the p_i's are functions of N and E: $p_i \equiv N_i(N, E)/N$. At fixed N we thus straightforwardly obtain $S = S(E)$. Numerical elaborations finally give the temperature

$$T \equiv T(E) = \left[\frac{\partial S}{\partial E}\right]_N^{-1}.$$

Once more this function is one-to-one, and from $T = T(E)$ we can obtain $E = E(T)$, whence the specific heat $C \equiv \frac{1}{N}\left[\frac{\partial E}{\partial T}\right]_N$ derives by (numerical) differentiation (see Fig. 1).

The result can be checked for $T > T_c$, where C is found to be constant, equal to $3k_B$ (as it should, since the system does not condense and classical theory works). The physics is completely different for $T < T_c$: C is almost zero (as entropy and energy decrease with quite different slopes because only a few atoms progressively migrate from the warmer tail of the spectrum into the fundamental level) for $T_d < T < T_c$, and it exhibits a peak for $T \approx T_d$, where condensation becomes a global effect. Note that after the spike, below T_d, the specific heat goes to zero as T goes to zero, as required by Nernst's theorem. These results (as well as the complete description of the physics of BEC given in [8]) are quantitatively correct up to the detailed structure of the ground state, and this approach can perhaps be the appropriate background in which to introduce a collective perturbation [9].

In conclusion, the relevance of the algebraic structure of quantum statistics has been stressed. An important ingredient of the latter is analyticity that has been explicitly introduced in the scheme. This leads to a unified vision of the problem of bosons. Thanks to such approach, apparent differences of bosons in various contexts (particle physics, non-interacting gas, condensate) have been shown to be related only to the role of the environment. The notion of boson acquires thus a primary fundamental role in a unified vision which encompasses different definitions and embraces different fields of physics.

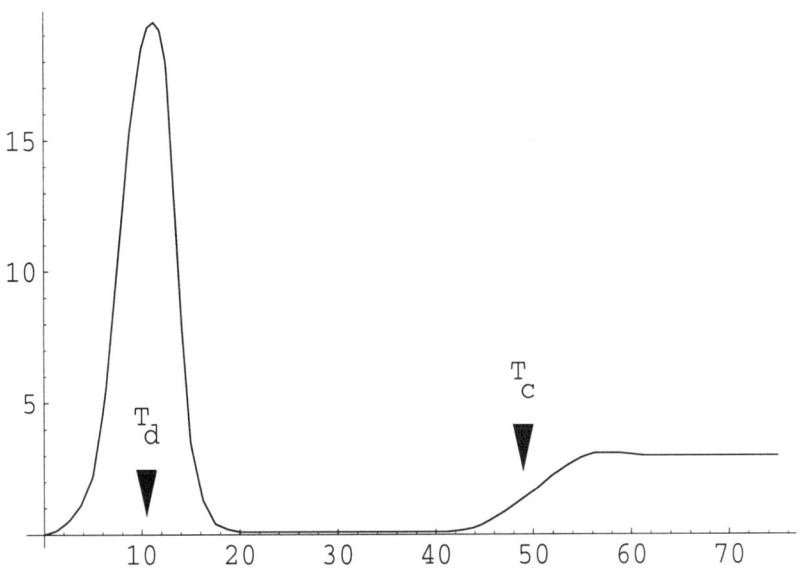

FIGURE 1. Specific heat C vs. T in units $k_B = h\nu = 1$ for $N = 10^6$ bosons in an isotropic harmonic trap, as predicted by discrete quantum statistics.

REFERENCES

1. Celeghini E., and Rasetti M., *Phys. Rev. Lett.* **80**, 3424-3427 (1998).
2. Perelomov A. M., *Generalized Coherent States and their Applications*, Springer-Verlag, Berlin, 1986.
3. Huang K., *Statistical Mechanics*, J.Wiley & Sons, New York, 1987.
4. Abe E., *Hopf Algebras*, Cambridge Tracts in Math. 74, Cambridge University Press, Cambridge, 1980.
5. Abramowitz M., and Stegun I., *Handbook of Mathematical Functions*, Dover, New York, 1965.
6. Lörstad B., *Int. J. Mod. Phys.* **A4**, 2861-2896 (1989).
7. ALEPH Collaboration, *Phys. Lett.* **B478**, 50-64 (2000).
8. Celeghini E., and Rasetti M., *arXiv:cond-mat*/0004096, 14 Apr 2000, submitted to *Phys. Rev. Lett.*
9. Dalfovo F., Georgini S., Pitaevskii L.P. and Stringari S., *Rev. Mod. Phys.* **71**, 463-512 (1999).

A Group-Theoretic Approach To Constructions Of Non-Relativistic Spin-Statistics

J. M. Harrison[a] & J. M. Robbins[a,b]

[a]*Department of Mathematics, University of Bristol, University Walk, Bristol, BS8 1TW, U.K.*
[b]*Basic Research Institute in the Mathematical Sciences, Hewlett-Packard Laboratories Bristol, Filton Road, Stoke Gifford, Bristol BS12 6QZ, U.K.*

Abstract. We give a group-theoretical generalization of Berry and Robbins' treatment of identical particles with spin. The original construction, which leads to the correct spin-statistics relation, is seen to arise from particular irreducible representations – the totally symmetric representations - of the group SU(4). Here we calculate the exchange signs and corresponding statistics for all irreducible representations of SU(4).

INTRODUCTION

Berry and Robbins [1 - 3] formulate quantum mechanics for identical spin-s particles – here we suppose there are two - on a configuration space in which permuted configurations are identified. The two-particle wavefunctions are expanded in a basis of transported spin states $|m_1 m_2(r)\rangle\rangle$ which are made to vary smoothly with the relative coordinate $r = r_1 - r_2$ so that spins are exchanged with positions up to a sign $(-1)^k$, i.e.

$$|m_1 m_2(-r)\rangle\rangle = (-1)^k |m_2 m_1(r)\rangle\rangle. \qquad (1)$$

The position, momentum and spin operators are defined so as to satisfy the standard commutation relations, and the statistics are determined by the exchange sign: +1 for bosons and −1 for fermions.

The construction in [1] is based on the Schwinger model for spin, in which each spin is described in terms of a pair of harmonic oscillators. The spin angular momentum S is expressed in terms of the creation and annihilation operators, and the spin-s basis states $|m\rangle$ - eigenstates of S_z - correspond to number states with $s+m$ (resp. $s-m$) quanta in the oscillators. Two-spin eigenstates $|m_1 m_2\rangle$ are analogously represented in terms of two pairs of oscillators.

The transported basis $|m_1 m_2(r)\rangle\rangle$ is obtained from the Schwinger states $|m_1 m_2\rangle$ by a unitary *exchange rotation*,

$$|m_1 m_2(r)\rangle\rangle = U(r)|m_1 m_2\rangle, \quad \text{where} \quad U(r) = \exp(-i\theta \hat{n}(r).E). \qquad (2)$$

The *exchange angular momentum* E, which generates the exchange rotation, couples quanta between pairs of oscillators, just as spin angular momentum couples quanta within a single oscillator pair. An exchange rotation by π about an axis perpendicular to z takes $|m_1 m_2\rangle$, up to a phase factor, to $|m_2 m_1\rangle$ (in analogy with the fact that a spin rotation by π about an axis perpendicular to z takes $|m\rangle$, up to a phase factor, to $|-m\rangle$). Because the $|m_2 m_1\rangle$ states are null eigenstates of E_z, i.e.

$$E_z |m_1 m_2\rangle = 0, \qquad (3)$$

the phase factor accompanying exchange is an axis-independent sign. Calculation gives $\exp(-i\pi E_y)|m_1 m_2\rangle = (-1)^{2s}|m_2 m_1\rangle$. It follows that the exchange condition (1) is satisfied, with $k = 2s$, if and only if $U(-r)$ and $U(r)$ differ on the right by the exchange rotation $\exp(-i\pi E_y) \times \exp(-i\alpha(r) E_z)$. A simple choice which satisfies this condition is to let $U(r)$ be an exchange rotation corresponding to any spatial rotation which maps $\hat{z}$ to $\hat{r}$.

In [1] it was suggested that certain features of the transported basis (smooth exchange, parallel transport) might characterize the spin-statistics relation. In [2] it was pointed out that this is not so; there are other constructions of the transported basis which have these features and yet yield different statistics. While the Schwinger construction appears to be the simplest in certain respects, specific principles are needed to establish a relation between spin and statistics.

Here we examine in a systematic way a family of alternative constructions motivated by group-theoretical considerations. We observe that the spin rotations constitute a group, $SU(2) \times SU(2)$ (there is one $SU(2)$ factor for each spin), and the exchange rotations form another $SU(2)$ group. Together, the spin and exchange rotations generate, and are subgroups of, a larger group, $SU(4)$. The Schwinger representation corresponds to particular irreducible representations, one for each s, of $SU(4)$, but there are many others, and the construction (2) of the transported basis generalizes to them. What is required is that the states $|m_1 m_2\rangle$ transform under spin rotations as an (s,s)-multiplet, and that they are null eigenstates of E_z (cf. (3)). The exchange sign in

$$\exp(-i\pi E_y)|m_1 m_2\rangle = (-1)^k |m_2 m_1\rangle, \qquad (4)$$

which determines the particle statistics as above, depends on the chosen representation and multiplet. It turns out that not all irreducible representations support the construction, but for those that do the exchange sign need not be unique. A detailed account of this work will be given elsewhere [4].

REPRESENTATIONS OF $SU(4)$

The defining representation of the generators of $SU(4)$ is

$$S_{1i} = \tfrac{1}{2}\begin{pmatrix} \sigma_i & 0 \\ 0 & 0 \end{pmatrix} \quad S_{2i} = \tfrac{1}{2}\begin{pmatrix} 0 & 0 \\ 0 & \sigma_i \end{pmatrix}$$
$$E_x = \tfrac{1}{2}\begin{pmatrix} 0 & I \\ I & 0 \end{pmatrix} \quad E_y = \tfrac{1}{2}\begin{pmatrix} 0 & -iI \\ iI & 0 \end{pmatrix} \quad E_z = \tfrac{1}{2}\begin{pmatrix} I & 0 \\ 0 & -I \end{pmatrix}, \quad (5)$$

where σ_i are the Pauli matrices and I is the 2×2 identity matrix. The operators S_{1i}, S_{2i} are the usual spin operators for the two particles and generate the $SU(2)\times SU(2)$ spin subgroup. The irreducible representations of $SU(4)$ are constructed by applying maximal symmetry conditions to a tensor product of basis states of the defining representation [5]. The symmetry conditions applied to the tensor product are recorded in a Young tableau. A Young tableau, denoted by $(n_1, n_2, \ldots, n_r)$, is a series of rows of boxes where row i has length n_i and $n_1 \geq n_2 \geq \ldots \geq n_r$. The irreducible representations of $SU(4)$ are in one-to-one correspondence with Young tableaux containing up to three rows, and the representations of $SU(2)$, with tableaux of one row. We refer to the representations of $SU(4)$ by using the notation (n_1, n_2, n_3), where $n_1 + n_2 + n_3 = n$. $SU(2)$ tableaux can be augmented on the left by any number of columns of two; an $SU(2)$ tableau with two rows (l_1, l_2) is equivalent to one with a single row $(l_1 - l_2)$. This equivalence is used in $SU(2)\times SU(2)$ content of the irreducible representation of $SU(4)$.

CALCULATING THE EXCHANGE SIGN

An irreducible representation (n_1, n_2, n_3) of $SU(4)$, when restricted to the subgroup of spin rotations $SU(2)\times SU(2)$, is reducible, and can be decomposed into its irreducible spin components. These are labeled by two $SU(2)$ tableaux (k_1, k_2) and (l_1, l_2), corresponding to spins $(l_1 - l_2)/2$ and $(k_1 - k_2)/2$. Requiring the two spins to have the same value s and that the component be a nullspace of E_z, we find that (k_1, k_2) and (l_1, l_2) must coincide, with

$$s = (l_1 - l_2)/2. \quad (6)$$

Using the rules for decomposing the $SU(2)\times SU(2)$ subspace [6], we find that the number p of these (s,s) multiplets in a representation of $SU(4)$ is the minimum of the eleven integers

$$\begin{aligned}&1)\ l_2+1 &&7)\ n_1+n_2-2l_1+1\\&2)\ n_1-n_2+1 &&8)\ l_1+l_2-n_2+1\\&3)\ n_1-l_1+1 &&9)\ n_1+n_2-2l_2-l_1+1\\&4)\ n_1-2l_1+1 &&10)\ n_1+2n_2-2l_1-2l_2+1\\&5)\ n_2-l_2+1 &&11)\ 2l_1+2l_2-n_1-n_2+1\\&6)\ l_1-l_2+1&&\end{aligned} \quad (7)$$

(Interestingly, none of the quantities 1) – 11) are redundant). If the minimum of (7) is negative or zero, there are no allowed spin-s multiplets in the representation. If p is even, there are equal numbers of bosonic (exchange sign = +1) and fermionic (exchange sign = -1) multiplets. If p is odd, there is either an extra bosonic or fermionic multiplet; the exchange sign of the extra multiplet can be calculated explicitly.

Table 1 shows an example of the allowed (s,s) multipets in the $(8,5,3)$ representation of $SU(4)$, along with their exchange signs. For the case $s=1$, the representation of $SU(2)$ is described by a tableau with two rows, $l_1 = 5$ and $l_2 = 3$. There are three $s=1$ multiplets, one with exchange sign $+1$ (the physically correct sign) and two with exchange sign -1.

TABLE 1. Decomposing the $(8,5,3)$ representation of $SU(4)$ into its (s,s) multiplets.

Spin	Number of multiplets	Exchange signs
2	3	-1, +1, -1
1	3	-1, +1, -1
0	1	-1

The derivation of these results will be given elsewhere [4].

CONCLUSION

From the perspective of representation theory, the expression (2) for the transported basis for two spins generalizes to other irreducible representations of $SU(4)$, and the exchange signs for all admissible representations can be calculated. These generalized results do not yield any spin-statistics relation, much less the physically correct one. A typical irreducible representation gives rise to transported bases for several values of spin, and for each value may lead to transported bases with either exchange sign. The only systematic scheme for associating representations to spin to have suggested itself is the one provided implicitly by the Schwinger model of [1]. The completely symmetric representations, the simplest from the point of view of representation theory, give rise to a single transported basis for just one value of spin, and thus lead to a definite spin-statistics relation.

Within this representation-theoretic framework, the completely symmetric representations provide a natural mechanism for incorporating the indistinguishability of spins along with positions of identical particles in nonrelativistic quantum mechanics. But the framework itself requires justification. A compelling derivation of the spin-statistics relation should proceed from general principles motivated by considerations of physics and/or mathematical simplicity. The models introduced here should prove useful in formulating and testing such principles.

REFERENCES

1. Berry, M. V., and Robbins, J. M., *Proc. Roy. Soc.* **A453**, 1771-1790 (1997)
2. Berry, M. V., and Robbins, J. M., *J. Phys. A: Math. Gen.* **33**, L207-L214 (2000)
3. Berry, M. V., and Robbins, J. M., *(In these proceedings)*
4. Harrison, J. M., and Robbins, J. M., *(In preparation)*
5. Georgi, H., *Lie Algebras in Particle Physics*, Benjamin/Cummings, Reading, 1982, pp.109-112.
6. Itzykson, C., and Nauenberg, M., *Reviews of Modern Physics* **38**, 95-120 (1966)

Pauli Spin-Statistics Theorem and Statistics of Quasiparticles in a Periodical Lattice

Ilya G. Kaplan

Instituto de Investigación en Materiales, UNAM, Apdo. Postal 20-360, 04510 México, D.F., MEXICO

Abstract. The connection between the Pauli spin-statistics theorem, parastatistics, and quasiparticles is discussed. Although all elementary particles known at present are bosons or fermions, the parastatistics can be realized for quasiparticles. As was shown by author [13], the quasiparticles in a periodical lattice (the Frenkel excitons and magnons) obey the modified parafermi statistics. This result was extended recently to delocalized hole [17] and hole pairs in high T_c superconducting ceramics [18]. The properties of quasiparticle system obeying the modified parafermi statistics are analyzed.

SPIN-STATISTICS THEOREM, PARASTATISTICS AND QUASIPARTICLES

In his famous spin-statistics theorem [1], Pauli did not give a direct proof. He showed that due to some physical contradictions, the second quantization operators for particles with integral spins cannot obey the fermion commutation relations, while for particles with half-integral spins, they cannot obey the boson commutation relations. From this Pauli concluded that particles with integral spin have to obey the Bose-Einstein statistics, while those with half-integral spin have to obey the Fermi-Dirac statistics. Thus, the connection between the value of spin and the permutation symmetry of many-particles wave function:

$$P_{ij}\Psi(1,\ldots,i,\ldots,j,\ldots,N) = (-1)^{2S}\Psi(1,\ldots,i,\ldots,j,\ldots,N) \tag{1}$$

follows if we assume that particles can obey only two types of commutation relations: boson or fermion relations. At that time it was believed that is really so. However, in 1953 Green [2] showed that the more general, parastatistics commutation relations, satisfying all physical requeriments and containing the boson and fermion commutation relations as particular cases, can be introduced. There is no direct connection between the parastatistics commutation relations and permutation symmetry of wave function. And so, the connection (1) was not rigorously established.

After 1940, numerous proofs of the spin-statistics theorem were published. All these proofs contain some additional assumptions, see book by Duck and Sudarshan [3] and critical review on it by Wightman [4]. On the whole, we should resume that till now we have not a rigorous proof of the spin-statistics theorem.

As follows from experiment, the Pauli principle is fulfilled also for composite particles: particles with odd number of fermions obey the Fermi-Dirac statistics and those with even number obey the Bose-Eisntein statistics. The well known example is the molecule $^{16}O_2$. The nucleus ^{16}O is a boson composite particle, so the total wave function of the molecule $^{16}O_2$ must be symmetrical according to the permutations of nuclei. From this follows that in the ground electronic state, which is antisymmetrical, the nuclear wavefunction should to be also antisymmetrical, so, only odd values of the rotational angular momentum are permitted. This was revealed many years ago [5,6]. Recent tests of possible violation of the Pauli principle for the $^{16}O_2$ molecule [7,8] give the probability of such violation $\leq (5-8) \times 10^{-7}$.

Although, the wave function of composite particle has boson (fermion) permutation symmetry, it is not true for its second quantization operators [9,10]. The most simple example is the Cooper pair operators [11]. They are constructed as a product of two fermion operators

$$b_k^\dagger = c_{k\alpha}^\dagger c_{-k\beta}^\dagger, \qquad b_k = c_{-k\beta} c_{k\alpha} . \qquad (2)$$

As can be easily verified,

$$[b_k, b_{k'}^\dagger]_- = [b_k^\dagger, b_{k'}^\dagger]_- = [b_k, b_{k'}]_- = 0, \qquad k \neq k' \qquad (3)$$

$$[b_k, b_k^\dagger]_- = 1 - (\hat{n}_k + \hat{n}_{-k}) \qquad (4)$$

where $\hat{n}_k = c_{k\alpha}^\dagger c_{k\alpha}$ is the electron number operator. The number of electrons in pairs is twice as large as the number of pairs: $\hat{n}_k + \hat{n}_{-k} = 2 b_k^\dagger b_k$. This allows to rewrite Eq.(4) in the fermion form:

$$[b_k, b_k^\dagger]_+ = 1 . \qquad (5)$$

Thus, for $k \neq k'$ Cooper's pairs are bosons, for $k = k'$ they are fermions. The particles (quasiparticles) of such type are named paulions. This is an additional (to Green's parastatistics) example that the variety of allowed commutation relations does not exhaust by the boson and fermion ones.

As was shown by Greenberg and Messiah [12], all known elementary particles are bosons or fermions. But in the quasiparticle case, the situation can be different. As was shown by Kaplan [13], the parastatistics is realized for quasiparticles in a periodical lattice. Namely: the Frenkel excitons and magnons obey a modified parafermi statistics of rank M, where M is the number of equivalent lattice sites within the delocalization region of collective excitation.

Later on, the results [13] for the Frenkel excitons and magnons were extended to polaritons [14], defectons in quantum crystals [15], and the Wannier-Mott excitons [16]. Recently, we proved that delocalized holes [17] and delocalized coupled hole pairs [18] also obey the modified parafermi statistics. Below I describe this statistics for a general system of quasiparticles in a periodical lattice.

STATISTICS OF QUASIPARTICLES IN A PERIODICAL LATTICE

Let us consider a system of spinless excitations of one type on the sites of a crystal lattice. The nature of excitations is not specified, it can be electron or spin excitations, holes or hole pairs on sites (in the last case it is sites of "superlattice" formed by centers of mass of hole pairs [18]). In the absence of dynamic interaction between excitations, the model Hamiltonian in the site representation can be written as follows:

$$H = \varepsilon_0 \sum_n b_n^\dagger b_n + \sum_{n,n'} M_{nn'} b_n^\dagger b_{n'} \qquad (6)$$

where $b_n^\dagger$, b_n are the creation and annihilation operators of excitation on site n, ε_0 is the creation energy of excitation in a lattice, $M_{nn'}$ is the so-called hopping integral, characterizing the efficienty of excitation transfer from site n to site n´.

The operators $b_n^\dagger$, $b_{n'}$ acting on different sites are independent and must commute. So, they obey the Bose commutation relations:

$$[b_n, b_{n'}^\dagger]_- = [b_n^\dagger, b_{n'}^\dagger]_- = [b_n, b_{n'}]_- = 0 \qquad \text{for } n \neq n'. \qquad (7)$$

In the Hamiltonian (6) is assumed that at a site only one excitation and of one type can be created, that is $(b_n^\dagger)^2 |0\rangle = 0$, where $|0\rangle$ is the vacuum state. So, the operators acting on one site satisfy the Pauli principle and the Fermi commutation relations:

$$[b_n, b_n^\dagger]_+ = 1 \quad , \quad [b_n^\dagger, b_n^\dagger]_+ = [b_n, b_n]_+ = 0 \qquad (8)$$

Thus, the excitation operators in site representation have to be attributed to the Pauli operators like the Cooper pair operators. The only difference: unlike the Cooper pair operators which are defined in impulse space and so are delocalized, the excitation operators are localized at lattice sites.

The introduced excitations do not belong to stationary states. They can move in crystal and occupy any sites. This is provided by the hopping term in the Hamiltonian (6). The stationary states can be created by some unitary transformation:

$$B_q = \frac{1}{\sqrt{M}} \sum_n u_{qn} b_n , \quad B_q^\dagger = \frac{1}{\sqrt{M}} \sum_n u_{qn}^* b_n^\dagger \qquad (9)$$

which transfer the Hamiltonian (6) to the diagonalized form in the quasi-impulse space,

$$H = \sum_q \varepsilon_q B_q^\dagger B_q . \qquad (10)$$

In the Eq.(9), M is the number of equivalent lattice sites. For simple lattices with one site per cell, the unitary transformation (9) is determined by the translation symmetry and $u_{qn} = \exp(-i\mathbf{q}\mathbf{r}_n)$. The self-energy of the diagonalized Hamiltonian (10) is equal to

$$\varepsilon_q = \varepsilon_0 + \sum_{n'(\neq n)} M_{nn'} \exp[i\mathbf{q}(\mathbf{r}_n - \mathbf{r}_{n'})] . \qquad (11)$$

The action of the operator $B_q^\dagger$ on the vacuum state $|0\rangle$ creates a quasiparticle in the quasi-impulse space. In the coordinate space the quasiparticle is delocalized among all equivalent sites. It can be exciton, magnon, delocalized hole or coupled hole pair, etc..

Since the operators $b_n^\dagger, b_n$ obey neither the boson nor the fermion commutation relations, the unitary transformation in general case is not canonical; this means that it does not preserve the commutation properties of the transformed operators and the operators (9) do not correspond to the paulion quasiparticles. As was shown by author [13], for lattices diagonalized by an exponential unitary transformation, such operator obey the modified parafermi statistics with trilinear commutation relations:

$$[[B_q^\dagger, B_{q'}]_- B_{q''}]_- = 2M^{-1} B_{\bar{q}} \qquad \bar{\mathbf{q}} = \mathbf{q'} + \mathbf{q''} - \mathbf{q} \qquad (12)$$

$$[[B_q^\dagger, B_{q'}]_- B_{q''}^\dagger]_- = 2M^{-1} B_{\bar{q}}^\dagger \qquad \bar{\mathbf{q}} = \mathbf{q} - \mathbf{q'} + \mathbf{q''} . \qquad (13)$$

Instead of the Kronecker symbols as in the parafermion relations, the value of $\bar{\mathbf{q}}$ in the right-hand side is determined by the quasimomentum conservation law. This is the consequence of the translation symmetry. The absense of the Kronecker symbols causes important physical properties which we will discuss below.

It can be shown [13,17] that one state may be occupied by up to M quasiparticles:

$$(B_q^\dagger)^N |0\rangle \neq 0 \quad \text{for } N \leq M ; \quad (B_q^\dagger)^{M+1} |0\rangle = 0 . \qquad (14)$$

This means that quasiparticles in the quasi-impulse space obey the modified parafermi statistics of rank M with the trilinear commutation relations (12) and (13).

The state with N quasiparticles, each with the same $\mathbf{q}$, has to be normalized:

$$|N_q\rangle = C_N (B_q^\dagger)^N |0\rangle , \qquad (15)$$

$$C_N = [N!(1-\frac{1}{M})(1-\frac{2}{M})\cdots(1-\frac{M-1}{M})]^{-\frac{1}{2}} . \qquad (16)$$

The expression for C_N is found by the induction method using the operator equation for shifting the operator B_q to the right, see [17], and then taking into account that

$$B_q |0\rangle = 0 \quad \text{and} \quad B_q \cdot B_q^\dagger |0\rangle = \delta_{qq'} |0\rangle . \qquad (17)$$

The second equation in (17) does not contain the rank of parastatistics because, in contrary to the Green Ansatz, our Eqs. (9) are normalized.

The results of applying the operators $B_q^\dagger$ and B_q to the state vector (15) are

$$B_q^\dagger |N_q\rangle = \sqrt{(N_q+1)\left(1-\frac{N_q}{M}\right)} |N_q+1\rangle \qquad (18)$$

$$B_q |N_q\rangle = \sqrt{N_q\left(1-\frac{N_q-1}{M}\right)} |N_q-1\rangle . \qquad (19)$$

Eq.(18) shows that the effect of applying $B_q^\dagger$ on a state with a maximum occupation number $N_q = M$ is equal to zero. As $M \to \infty$, relations (18) and (19) turn into the well known relations for bosons.

From Eqs. (18) and (19), it follows that

$$B_q^\dagger B_q |N_q\rangle = N_q \left(1-\frac{N_q-1}{M}\right) . \qquad (20)$$

So, the operator $B_q^\dagger B_q$ is not a particle number operator in a state **q**, as in the case of boson, fermion or paulion operators. It can be proved that for the introduced above parafermi statistics, the operator of particle number in a state **q** does not exist, see Ref.[13,17]. The mathematical reason for this is the absence of the Kronecker symbols in the trilinear commutation relations (12) and (13). The physical consequence: the absence of independent one-particles states.

SOME PROPERTIES OF QUASIPARTICLES GAS

From the principal impossibility of introducing the operator of particle number $\hat{N}_q$ for quasiparticles in crystal follows that the ideal gas of such quasiparticles does not exist. The Hamiltonian

$$H_0 = \sum_q \varepsilon_q B_q^\dagger B_q \neq \sum_q \varepsilon_q \hat{N}_q , \qquad (21)$$

because $\hat{N}_q$ cannot be defined. Let us calculate the mean value of the Hamiltonian (21) in the vector state $|N_q\rangle$ with N quasiparticles. The form of Hamiltonian means that quasiparticles do not interact (dynamically) with each other. We have to calculate

$$E(N_q) = \langle N_q | H_0 | N_q \rangle = C_N^2 \langle 0 | B_q^N \left(\sum_{q'} \varepsilon_{q'} B_{q'}^\dagger B_{q'} \right) \left(B_q^\dagger \right)^N | 0 \rangle . \qquad (22)$$

Using an operator equation for shifting B_q to the right, after cumbersome but streight-forward calculations (see Ref. [17]), we obtain

$$E(N_q) = N\varepsilon_q \left(1 - \frac{N-1}{M}\right) + \frac{N(N-1)}{M(M-1)} \sum_{q'(\neq q)} \varepsilon_{q'} = N\left[\varepsilon_q + \frac{N-1}{M}(\bar{\varepsilon} - \varepsilon_q)\right] \qquad (23)$$

where $\bar{\varepsilon} = \left(1/{M-1}\right) \sum_{q'(\neq q)} \varepsilon_{q'}$ is the mean energy of the quasiparticle band. The second term in the rigth-hand part of Eq.(23) is called the *kinematic* interaction. It is proportional to the concentration of quasiparticles. At small concentration it approaches to zero and gas of quasiparticles can be considered as a Bose gas. This kind of interaction depending on the deviation of quasiparticle statistics from the Bose (Fermi) statistics was firstly revealed by Dyson [19] in magnetic systems.

In high T_c ceramics, the maximum T_c is achieved for a hole concentration in the CuO_2 planes equals to 0.2-0.25 per CuO_2 units [20]. The same order of magnitude has to be for the coupled hole-pair concentration [18]. For excitons a much higher concentrations can be achieved in modern laser instalations. Thus, the deviations from the Bose statistics for the quasiparticle systems are not negligible and have to be taken into account.

The kinematic interaction is larger the larger is the difference between ε_q and the mean energy of the quasiparticle band. For state $\varepsilon_q < \bar{\varepsilon}$, the kinematic interaction is positive, so, in the ground state, the kinematic interaction leads to repulsion. For higher states with $\varepsilon_q > \bar{\varepsilon}$, the kinematic interaction is attractive.

According to the expression (23) for energy $E(N_q)$, the kinematic interaction mixes all states of the quasiparticle band. Therefore, we cannot define quasiparticles in some particular state. The ideal gas of quasiparticles does not exist fundamentally. So, the Gentile distribution function [21], derived for an ideal paragas, cannot be applied to a quasiparticle system in the quasi-impulse space. As shown above, the quasiparticles in the quasi-impulse space obey the modified parafermi statistics of rank M, so one state can be occupied up to M quasiparticles. The number of quasiparticles is equal to the number of created excitations. The latter cannot exceed the number M of lattice sites at which the excitation can be created. So, all created quasiparticles can occupy one state, for example, the ground state. This means that, in spite of non-bosonic behavior, there is no statistical prohibition of the Bose-Einstein condensation (BEC) phenomenon in quasiparticle systems. On the other hand, the gas of quasiparticles is always non-ideal. Thus, the BEC phenomenon has to be considered like in the Bose gas with weak repulsion, as in the ^{4}He liquid case, but using the proper trilinear commutation relations (the kinematic repulsion instead of dynamical repulsion in liquid helium).

This study was supported by the proyect of CONACyT (Mexico) 32227E.

REFERENCES

1. Pauli, W., *Phys. Rev.* **58**, 717-722 (1940).
2. Green, H.S., *Phys. Rev.* **90**, 270-273 (1953).
3. Duck, I., and Sudarshan, E.C.G., *Pauli and the Spin-Statistics Theorem*, World Scientific, Singapore, 1997.
4. Wightman, A.S., *Amer. J. Phys.* **67**, 742-746 (1999).
5. Dieke, G.H., and Babcock, H.D., *Proc. Natl. Acad. Sci. U.S.A.* **13**, 670-678 (1927).
6. Heisenberg, W., *Z. Phys.* **41**, 239-267 (1927).
7. Angelis, M., Gagliardi, G., Gianfrani, L., and Tino, G.M., *Phys Rev. Lett.* **76**, 2840-2843 (1996).
8. Hilborn, R.C., and Yuca, C.L., *Phys. Rev. Lett.* **76**, 2844-2847 (1996).
9. Gilbert, J.D., *J. Math. Phys.* **18**, 791-805 (1977)
10. Kaplan, I.G. *Problems in Theor. and Nucl. Phys.*, Saratov Univ. (in Russian), N7, 66-77 (1980).
11. Schrieffer, J.R., *Theory of Superconductivity*, Addison-Wesley, Redwood City, CA, 1988.
12. Greenberg, O.W, and Messiah, A.M., *Phys.Rev.* **138**, B1155-B1167 (1965).
13. Kaplan, I.G., *Theor. Math. Phys.* **27**, 466-474 (1976).
14. Avdyugin, A.N., Zavorotnev, Yu. D., and Ovander, L.N., *Sov. Phys. Solid State.* **25**, 1437-1438 (1983).
15. Pushkarov, D.I., *Phys. Status Solidi* **(b) 133**, 525-531 (1986).
16. Nguen, B.A., and Hoang, N.C., *J. Phys.: Condens. Matter* **2**, 4127-4136 (1990).
17. Kaplan, I.G., and Navarro, O., *J. Phys.: Condens. Matter* **11**, 6187-6195 (1999).
18. Kaplan, I.G., and Navarro, O., *Physica C* (in Press).
19. Dyson, F., *Phys.Rev.* **102**, 1217-1230 (1956).
20. Zhang, H., and Sato, H., *Phys. Rev. Lett.* **70**, 1697-1700 (1993).
21. Gentile, G., *Nuovo Cimento* **17**, 493-497 (1940).

Permutational Symmetry of Many-Particle Spin-Functions: The Dirac Identities

J. Katriel

Department of Chemistry, Technion - Israel Institute of Technology
Haifa 32000, Israel.

Abstract. Explicit expressions are derived for the action of all conjugacy class-sums of the symmetric group within the space spanned by n identical spin-$\frac{1}{2}$ particles, in terms of the total spin operator. For identical particles with elementary spin σ all conjugacy class-sums are presented in terms of the lowest 2σ single cycle conjugacy class-sums.

1. INTRODUCTION

The indistinguishability of quantum-mechanical identical particles implies that permutations of such particles are symmetries of the system hamiltonian. Consequently, the corresponding eigenstates belong to well defined irreducible representations (irreps) of the symmetric group. For systems in which the hamiltonian is spin-independent the total wavefunction can be constructed by coupling a set of spin functions that span an irrep of the symmetric group with a set of spatial functions that span some other irrep, into a total wavefunction that belongs to yet another irrep. The elementary spin of the particles of interest specifies the maximum number of rows in the Young diagrams that designate the feasible total spin irreps. The spin-statistics connection is invoked at this point to specify the allowed irrep of the total wavefunction, thereby restricting the irreps of the spatial wavefunctions that can couple with a given spin irrep.

The present paper reviews both old and new results concerning the implications of the restrictions imposed by the value of the elementary spin on the allowed irreps of the symmetric group. The original Dirac identity involves the representation of the transposition of two spin-$\frac{1}{2}$ particles in terms of their spin operators. One type of generalization involves the representation of the action of all conjugacy class-sums of the symmetric group, within the space spanned by n spin-$\frac{1}{2}$ particles, in terms of the total spin operator. This generalization is fully achieved. A further generalization concerns the case of higher elementary spins, for which a larger number of independent central operators is required. The representation of all the conjugacy class-sums in terms of the minimal set, that for particles of spin

σ consists of the conjugacy class-sums $[(k)]_n$, $k = 2, 3, \cdots, 2\sigma + 1$, is explicitly constructed.

2. THE DIRAC IDENTITIES FOR SPIN-$\frac{1}{2}$ PARTICLES

Two spin-$\frac{1}{2}$ particles can be coupled into two total spin states, a triplet that consists of the states $\alpha\beta$, $\frac{1}{\sqrt{2}}(\alpha\beta + \beta\alpha)$, and $\beta\beta$, with a total spin quantum number $S = 1$, and a singlet, $\frac{1}{\sqrt{2}}(\alpha\beta - \beta\alpha)$, with a total spin $S = 0$. Since the triplet is symmetric and the singlet antisymmetric with respect to transpositions of the two particles, one can write, within the four dimensional space spanned by the two particle spin functions, the Dirac identity [1]

$$(1,2) = \hat{S}^2 - 1 = \frac{1}{2} + 2\hat{s}_1 \cdot \hat{s}_2, \tag{1}$$

in which the relation $\hat{S}^2 = \hat{s}_1^2 + \hat{s}_2^2 + 2\hat{s}_1 \cdot \hat{s}_2 = \frac{3}{2} + 2\hat{s}_1 \cdot \hat{s}_2$ was used. Moreover, the values of $\hat{s}_1 \cdot \hat{s}_2$ for the two total spin states imply that $(\hat{s}_1 \cdot \hat{s}_2)^2 = -\frac{1}{2}(\hat{s}_1 \cdot \hat{s}_2) + \frac{3}{16}$, which means that any integral power of $\hat{s}_1 \cdot \hat{s}_2$ can be expressed linearly in terms of $\hat{s}_1 \cdot \hat{s}_2$.

The expression for the transposition of two particles in terms of the scalar product of their spin operators, eq. 1, allows the presentation of the transposition conjugacy class-sum $[(2)]_n = \sum_{i<j}^n (i,j)$ in terms of the spin operators, in the form

$$[(2)]_n = \sum_{i<j}(\frac{1}{2} + 2\hat{s}_i \cdot \hat{s}_j) = \frac{n(n-1)}{4} + 2\sum_{i<j} \hat{s}_i \cdot \hat{s}_j. \tag{2}$$

Since $\hat{S}^2_{(n)} = \frac{3}{4}n + 2\sum_{i<j} \hat{s}_i \cdot \hat{s}_j$, one obtains $[(2)]_n = \hat{S}^2_{(n)} + \frac{n(n-4)}{4}$.

The two-particle Dirac identity, eq. 1, yields

$$(123) + (132) = (13)(12) + (12)(13)$$
$$= (\frac{1}{2} + 2\hat{s}_1 \cdot \hat{s}_3)(\frac{1}{2} + 2\hat{s}_1 \cdot \hat{s}_2) + (\frac{1}{2} + 2\hat{s}_1 \cdot \hat{s}_2)(\frac{1}{2} + 2\hat{s}_1 \cdot \hat{s}_3).$$

The identity $(\hat{s}_1 \cdot \hat{s}_2)(\hat{s}_1 \cdot \hat{s}_3) + (\hat{s}_1 \cdot \hat{s}_3)(\hat{s}_1 \cdot \hat{s}_2) = \frac{1}{2}(\hat{s}_2 \cdot \hat{s}_3)$ holds for spin-$\frac{1}{2}$ particles. Using this identity we obtain

$$(123) + (132) = \frac{1}{2} + 2(\hat{s}_1 \cdot \hat{s}_2 + \hat{s}_1 \cdot \hat{s}_3) + 4[(\hat{s}_1 \cdot \hat{s}_2)(\hat{s}_1 \cdot \hat{s}_3) + (\hat{s}_1 \cdot \hat{s}_3)(\hat{s}_1 \cdot \hat{s}_2)]$$
$$= \frac{1}{2} + 2(\hat{s}_1 \cdot \hat{s}_2 + \hat{s}_1 \cdot \hat{s}_3 + \hat{s}_2 \cdot \hat{s}_3).$$

It follows that [2]

$$[(3)]_n = \frac{1}{2}\binom{n}{3} + 2(n-2)\sum_{i<j}^n \hat{s}_i \cdot \hat{s}_j = (n-2)\hat{S}^2_{(n)} + \frac{n(n-2)(n-10)}{12}.$$

Similarly,

$$[(4)]_n = \frac{1}{2}(\hat{S}^2_{(n)})^2 + [\frac{1}{4}(n-4)(3n-8) - 2]\hat{S}^2_{(n)} + \frac{1}{32}n(n-2)(n-4)(n-18),$$

$$[(2)^2]_n = \frac{1}{2}(\hat{S}^2_{(n)})^2 + \frac{1}{4}(n^2 - 10n + 12)\hat{S}^2_{(n)} + \frac{1}{8}n(\frac{1}{3}n^3 - 3n^2 + 14n - 18),$$

etc. Proceeding in this way very quickly becomes prohibitively complex [3]. An effective procedure that allows the treatment of arbitrarily high conjugacy class-sums is presented in section 6.

3. TWO SPIN-σ PARTICLES

For two spin-σ particles the total spin obtains the values $S = 2\sigma, 2\sigma - 1, 2\sigma - 2, \cdots, 0$. The states $S = 2\sigma, 2\sigma - 2, \cdots$ are even, and the states $S = 2\sigma - 1, 2\sigma - 3, \cdots$ are odd with respect to particle transposition.

Using this observation, Schroedinger [4] wrote the transposition in the form

$$(1,2) = (P_{2\sigma} + P_{2\sigma-2} + \cdots) - (P_{2\sigma-1} + P_{2\sigma-3} + \cdots) \tag{3}$$

where P_S is a projection operator of the form

$$P_S = \prod_{\substack{S'=0 \\ (S' \neq S)}}^{2\sigma} \frac{\hat{S}^2 - S'(S'+1)}{S(S+1) - S'(S'+1)} \quad ; \quad S = 2\sigma, 2\sigma - 1, \cdots, 0.$$

Thus, P_S is a polynomial in $\hat{S}^2$ (hence $\hat{s}_1 \cdot \hat{s}_2$) of order 2σ. For $\sigma = \frac{1}{2}$ eq. 3 reduces to the Dirac identity, eq. 1. Very little progress has so far been achieved in formulating an extension of this relation to more than two particles. The results presented in section 6 can be viewed as a step towards this generalization.

4. AN APPLICATION: THE EXCHNAGE INTERACTION MODEL

The Heisenberg hamiltonian

$$H = -\frac{J}{2\nu} \sum_{(i,j)}^{N} \hat{s}_i \cdot \hat{s}_j, \tag{4}$$

where (i, j) stands for a neighboring pair of spins on an appropriate lattice, and ν is the number of nearest neighbors of each spin, is the most common model of ferromagnetism. The infinite range counterpart of this hamiltonian

$$H = -\frac{J}{N} \sum_{i<j}^{N} \hat{s}_i \cdot \hat{s}_j. \tag{5}$$

has a phase diagram which is identical with that of the mean-field approximation of (4). For spin-$\frac{1}{2}$ particles use of eq. 2 allows writing eq. 5, up to irrelevant constants, in the form $H = -\frac{J}{N}[(2)]_N$. The infinite range exchange interaction model [5] amounts to taking the latter form as the hamiltonian for particles with arbitrary elementary spins. For particles with an elementary spin σ the eigenvalues of the transposition conjugacy class-sums, $[(2)]_N$, are $\Lambda_\Gamma = \frac{1}{2}\sum_{i=1}^{k} \mu_i(\mu_i - 2i)$, where $\mu_1, \mu_2, \cdots, \mu_k$ are the row lengths of the corresponding Young diagram, Γ, and $k = 2\sigma + 1$. The dimension of the irrep Γ is $\Omega_\Gamma = N! \frac{\prod_{i<j}^{k}(\mu_i - \mu_j + j - i)}{\prod_{i=1}^{k}(\mu_i + n - i)!}$. This irrep appears in the space spanned by N identical particles with an elementary spin σ

$$g_\sigma(N, \Gamma) = \frac{\prod_{i<j}^{k}(\mu_i - \mu_j + j - i)}{\prod_{i<j}^{k}(j-i)} \text{ times.}$$

The spectral and combinatorial data presented above allows the exact statistical mechanical treatment of the infinite range exchange interaction hamiltonian. The order parameters are just the thermal averages of the row lengths of the Young diagram that specifies the state of the system. One obtains $\lambda_\ell = \frac{1}{q}\exp(\beta J \lambda_\ell)$ where $\lambda_\ell = \frac{\mu_\ell}{N}$, $\ell = 1, 2, \cdots, k$, are the reduced order parameters, and where $q = \sum_{i=1}^{2\sigma+1} \exp(\beta J \lambda_i)$ is the corresponding canonical partition function.

The order parameters are found to be coupled in such a way that the only consistent solutions of the set of equations are of the form $\lambda_1 = \lambda_2 = \cdots = \lambda_m \equiv \lambda_+$ and $\lambda_{m+1} = \lambda_{m+2} = \cdots = \lambda_{2\sigma+1} \equiv \lambda_-$.

For $\sigma = \frac{1}{2}$ the mean-field equation reduces to that of the Heisenberg hamiltonian, i.e., $\lambda_\pm = \frac{1}{q}\exp(\beta J \lambda_\pm)$ or $S = \frac{\lambda_+ - \lambda_-}{2} = \frac{1}{2}\tanh(\beta J S)$. Further details are presented in ref. [6].

5. EIGENVALUES OF THE CONJUGACY CLASS-SUMS OF THE SYMMETRIC GROUP

The eigenvalues of the conjugacy class-sums of the symmetric group, which are sometimes referred to as the central characters, are in fact closely related to the ordinary characters, from which they can be obtained by simple normalization. The procedure presented in section 6 for the derivation of generalizations of the Dirac identity is based on the availability of explicit expressions for these eigenvalues in a particularly convenient form, as follows.

For a Young diagram Γ that is specified in terms of the k row lengths $\lambda_1, \lambda_2, \cdots, \lambda_k$ (that form a partition of n) we define

$$\sigma_m = \sum_{i=1}^{k}\sum_{j=1}^{\lambda_i}(j-i)^m.$$

In terms of these symmetric power sums the eigenvalues of the single-cycle conjugacy class-sums are given by the expression

$$\lambda^\Gamma_{[(p)]_n} = \sum_{(2^{p_2} 3^{p_3}\cdots)} f_{p_2 p_3 \cdots}(n) \prod_{i \geq 3} \sigma_{i-2}^{p_i}, \qquad (6)$$

where $\sum_{i\geq 2} i p_i = p+1$ and $f_{p_2 p_3 \cdots}(n)$ is a polynomial in n of degree p_2. These polynomials are determined by noting that the coefficient of the leading term, σ_{p-1}, should be equal to unity and that the expression on the right-hand-side of eq. 6 should vanish when the symmetric power sums $\sigma_1, \sigma_2, \cdots, \sigma_{p-1}$ are evaluated for $n < p$. The explicit expressions for the eigenvalues of the first four single-cycle conjugacy class-sums are

$$\lambda^\Gamma_{[(2)]_n} = \sigma_1$$
$$\lambda^\Gamma_{[(3)]_n} = \sigma_2 - \frac{1}{2}n(n-1) \qquad (7)$$
$$\lambda^\Gamma_{[(4)]_n} = \sigma_3 - (2n-3)\sigma_1$$
$$\lambda^\Gamma_{[(5)]_n} = \sigma_4 - (3n-10)\sigma_2 - 2\sigma_1^2 + \frac{1}{6}n(n-1)(5n-19)$$

An algorithm applicable to arbitrary (multi cycle) conjugacy class-sums was formulated in [7].

One easily notes that σ_m is a symmetric polynomial in the "reduced row lengths" $\bar{\lambda}_i = \lambda_i - i$, $i = 1, 2, \cdots, n$. This turns out to be a key feature in the algorithm that allows the generation of Dirac-type identities for arbitrary elementary spins.

6. IRREPS WITH A RESTRICTED NUMBER OF ROWS

The crucial property of the expressions for the eigenvalues of the conjugacy class-sums of S_n in terms of the symmetric power-sums over the contents of the Young diagram is the following: For all $k \leq n$ the eigenvalues of the conjugacy class-sum $[(k)]_n$ are given in terms of a polynomial in the symmetric power-sums $\sigma_1, \sigma_2, \cdots, \sigma_{k-1}$. Thus, the set of polynomials that represent $[(2)]_n, [(3)]_n, \cdots, [(k)]_n$ can be inverted to yield expressions for $\sigma_1, \sigma_2, \cdots, \sigma_{k-1}$. If the allowed Young diagrams are restricted to have at most $k = 2\sigma + 1$ rows, these symmetric power-sums determine the higher symmetric power-sums $\sigma_k, \sigma_{k+1}, \cdots$. Thus, the latter can be expressed in terms of $\lambda_{[(2)]_n}, \lambda_{[(3)]_n}, \cdots, \lambda_{[(k)]_n}$. A detailed algorithm is presented in ref. [8]. Using that algorithm the following relations were obtained.

TWO-ROW YOUNG DIAGRAMS ($\sigma = \frac{1}{2}$):

$$\lambda_{[(3)]_n} = (n-2)\lambda_{[(2)]_n} - \frac{1}{6}n(n-1)(n-2)$$
$$\lambda_{[(4)]_n} = \frac{1}{2}\lambda^2_{[(2)]_n} + \frac{1}{2}(n-2)(n-6)\lambda_{[(2)]_n} - \frac{1}{8}n(n-1)(n-3)(n-4)$$
$$\lambda_{[(5)]_n} = (n-4)\lambda^2_{[(2)]_n} - 3(n-2)(n-4)\lambda_{[(2)]_n}$$
$$\qquad - \frac{1}{20}n(n-1)(n-3)(n-4)(n-12)$$

These identities are equivalent to those presented in section 2. However, the algorithm referred to above allows the systematic generation of relations of this type for arbitrary conjugacy class-sums.

THREE-ROW YOUNG DIAGRAMS ($\sigma = 1$):

$$\lambda_{[(4)]_n} = \left(n - \frac{9}{2}\right)\lambda_{[(3)]_n} + \frac{1}{2}\lambda_{[(2)]_n}^2 - \frac{1}{2}(n-2)(n-3)\lambda_{[(2)]_n}$$
$$+ \frac{1}{24}n^2(n-1)(n-5)$$

$$\lambda_{[(5)]_n} = \lambda_{[(3)]_n}\lambda_{[(2)]_n} + \frac{1}{2}(n^2 - 15n + 48)\lambda_{[(3)]_n} - 2\lambda_{[(2)]_n}^2$$
$$- \frac{1}{3}(n-2)(n-4)(n-9)\lambda_{[(2)]_n} + \frac{1}{30}n(n-1)(n-8)(n^2 - 6n + 3)$$

FOUR-ROW YOUNG DIAGRAMS ($\sigma = \frac{3}{2}$):

$$\lambda_{[(5)]_n} = (n-8)\lambda_{[(4)]_n} + \lambda_{[(3)]_n}\lambda_{[(2)]_n} - \frac{1}{2}(n-4)(n-6)\lambda_{[(3)]_n} - \frac{1}{2}(n-4)\lambda_{[(2)]_n}^2$$
$$+ \frac{1}{6}n(n-2)(n-7)\lambda_{[(2)]_n} - \frac{1}{120}n(n-1)(n-4)(n-8)(n+3)$$

These expressions for the eigenvalues of arbitrary conjugacy class-sums in terms of the minimal set that consists of the single-cycle conjugacy class-sums corresponding to cycles of lengths $2, 3, \cdots, 2\sigma + 1$, provide the full group-theoretical content of the generalized Dirac identities.

REFERENCES

1. Dirac, P. A. M., *Proc. Roy. Soc. London A*, **123**, 714-733 (1929).
2. Klein, D. J., *J. Phys. A*, **15**, 661-671 (1982).
3. Katriel, J., Paldus, J. and Pauncz, R., *Intern. J. Quantum Chem.*, **28**, 181-202 (1985).
4. Schroedinger, E., *Proc. Roy. Irish Acad. A*, **48**, 39-52 (1941).
5. Chen, Y. C., *Phys. Rev. B* **57**, 5009-5012 (1998), and references therein.
6. Katriel, J. and Kventsel, G. F., *Phys. Rev. B* (in press).
7. Katriel, J., *Discrete Appl. Math.* **67**, 149-156 (1996).
8. Katriel, J., *Intern. J. Quantum Chem.* **78**, 407-411 (2000).

Commutation Relations in Mesoscopic Electric Circuits

You-Quan Li

*Institut für Physik, Universität Augsburg, D-86135 Augsburg, Germany
and Department of Physics, Zhejiang University, Hangzhou 310027, China
email: yqli@physik.uni-augsburg.de*

Abstract. In the talk, I briefly demonstrate the quantum theory for mesoscopic electric circuits and its applications. In the theory, the importance of the charge discreteness in a mesoscopic electric circuit is addressed. As a result, a new kind of commutation relation for electric charge and current occurred inevitably. The charge representation, canonical current representation and pseudo-current representation are discussed extensively. It not only provides a concrete realization of mathematical models which discuss the space quantization in high energy physics and quantum gravity but also presents a sequence of applications in condensed matter physics from a different point of view. A possible generalization to coupled circuits is also proposed.

INTRODUCTION

The dramatic achievement in nanotechnology has aroused tremendous developments in experimental physics in mesoscopic scale. Miniaturization of integrated circuits is undoubtedly a persistent trend for electronic device community. A theory for mesoscopic circuits was proposed by Li and Chen, in which the charge discreteness is first introduced in the quantization of electric circuits [1]. The possibility of space-time discreteness was early considered by Snyder [2] who indicated that the Lorentz invariance do not exclude quantized (discrete) space-time, it was also argued by Li [3] from the finiteness of the observed universe. S. Mantecinos, I. Saavedra, and O. Kunstmann [4] discussed the commutation relations of [2] arguing that it may be related to physics in high energy scale ($10^9 - 10^{12}$eV). Numerous attempts to the argument of exiting minimal position uncertainty were made [5] on the basis of various considerations in string theory as well as in quantum gravity. Actually, the approach of [1] not only provides a concrete realization of mathematical models for exploring the space quantization in high energy physics and quantum gravity but also presents a sequence of applications in condensed matter physics from a different point of view. For example, the persistent current is solved by regarding the mesoscopic metal ring as the circuit of a pure L-design. Application of the theory to a pure C-design gives rise to the Coulomb blockade solution [6].

BASIC DEFINITIONS AND COMMUTATION RELATIONS

Let $\hat{q}$ denote for the *charge* operator, and $\hat{p}$ for the canonical conjugation of the charge satisfying $[\hat{q}, \hat{p}] = i\hbar$. We call $\hat{p}$ the *canonical current* operator since it is not only the canonical conjugation of charge but also the current operator in the quantization approach [7] where the charge was considered as a continuous variable. Taking into account of the discreteness of electronic charge in quantization procedure, we must impose that the eigenvalues of the self-adjoint operator $\hat{q}$ take discrete values [1], i.e. $\hat{q}|n\rangle = nq_e|n\rangle$ where $n \in \mathbf{Z}$ (set of integers) and $q_e = 1.602 \times 10^{-19}$ coulomb, the elementary electric charge. We therefore introduce a minimum 'shift operator' $\hat{Q} := e^{iq_e \hat{p}/\hbar}$ in charge space, which satisfies [1]

$$[\hat{q}, \hat{Q}] = -q_e \hat{Q}, \quad \hat{Q}^{-1} = \hat{Q}^\dagger. \tag{1}$$

These relations determine the structure of the whole Fock space, accordingly, $\hat{Q}^+|n\rangle = e^{i\alpha_{n+1}}|n+1\rangle$, $\hat{Q}|n\rangle = e^{-i\alpha_n}|n-1\rangle$ where α_n's being undetermined phases. The Fock space for the algebra (1) differs from the well known Fock space for the Heisenberg-Weyl algebra because the spectrum of the former is isomorphic to the set of integers $\mathbf{Z}$ but that of the later is isomorphic to the set of non-negative integers $\mathbf{Z}^+ + \{0\}$. Since $\{|n\rangle | n \in \mathbf{Z}\}$ spans a Hilbert space and $\hat{q}$ is self-adjoint, both the completeness $\sum_{n \in \mathbf{Z}} |n\rangle\langle n| = 1$ and the orthogonality $\langle n|m\rangle = \delta_{nm}$ faithful.

The *quasi-current* $\hat{J}$ for a mesoscopic circuit is defined by $\hat{J} = -i\hbar(\hat{Q}^{1/2} - \hat{Q}^{-1/2})/q_e$ which reduces to the canonical current in the limit $q_e \to 0$. The Hamiltonian of a mesoscopic LC-circuit is given by [1]

$$\hat{H} = -\frac{\hbar^2}{2L}\hat{J}^2 + \frac{1}{2C}\hat{q}^2 + \varepsilon \hat{q}, \tag{2}$$

where ε stands for the voltage source, L for inductance, and C for capacity of the circuit. Using eq.(1) we easily obtain the new commutation relations for the quasi-current operator,

$$[\hat{q}, \hat{J}] = i\frac{\hbar}{2}\hat{K}, \quad [\hat{q}, \hat{K}] = -i\frac{q_e^2}{2\hbar}\hat{J}, \tag{3}$$

where an auxiliary operator $\hat{K} = \hat{Q}^{1/2} + \hat{Q}^{-1/2}$ is introduced. Obviously eq.(3) obeys the SU(2) algebra after rescaling the operators. In terms of $\hat{K}$ and $\hat{J}$ we can define a useful operator $\hat{P} = \hat{J}\hat{K}^{-1}$ which we call the *pseudo-current* operator. Obviously the pseudo-current also reduces to canonical current in the limit $q_e \to 0$. With the help of (3), we obtain the following commutation relations,

$$[\hat{q}, \hat{P}] = i\hbar\left(1 + (\frac{q_e}{2\hbar})^2 \hat{P}^2\right). \tag{4}$$

Similar kind of commutation relation was considered earlier in [2] in searching the possibility of space-time discreteness. From the commutation relation (4) one will have a uncertainty relation [4,8] for the charge and pseudo-current [1], which is different from the conventional Heisenberg uncertainty relation.

The definition of *physical current* $\hat{I}$ arises from the Heisenberg equation $\hat{I} = d\hat{q}/dt = (1/i\hbar)[\hat{q}, \hat{H}]$. For the *LC*-design circuit, one can immediately obtain [9],

$$\hat{I} = -i\frac{\hbar}{2q_e L}(\hat{Q} - \hat{Q}^\dagger). \tag{5}$$

PSEUDO-CURRENT REPRESENTATION

We consider the pseudo-current representation $\hat{P}|\eta\rangle = \eta|\eta\rangle$. The differential realization of commutation relation (4) is given by [2]

$$\hat{P} = \eta, \quad \hat{q} = i\hbar\left(1 + (\frac{q_e}{2\hbar}\eta)^2\right)\frac{\partial}{\partial \eta}. \tag{6}$$

Obviously, $\int d\eta \psi^*(\eta) \hat{q} \phi(\eta)$ fails in guaranteeing the charge operator being self-adjoint. The factor $(1+(\eta q_e/2\hbar)^2)^{-1}$ in the measure on the pseudo-current space is therefore required [8] to cancel the corresponding factor of $\hat{q}$ in this representation. The inner product must be so defined,

$$\langle \psi | \phi \rangle = \int_{-\infty}^{\infty} \frac{d\eta}{1 + (\frac{q_e}{2\hbar}\eta)^2} \psi^*(\eta) \phi(\eta) \tag{7}$$

that both $\hat{q}$ and $\hat{P}$ could be self-adjoint.

The completeness is given by

$$\int_{-\infty}^{\infty} \frac{d\eta}{1 + (\frac{q_e}{2\hbar}\eta)^2} |\eta\rangle\langle\eta| = 1. \tag{8}$$

Consequently, the inner product of two eigenstates of the pseudo-current operator yields

$$\langle \eta' | \eta \rangle = \left(1 + (\frac{q_e}{2\hbar}\eta)^2\right) \delta(\eta - \eta'). \tag{9}$$

The eigen-equation of charge operator in this representation reads

$$i\hbar\left(1 + (\frac{q_e}{2\hbar}\eta)^2\right)\frac{\partial}{\partial \eta}\psi_q(\eta) = q\psi_q(\eta), \tag{10}$$

where $\psi_q(\eta) := \langle \eta | \psi_q \rangle$. The deferential equation (10) is solved by

$$\psi_q(\eta) = (\frac{q_e}{2\pi\hbar})^{1/2} \exp(-iq\frac{2}{q_e}\tan^{-1}(\frac{q_e}{2\hbar}\eta)), \tag{11}$$

which has been normalized. It is interesting to evaluate their inner product

$$\langle\psi_q|\psi_{q'}\rangle = \frac{q_e}{(q'-q)\pi}\sin(\frac{q'-q}{q_e}\pi). \tag{12}$$

This clearly brings about a orthogonal catastrophe because the eigenstate of a self-adjoint operator with different eigenvalues must be mutually orthogonal. Actually, it can be avoided provided that $q' - q = nq_e$. We conclude that the electric charge must be quantized (other eigenvalues are not physically permitted).

A natural choice is $q = nq_e$, then the transformation from charge representation to pseudo-current representation is easily derived,

$$\begin{aligned}\langle\eta|\psi\rangle &= \sum_{n=-\infty}^{\infty}\langle\eta|n\rangle\langle n|\psi\rangle \\ &= (\frac{q_e}{2\pi\hbar})^{1/2}\sum_{n=-\infty}^{\infty}\langle n|\psi\rangle e^{-in\Theta(\eta)},\end{aligned} \tag{13}$$

where $\Theta(x) = 2\tan^{-1}(xq_e/2\hbar)$. Multiplying (13) by $e^{in'\Theta(\eta)}/[1+(\eta q_e/2\hbar)^2]$ and integrating with respect to η give rise to the inverse transformation:

$$\langle n|\psi\rangle = (\frac{q_e}{2\pi\hbar})^{1/2}\int_{-\infty}^{\infty}\frac{d\eta}{1+(\frac{q_e}{2\hbar}\eta)^2}e^{in\Theta(\eta)}\langle\eta|\psi\rangle. \tag{14}$$

CANONICAL CURRENT REPRESENTATION

In the canonical current space $\hat{p}|p\rangle = p|p\rangle$, the $\hat{q}$ and $\hat{p}$ are realized by $\hat{p} = p$, $\hat{q} = i\hbar\partial/\partial p$. The eigen-equation of charge operator is,

$$i\hbar\frac{\partial}{\partial p}\psi_q(p) = q\psi_q(p),$$

which is solved by plane waves $\psi_q(p) = e^{-iqp/\hbar}$. Obviously, the periodic condition in p-space, $\psi_q(p+2\pi\hbar/q_e) = \psi_q(p)$ should be imposed so that the charge is quantized (discrete) $q/q_e = n$, consequently,

$$\psi_n(p) = \langle p|n\rangle = e^{-inpq_e/\hbar} \tag{15}$$

The transformation from charge representation to canonical current representation is easily obtained,

$$\begin{aligned}\langle p|\psi\rangle &= \sum_{n=-\infty}^{\infty}\langle p|n\rangle\langle n|\psi\rangle \\ &= \sum_{n=-\infty}^{\infty}e^{-inpq_e/\hbar}\langle n|\psi\rangle.\end{aligned} \tag{16}$$

Multiplying eq.(16) with $e^{in'pq_e/\hbar}$ and integrating with respect to p, we get the inverse transformation, canonical current representation to charge representation:

$$\langle n|\psi\rangle = \frac{q_e}{2\pi\hbar}\int_{-\pi\hbar/q_e}^{\pi\hbar/q_e} dp\langle p|\psi\rangle e^{inpq_e/\hbar}. \tag{17}$$

In the p-space, the Hamiltonian for a mesoscopic LC-design circuit becomes

$$\hat{H} = -\frac{\hbar^2}{q_e^2 L}\left[\cos(\frac{q_e}{\hbar}p) - 1\right] - \frac{\hbar^2}{2C}(\frac{\partial}{\partial p} + i\frac{C}{\hbar}\varepsilon)^2 - \frac{C}{2}\varepsilon^2. \tag{18}$$

The advantage of the canonical current representation is that the Schrödinger equation for the Hamiltonian (18) becomes the standard Mathieu equation after a unitary transformation. The wave function was solved in terms of periodic Mathieu functions, and the energy spectrum was expressed by the eigenvalues of Mathieu equation. The details can be find in [1].

APPLICATIONS

In Coulomb blockade experiments, the mesoscopic capacity may be relatively very small (about $10^{-8}F$) but the inductance of a macroscopic circuit connecting to a source is relatively large because it is proportional to the area which the circuit spans. We can neglect the term reversely proportional to L in (2), and study the equation for a pure C-design. Because a mesoscopic metal ring can be regarded as a pure L-design, we can also study the persistent current on a mesoscopic ring.

Coulomb blockade

The Schrödinger equation for a pure C-design reads

$$\left(\frac{1}{2C}\hat{q}^2 - \varepsilon\hat{q}\right)|\psi\rangle = E|\psi\rangle, \tag{19}$$

where ε is an adiabatic voltage source. The Hamiltonian and charge operator commute each other, so $|n\rangle$ is the eigenstate with energy $E = (nq_e - C\varepsilon)^2/2C - C\varepsilon^2/2$, where both the charge quantum number and the voltage source are involved. The relation between charge q and the voltage ε for the *ground* state is given by

$$q = \sum_{m=0}^{\infty}\left\{\theta[\varepsilon - (m+\frac{1}{2})\frac{q_e}{C}] - \theta[-\varepsilon - (m+\frac{1}{2})\frac{q_e}{C}]\right\}q_e \tag{20}$$

where $\theta(x)$ is the step function. The corresponding eigenstate is

$$|\psi(\varepsilon)\rangle_{ground} = \sum_{m=-\infty}^{\infty}\left\{\theta[\varepsilon - (m-\frac{1}{2})\frac{q_e}{C}] - \theta[\varepsilon - (m+\frac{1}{2})\frac{q_e}{C}]\right\}|m\rangle. \tag{21}$$

The dependence of the current on time is obtained by taking derivative

$$\frac{dq}{dt} = \sum_{m=0}^{\infty} q_e \left\{ \delta[\varepsilon - (m + \frac{1}{2})\frac{q_e}{C}] + \delta[\varepsilon + (m + \frac{1}{2})\frac{q_e}{C}] \right\} \frac{d\varepsilon}{dt}. \tag{22}$$

Clearly, the current is of a form of sharp pulses which occurs periodically (with periodicity q_e/C) according to the changes of voltage.

Persistent current

The Schrödinger equation for a pure L-design in the presence of magnetic flux is given by,

$$-\frac{\hbar^2}{2q_e^2 L}(e^{-i\frac{q_e}{\hbar}\phi}\hat{Q} + e^{i\frac{q_e}{\hbar}\phi}\hat{Q}^+ - 2)|\psi\rangle = E|\psi\rangle. \tag{23}$$

It is obtained on the basis of gauge covariance [1]. The eigenstates can be simultaneous eigenstates of $\hat{p}$, eq.(23) is solved by the eigenstate $|p\rangle = \sum_{n \in \mathbf{Z}} \kappa_n e^{inq_e p/\hbar}|n\rangle$ ($\kappa_n := \exp(i \sum_{j=1}^{n} \alpha_j)$) with the energy spectrum:

$$E(p, \phi) = \frac{2\hbar^2}{q_e^2 L} \sin^2\left(\frac{q_e}{2\hbar}(p - \phi)\right). \tag{24}$$

It oscillates with respect to ϕ or p. Differing from the usual classical pure L-design, the energy of a mesoscopic quantum pure L-design can not be large than $2\hbar^2/q_e^2 L$. Clearly, the lowest energy states are those states with $p = \phi + nh/q_e$, the eigenvalues of the electric current $\hat{I}$ of ground state can be obtained [1]. The electric current on a mesoscopic circuit of pure L-design is not null in the presence of a magnetic flux (except $\phi = nh/q_e$). This is a quantum characteristic property. The persistent current in a mesoscopic L-design is an observable quantity periodically depending on the flux ϕ. In terms of the inductance of mesoscopic metal ring, $L = 8\pi r(\frac{1}{2} \ln \frac{8r}{a} - 1)$ where r is the radius of the ring and a is the radius of the metal wire, the formula for persistent current on a mesoscopic ring is obtained

$$I(\phi) = \frac{\hbar}{8\pi r(\frac{1}{2} \ln \frac{8r}{a} - 1)q_e} \sin(\frac{q_e}{\hbar}\phi). \tag{25}$$

Differing from the conventional formulation of the persistent current on the basis of quantum dynamics for electrons, This formulation presented a method from a new point of view. Formally, the $I(\phi)$ is a sine function with periodicity of $\phi_0 = h/q_e$. But either the model that the electrons move freely in an ideal ring [10], or the model that the electrons have hard-core interactions between them [11] can only give the sawtooth-type periodicity. Obviously, the sawtooth-type function is only the limit case for $q_e/\hbar \to 0$.

COUPLED CIRCUITS

The above discussions are based on a single mesoscopic circuit. Let us consider the case that several circuits coupled to each other by mutual inductances, associated capacities or any other kind of coupling. Because of quantum tunneling and quantum fluctuations, the charges in individual circuit is no longer precisely measurable. The charge on each circuit are not good quantum numbers, and therefore the charge operators $\hat{q}_j$ are expected to be noncommutative. A natural generalization of the formulation for single circuit is easily carried out in the pseudo-current representation, namely,

$$\hat{P}_j = \eta_j, \quad \hat{q}_j = i\hbar\left(1 + (\frac{q_e}{2\hbar})^2 \vec{\eta}^2\right)\frac{\partial}{\partial \eta_j}. \tag{26}$$

It is easily to obtain the following commutation relation,

$$[\hat{q}_i, \hat{q}_j] = i\frac{q_e^2}{2\hbar}M_{ij}, \tag{27}$$

where $M_{ij} := \hat{P}_i\hat{q}_j - \hat{P}_j\hat{q}_i$. The charge operators for individual circuit are generally noncommutative for the non-vanishing $\langle M_{ij} \rangle$. This provides a concrete physical example of noncommutative geometry. Further studies are in progress.

ACKNOWLEDGMENT

The work is supported by AvH Stiftung, NSFC-19975040 and EYFC98. I would like to thank J.C. Flores for communication and reference. Interesting discussions with O.W. Greenburg, M.S. Plyushchay, A. Solomon, E.C.G. Sudarshan, A. Zee et.al. are also acknowledged.

REFERENCES

1. Li Y.Q. and Chen B., *Phys. Rev. B* **53**, 4027-4032 (1996).
2. Snyder H.S., *Phys. Rev.* **71**, 38-41 (1947).
3. Li Y.Q., *Commun. Theor. Phys.* **24**, 255-256 (1995).
4. Mantecinos S., Saavedra I., and Kunstmann O., *Phys. Lett. A* **109**, 139-142 (1985).
5. Garay L.J., *Int. J. Mod. Phys. A* **10**, 145-166 (1995) and references therein.
6. Li Y.Q., "Mesoscopic quantum circuit theory to persistent current and Coulomb blockade" in *Proceedings of the 5th Wigner Symposium*, edited by P. Kasperkovitz & D. Grau, World. Sci., Singpore 1998 pp.307-310.
7. Louisell W.H., *Quantum Statistical Properties of Radiation*, John Wiley, New York 1973.
8. Kempf A., Mangano G., and Mann K.B., *Phys. Rev. D* **52**, 1108-1118 (1995).
9. Flores J.C., *e-print* cond-mat/9908012.
10. Cheung H.F., Gefen Y., Riedel E.K., et. al., *Phys. Rev. B*, **37**, 6050-6062 (1988).
11. Li Y.Q. and Ma Z.S., *J. Phys. Soc. Jpn.*, **65**, 1519-1522 (1996).

Inner Composition Law of Pure-Spin States

Vladimir I. Man'ko,[*] Giuseppe Marmo,[+] E. C. George Sudarshan,[†] and Francesco Zaccaria[+]

[*] *P.N. Lebedev Physical Institute, Leninskii Prospect 53, Moscow 117924, Russia*
[+] *Dipartimento di Scienze Fisiche, Universitá "Federico II" di Napoli and Istituto Nazionale di Fisica Nucleare, Sezione di Napoli, Complesso Universitario di Monte S. Angelo, Via Cintia, Napoli 80126, Italia*
[†] *Physics Department, Center for Particle Physics, University of Texas, Austin, Texas 78712, USA*

Abstract. Superposition principle for spin degrees of freedom is described in terms of density operators only using a formulated composition law of pure-state density operators. Decoherence phenomenon and visibility of the interference pattern are discussed.

INTRODUCTION

The pure states in quantum mechanics are associated with vectors $\mid \psi \rangle$ in Hilbert space [1] which correspond to the state wave functions [2]. The pure state can be also described by the projector, which is the density operator introduced in [3]. Recently [4–6] the composition of the pure-state density operators was discussed in the connection with the interference problem in quantum mechanics.

The discrete spin degrees of freedom have no classical limit. Due to this, it is worthy to consider the composition law of the spin-state projectors which provides the pure-state projector of another spin state.

Let us consider spin state of spin $j = 1/2$.

The generic normalized spin state is described by the vector $\mid \psi \rangle$ in two-dimensional Hilbert space

$$\mid \psi \rangle = \begin{pmatrix} a \\ b \end{pmatrix}, \qquad a = \left\langle \frac{1}{2}, \frac{1}{2} \middle| \psi \right\rangle, \qquad b = \left\langle \frac{1}{2}, -\frac{1}{2} \middle| \psi \right\rangle. \tag{1}$$

We used base vectors $\mid j, m \rangle$ with spin projection $m = \pm 1/2$. In (1), the complex numbers

$$a = |a|\, e^{i\varphi_a}, \qquad b = |b|\, e^{i\varphi_b}$$

satisfy the normalization condition

$$|a|^2 + |b|^2 = 1. \tag{2}$$

For the pure state $|\psi\rangle$, the density operator reads

$$\rho_\psi = |\psi\rangle\langle\psi|. \tag{3}$$

The positive densiy operator ρ_ψ of the pure spin-state satisfies the relations determining projector operator

$$\rho_\psi^\dagger = \rho_\psi, \qquad \rho_\psi^2 = \rho_\psi, \qquad \operatorname{Tr} \rho_\psi = 1. \tag{4}$$

The density matrix of pure state has rank equal unity and the unique nonzero eigenvalue equals to unity.

The density operator ρ of impure state is characterized by the impurity parameter

$$\mu_0 = \operatorname{Tr} \rho^2 < 1. \tag{5}$$

The example of density matrix of the pure spin state is the Hermitian matrix with two angle parameters

$$0 \leq \theta < \pi, \qquad 0 \leq \gamma < 2\pi,$$

which determine the point on the sphere S^2

$$\rho_\psi = \begin{pmatrix} \cos^2\theta & e^{i\gamma}\cos\theta\sin\theta \\ e^{-i\gamma}\cos\theta\sin\theta & \sin^2\theta \end{pmatrix}, \tag{6}$$

where

$$|a|^2 = \cos^2\theta, \qquad \gamma = \varphi_a - \varphi_b. \tag{7}$$

COMPOSITION LAW

In order to formulate a composition law of two density operators

$$\rho_1 = |\psi_1\rangle\langle\psi_1| \tag{8}$$

and

$$\rho_2 = |\psi_2\rangle\langle\psi_2|, \tag{9}$$

which corresponds to the superposition of two orthogonal state vectors $|\psi_1\rangle$ and $|\psi_2\rangle$ with complex coefficients c_1 and c_2 satisfying the condition

$$|c_1|^2 + |c_2|^2 = 1,$$

we use a normalized fiducial vector $|\psi_0\rangle$.

Two orthogonal pure spin-states have density matrices

$$\rho_{1,2} = \begin{pmatrix} \cos^2 \theta_{1,2} & e^{i\gamma_{1,2}} \cos \theta_{1,2} \sin \theta_{1,2} \\ e^{-i\gamma_{1,2}} \cos \theta_{1,2} \sin \theta_{1,2} & \sin^2 \theta_{1,2} \end{pmatrix}, \quad (10)$$

where

$$\theta_2 = \frac{\pi}{2} - \theta_1, \qquad \gamma_2 = \pi + \gamma_1. \quad (11)$$

Let us associate with the density operators ρ_1 and ρ_2 two normalized vectors

$$|\psi_1\rangle = \frac{\rho_1 |\psi_0\rangle}{\sqrt{\langle\psi_0|\rho_1|\psi_0\rangle}} \quad (12)$$

and

$$|\psi_2\rangle = \frac{\rho_2 |\psi_0\rangle}{\sqrt{\langle\psi_0|\rho_2|\psi_0\rangle}}. \quad (13)$$

If one considers a superposition of two arbitrary state vectors $|\psi_1\rangle$ and $|\psi_2\rangle$

$$|\psi\rangle = c_1 |\psi_1\rangle + c_2 |\psi_2\rangle, \quad (14)$$

one obtains a corresponding density operator which reads

$$\rho = |c_1|^2 |\psi_1\rangle\langle\psi_1| + |c_2|^2 |\psi_2\rangle\langle\psi_2| + c_1 c_2^* |\psi_1\rangle\langle\psi_2| + c_2 c_1^* |\psi_2\rangle\langle\psi_1|. \quad (15)$$

Each of the state vectors $|\psi_1\rangle$ and $|\psi_2\rangle$ can be multiplied by constant phase factors $e^{i\Theta_1}$ and $e^{i\Theta_2}$, respectively. The density operators

$$|\psi_1\rangle\langle\psi_1| = \rho_1 \quad \text{and} \quad |\psi_2\rangle\langle\psi_2| = \rho_2$$

do not depend on these phase factors. But the density operator of superposition (14) given by (15) depends on the relative phase

$$\varphi = \Theta_1 - \Theta_2.$$

To obtain the composition law of two density operators ρ_1 and ρ_2 corresponding to the superposition of state vectors (14), we define it using the superposition of vectors (12) and (13)

$$\rho = \left(c_1 |\psi_1\rangle + c_2 |\psi_2\rangle\right)\left(c_1^* \langle\psi_1| + c_2^* \langle\psi_2|\right). \quad (16)$$

In terms of density operators of pure states ρ_1 and ρ_2, the density operator (16) reads

$$\rho = |c_1|^2 \rho_1 + |c_2|^2 \rho_2 + [c_1 c_2^* \rho_1 P_0 \rho_2 + \text{h.c.}] \left[\text{Tr}(\rho_1 P_0) \text{Tr}(\rho_2 P_0)\right]^{-1/2}, \quad (17)$$

where
$$P_0 = |\psi_0\rangle\langle\psi_0|.$$

The first two terms in (17) provide only impure density operator. The rest interference term provides a purification of the impure density operator. One can see that

$$\frac{\rho_1 P_0 \rho_2}{\left[\text{Tr}(\rho_1 P_0)\text{Tr}(\rho_2 P_0)\right]^{1/2}} = |\psi_1\rangle\langle\psi_2| \, e^{i\varphi}, \qquad (18)$$

where

$$e^{i\varphi} = \frac{\langle\psi_1|\psi_0\rangle\langle\psi_0|\psi_2\rangle}{|\langle\psi_1|\psi_0\rangle\langle\psi_0|\psi_2\rangle|}. \qquad (19)$$

This means that the introduction of a fiducial vector gives us the possibility to describe the relative phase of the state vectors $|\psi_1\rangle$ and $|\psi_2\rangle$ in the composition law of the density operators. The role of the fiducial vector $|\psi_0\rangle$ is to select a vector each within the one-dimensional vector space determined, respectively, by the density operators ρ_1 and ρ_2.

To describe the phenomenon of a partial decoherence, one can generalize formula (17). The trace of the last interference term in (17) equals zero. By elaborating an additional parameter, visibility of the interference pattern in terms of the projectors can be associated with the extended composition rule

$$\rho = |c_1|^2 \rho_1 + |c_2|^2 \rho_2 + \gamma\left[c_1 c_2^* \rho_1 P_0 \rho_2 + \text{h.c.}\right]\left[\text{Tr}(\rho_1 P_0)\text{Tr}(\rho_2 P_0)\right]^{-1/2}. \qquad (20)$$

If the coefficient $\gamma = 1$, one has the pure state (17). If the coefficient $\gamma = 0$, one has the mixed state (20) without the interference pattern. The purity parameter of the mixed state (20) reads

$$\mu_0 = 1 - 2\left(1 - \gamma^2\right)|c_1 c_2|^2. \qquad (21)$$

The parameter γ can be incorporated by using complex values of the angle in Eq. (19).

COMPOSITION LAW AND DEFORMED ASSOCIATIVE PRODUCT

Now we may incorporate the factor P_0 by using a K-deformed associative product on the space of operators by setting

$$A \cdot_K B := AKB; \qquad K = P_0,$$

as considered in [7]. In this more compact form, the extension to many projection operators can be easily written down and, for composition of n density operators, it reads [6]

$$\rho = \sum_{i,j=1}^{n} c_i c_j^* \frac{\rho_i \cdot_K \rho_j}{\left[\text{Tr} \left(\rho_i \cdot_K \rho_j \cdot_K P_0 \right) \right]^{1/2}}. \tag{22}$$

We used the relation:

$$\text{Tr}\left(\rho_i P_0\right) \text{Tr}\left(\rho_j P_0\right) = \text{Tr}\left(\rho_i P_0 \rho_j P_0\right). \tag{23}$$

The above formulas are written for generic spin j. The relative phases of the pure spin states can be included into relative phases of complex numbers c_i, c_j^*.

For two j-spin states, the composition law of nonorthogonal density operators holds by the replacement of the operator P_0 with the projector

$$P_j = |j,j\rangle\langle j,j| \tag{24}$$

and using angle φ given by Eq. (19). Thus, for the j-spin case, one has [4]

$$\rho_\varphi = \Big\{ |c_1|^2 \rho_1 + |c_2|^2 \rho_2 + \left(c_1 c_2^* e^{i\varphi} \rho_1 |j,j\rangle\langle j,j| \rho_2 + \text{h.c.} \right)$$
$$\times \left[\text{Tr}\left(\rho_1 P_j\right) \text{Tr}\left(\rho_2 P_j\right) \right]^{-1/2} \Big\}$$
$$\times \Big\{ 1 + \left[c_1 c_2^* e^{i\varphi} \text{Tr}\, \rho_1 |j,j\rangle\langle j,j| \rho_2 + c_1^* c_2 e^{-i\varphi} \text{Tr}\, \rho_2 |j,j\rangle\langle j,j| \rho_1 \right]$$
$$\times \left[\text{Tr}\left(\rho_1 P_j\right) \text{Tr}\left(\rho_2 P_j\right) \right]^{-1/2} \Big\}^{-1}. \tag{25}$$

One can check that, for arbitrary projectors ρ_1 and ρ_2, the Hermitian operator ρ_φ given by (25) satisfies the relations

$$\rho_\varphi^2 = \rho_\varphi; \qquad \text{Tr}\, \rho_\varphi^2 = 1.$$

Thus, for the spin superposition state in the case

$$|c_1|^2 + |c_2|^2 = 1,$$

we obtain the density matrix with parameters θ and γ given by the relations

$$\cos^2\theta = \frac{|c_1|^2 \cos^2\theta_1 + |c_2|^2 \cos^2\theta_2 + \left(c_1 c_2^* e^{i\varphi} + c_1^* c_2 e^{-i\varphi} \right) \cos\theta_1 \cos\theta_2}{1 + \left(c_1 c_2^* e^{i\varphi} + c_1^* c_2 e^{-i\varphi} \right) \cos(\theta_1 - \theta_2)} \tag{26}$$

and

$$\cos\theta \sin\theta\, e^{i\gamma} = \frac{|c_1|^2 \cos\theta_1 \sin\theta_1\, e^{i\gamma_1} + |c_2|^2 \cos\theta_2 \sin\theta_2\, e^{i\gamma_2}}{1 + \left(c_1 c_2^* e^{i\varphi} + c_1^* c_2 e^{-i\varphi} \right) \cos(\theta_1 - \theta_2)}$$
$$+ \frac{c_1 c_2^* e^{i\varphi} \cos\theta_1 \sin\theta_2\, e^{i\gamma_2} + c_1^* c_2 e^{-i\varphi} \sin\theta_1 \cos\theta_2\, e^{i\gamma_1}}{1 + \left(c_1 c_2^* e^{i\varphi} + c_1^* c_2 e^{-i\varphi} \right) \cos(\theta_1 - \theta_2)}. \tag{27}$$

The main result of our paper is the formulation of the composition law for density operators of pure spin states, which yields the rule of purification of the impure mixture of quantum states. This composition law connects the Schrödinger and von Neumann descriptions of pure spin states and provides a rule of the pure-state superposition in terms of density operators.

ACKNOWLEDGMENTS

V.I.M. and E.C.G.S. thank Dipartimento di Scienze Fisiche Universitá "Federico II" di Napoli and Istituto Nazionale di Fisica Nucleare, Sezione di Napoli for kind hospitality.

V.I.M. was partially supported by the Russian Foundation for Basic Research under Project No. 99-02-17753 and the Ministry for Science and Technology of the Russian Federation within the framework of the Program "Fundamental Nuclear Physics."

REFERENCES

1. Dirac, P.A.M., *The Principles of Quantum Mechanics*, Pergamon, Oxford, 1958, 4th ed.
2. Schrödinger, E., *Ann. Physik* **79**, 489 (1926).
3. Von Neumann, J., *Mathematische Grundlagen der Quantummechanik*, Springer, Berlin, 1932; *Göttingenische Nachrichten*, 11 (Nov. 1927) S. 245–272.
4. Man'ko, V.I., Marmo, G., Sudarshan, E.C.G., and Zaccaria, F., *J. Russ. Laser Res.* (Kluwer Academic/Plenum Publishers) **20**, 421–437 (1999).
5. Man'ko, V.I., Marmo, G., Sudarshan, E.C.G., and Zaccaria, F., "Purification of impure density operators and the recovery of entanglements," E-print quant/ph-9910080.
6. Man'ko, V.I., Marmo, G., Sudarshan, E.C.G., and Zaccaria, F., "Inner composition law of pure states as a purification of impure state," *Phys. Lett. A* (2000, to appear).
7. Man'ko, V.I., Marmo, G., Sudarshan, E.C.G., and Zaccaria, F., *Int. J. Mod. Phys. A* **11**, 1281–1296 (1997).

Symmetry Considerations in Quantum Computing

A. Otte and G. Mahler

*University of Stuttgart, Institute for Theoretical Physics I
Pfaffenwaldring 57/IV, 70550 Stuttgart, Germany
email: al@theo.physik.uni-stuttgart.de*

Abstract. A quantum computer is, if realized, the ultimate machine for completely controlling a quantum-mechanical system. This control enables one to prepare any arbitrary state of the Hilbert space and to determine by will its time evolution. From this point of view, the realization of quantum states with unusual symmetries seems to be possible by the use of such quantum information processing devices. On the other hand, symmetry arguments can be used to gain more insight in such a quantum network consisting of identical qubits and they can even further lead to error prevention codes by identifying decoherence free subspaces. By introducing a collective operator description for quantum networks, we show how to describe ensemble systems using a reduced set of parameters. Further we want to point out some connections between quantum computers and symmetry considerations.

COLLECTIVE OPERATORS

A complete description of large quantum networks turns out to be virtually impossible, as the Hilbert space dimension grows exponentially with the number of particles used. However, symmetry can reduce the complexity of the total system by a significant amount: For a perfect permutation-symmetric system we will show that the reduction in the number of parameters needed is so enormous, that a polynomial increasing number of parameters is enough to describe networks of arbitrary size.

For now we will stick to quantum networks (and therefore quantum computers) built out of N two-level systems, knowing that a generalization to n-level systems is straight forward. First one needs to specify which single particle operator basis is used, e.g. $\hat{\sigma}_x$, $\hat{\sigma}_y$, $\hat{\sigma}_z$ (which is hermitian and unitary) or $\hat{\sigma}_\pm = \hat{\sigma}_x \pm i\hat{\sigma}_y$, $\hat{\sigma}_z$ or any other complete basis. We then introduce cluster operators $\hat{C}_{\alpha\beta\gamma,p}$, meaning an operator of dimension $2^N \times 2^N$, where α, β and γ specify the multiplicity of $\hat{\sigma}_x$, $\hat{\sigma}_y$, and $\hat{\sigma}_z$, respectively ($\alpha + \beta + \gamma \leq N$). Index p specifies a permutation of these among the N subsystems. The number of such permutations and hence the index range for $p \in [0, \Omega - 1]$ is

$$\Omega(\alpha, \beta, \gamma) = \frac{N!}{\alpha!\beta!\gamma!(N-\alpha-\beta-\gamma)!} . \tag{1}$$

These operators $\hat{C}_{\alpha\beta\gamma,p}$, again, span the whole Liouville-space by defining all subsystem specific properties. We now go on to define collective operators $\hat{E}$ by

$$\hat{E}_{\alpha\beta\gamma,b} = \sum_{p=0}^{\Omega-1} \omega_\Omega^{pb} \hat{C}_{\alpha\beta\gamma,p} ; \qquad \omega_\Omega = e^{\frac{2\pi i}{\Omega}} . \tag{2}$$

To ensure that all subsystems are treated on equal footing, the sum extends over all permutations p, weighted only with pure phase fators, where $b \in [0, \Omega-1]$ labels the phase shift between "neighbouring" p. (Here the numbering of permutations is a matter of choice and the phase has no physical meaning.) The set of collective operators is orthonormal

$$\frac{1}{\Omega\, 2^N} \operatorname{tr} \left\{ \hat{E}_{\alpha\beta\gamma,b} \hat{E}^\dagger_{\alpha'\beta'\gamma',b'} \right\} = \delta_{\alpha\alpha'} \delta_{\beta\beta'} \delta_{\gamma\gamma'} \delta_{bb'} \tag{3}$$

and complete, so that the density operator $\hat{\rho}$ of the network can be decomposed as

$$\hat{\rho} = \frac{1}{2^N} \sum_{\{\alpha\beta\gamma\}} \sum_b E_{\alpha\beta\gamma,b} \hat{E}_{\alpha\beta\gamma,b} , \tag{4}$$

$$E_{\alpha\beta\gamma,b} = \frac{1}{\Omega} \operatorname{tr} \left\{ \hat{\rho} \cdot \hat{E}^\dagger_{\alpha\beta\gamma,b} \right\} . \tag{5}$$

The expectation values $E_{\alpha\beta\gamma,b}$ are collective in the sense that they do *not* refer to specific subsystem-indices.

PERMUTATION SYMMETRY

Within a perfect permutation-symmetric network no subsystem can be distinguished, neither in preparation nor in detection. Two possible setups can be thought of: Fundamental indistinguishable and operational indistinguishable subsystems. A typical setup of the former type would be several electrons in a box. As the electrons are fermions and their location is not fixed, there is no way to act independently on a specific electron. All controlling and measurement procedures act on the system as a whole. Therefore no information loss can occur when reducing the system-description to permutation-symmetric operators.

Opposed to that, in the case of operational indistinguishability, the design of the experimental setup is the source of reduction. A linear ion trap with a laser beam acting on the ions could be used as an example. Only if the beam waist is less than the spatial separation of the ions, the particles become distinguishable, so by controlling the laser beam the experimentalist can choose which case he is in.

Collective operators $\hat{E}_{\alpha\beta\gamma,b}$ with $b = 0$ are invariant under any arbitrary subsystem permutation $\hat{P}$

$$\hat{P}^\dagger \hat{E}_{\alpha\beta\gamma,0} \hat{P} = \hat{E}_{\alpha\beta\gamma,0} \ . \tag{6}$$

For indistinguishable subsystems, these are the only operators allowed: Any control or measurement operator has to be part of the Liouville-space spanned by $\hat{E}_{\alpha\beta\gamma,0}$. The number of such operators is (for $n = 2$)

$$\xi_0 = \frac{(N+1)(N+2)(N+3)}{6} \ , \tag{7}$$

and thus scales polynomially of order $\mathcal{O}(N^3)$ with the number of subsystems involved. Therefore, such a highly symmetric system constitutes a class of reduced complexity as compared with a general quantum network of the same size N. This is somewhat surprising as indistinguishability might have been expected to enhance non-classical features.

Alternatively, if all expectation-values happen to be permutation-symmetric (no "structure"), then

$$\langle \hat{C}_{\alpha\beta\gamma,p} \rangle = \langle \hat{C}_{\alpha\beta\gamma} \rangle \tag{8}$$

for all p-permutations, and the definition eq. (2) leads to

$$\langle \hat{E}_{\alpha\beta\gamma,b} \rangle = \Omega \langle \hat{C}_{\alpha\beta\gamma} \rangle \delta_{0b} \tag{9}$$

i.e. we need to consider only the $\hat{E}_{\alpha\beta\gamma,0}$ operators.

Symmetry classes and unusual symmetries

Permutation symmetry defines a kind of operational indistinguishability between the subsystems of a given network. This symmetry alone, however, would allow for more symmetry classes than realized in nature by the fundamentally indistinguishable particles, Fermions and Bosons, respectively. The spin-statistics-relation going back to Pauli [1,3] might be relaxed, if the particles or subsystems are localized in different areas of real space [2]. The location index would, in principle, render these subsystems distinguishable; however, for the following we assume that the actual operators describing the network and its coupling to the outside world are still permutation-symmetric, so that the corresponding super-selection rules apply, as will be discussed below. This would imply, e.g. that electrons, localized in different semiconductor quantum dots, could live in the state-subspace of Bose-symmetry (or of any other "para-boson" [3] symmetry class as well, provided we are able to prepare such states from some standard initial state: Directed transient symmetry breaking could do this job.

Any symmetry type can be characterized by a Young diagram and is equivalent to a irreducible representation of the permutation group S_N spanned by the basis vectors. Since a definite angular momentum quantum number j is assigned to every Young diagram, there are $2j+1$ states of equal symmetry type but different energy (i.e. different configuration). If the system is subjected to permutation-symmetric operators only, the super-selection rules prohibit any transition between different Young tables. In case the system under consideration is in a specific state with angular momentum j, the state space reached by applying any collective operator $\hat{E}_{ijk,0}$ is of dimension $2j+1$. Therefore $(2j+1)^2$ parameters (expectation values) are needed to describe the system. The total number of parameters, $\sum_{j=0}^{N/2}(2j+1)^2 = \frac{1}{6}(N+1)(N+2)(N+3)$ is exactly the number ξ_0 of collective operators $\hat{E}_{ijk,0}$ (see eq. (7)).

Structure and Hamilton-models

For fundamentally indistinguishable particles, the subsystem index μ has no physical meaning. However, it may happen that a specific property is not only a good quantum number but a unique constant of motion for any subsystem. Localization in real space is a pertinent example; the subsystem-index μ is then mapped onto a spatial position-index R_μ. By this position any subsystem becomes distinguishable, in principle. The phases (entering the collective operators $\hat{E}$) get a physical meaning ("wave length"). An "operational" indistinguishability remains, if the Hamiltonian describing the network does still contain collective-operators of the type $\hat{E}_{\alpha\beta\gamma,0}$ only.

Typical Hamilton-models include $m = 1, 2$ -particle operators. The structure tends to break permutation-symmetry for the localized states as the coupling usually depends on the distance $|R_\mu - R_\nu|$. This partial selectivity can be described as a perturbation via collective operators $\hat{E}_{\alpha\beta\gamma,b}$ with $b \neq 0$.

If all pertinent distances could be made equal, the breaking of the permutation symmetry would go to zero. For $N > D+1$ (in D dimensions) however, the interaction distances cannot all be the same. A partial remedy is the introduction of a "quantum bus": In this case the nodes do not interact directly but only indirectly via a common degree of freedom. This degree of freedom could be a central spin, but typically is implemented as a collective mode (like the phonon mode of a cold ion trap [4]).

SYMMETRY BREAKING AND IRREVERSIBILITY

Constrained operations will lead, quantum mechanically, to selection rules accompanied by a tremendous reduction of the state space available to the system dynamics starting from a given initial state. This situation needs to be distinguished from lack of control (measurement data) implying lack of information (entropy $S > 0$). In so far as this lack of control refers to expectation values $b \neq 0$ (which would be absent for strict permutation symmetry), "uncontrolled" symmetry breaking may be said to lead to an ensemble description and irreversibility: The true state space has been reduced to the smaller one defined by the assumed permutation symmetry. In this case, however, the "decoherence" does not reflect the influence of an external bath but is rather of "internal" origin. Phenomenologically ony may try to model these effects, as usual, via some decoherence times; however, it is not clear yet under what conditions such a procedure would be appropriate. Clearly it should, at most, work for sufficiently large networks and only to the extent that the reduction is really substantial (cf. ref. [5]). The decoherence time would then, in turn, allow to assess symmetry breaking effects in a global way.

INTERPLAY BETWEEN SELECTIVE AND COLLECTIVE INTERACTIONS

Selective interactions violate the selection rules implicit in permutation-symmetric interactions. This qualitatively different dynamical behaviour can be exploited to implement specific functionalities: One possibility is to address different symmetry-classes via selective coupling (i.e. controlled symmetry breaking); another possibility is to suppress transition due to permutation-symmetric interactions by using states of different symmetry, which, nevertheless, could all be prepared selectively ("decoherence-free subspace").

To be specific, let us consider a network of N pseudo-spins without mutual interactions, but in the presence of a quantum bus (i.e. a collective mode to which all spins are coupled in the same way). We further assume that the coupling to a larger field can be made at will either selective or collective. We start (for $N = 4$, $n = 2$) with the permutation-symmetric ground state $|0000\rangle$, ($j = 2$). By applying selective laser pulses and exploiting the coupling to the quantum-bus we can generate the EPR-state $\frac{1}{\sqrt{2}}(|01\rangle - |10\rangle)$ within the pair $\nu = 1, 2$. The total 4-particle state then becomes a member of the symmetry class $j = 1$. If permutation symmetry is restored now, transitions are possible only within the $(2j + 1)$-dimensional subspace of this symmetry class (on a time-scale less than the decoherence time). This would allow to study finite systems of (operationally) indistinguishable subsystems of a symmetry not realized in nature by fundamentally indistinguishable particles! It is straightforward to extend this scheme to $N > 4$.

Alternatively, the selection rules can readily be exploited as a means for stabilization: For this purpose we assume that the "unwanted" dynamics (coupling to the bath) is permutation-symmetric, while the control dynamics to be used is selective (i.e. not subject to the selection rules). For $N = 4$ we could take the three lowest energy states $j = 2, 1, 0$. Unitary dynamics within this subspace would (in the ideal case) not be perturbed by dissipation, if the bath was kept at zero temperature. (There are higher energy levels with the same symmetry, though.)

It is thus preferable to use the multiplicity of the states $j = 0$ (for N even) which would decouple from the bath exactly, as each of those states is the only member of its symmetry class: There is nothing to connect to under the action of a permutation-symmetric coupling (cf. ref [6]). Such schemes have been investigated by a number of authors [7–11]. The resulting stabilization is limited by the fact that the symmetry selection rules will, in practice, not hold strictly.

REFERENCES

1. See, e.g. Streater, R. F., and Wightman, A. S., *PCT, Spin and Statistics, and All That*, B. Cummings, New York, 1964.
2. Alexandrov, A. S., and Giles, R. T., cond-mat/9704164 (1997).
3. Greenberg, O. W., and Messiah, A. M. L., *Phys. Rev.* **138**B, 1155 (1965).
4. Cirac, J. I., and Zoller, P., *Phys. Rev. Lett.* **74**, 4091 (1995).
5. Munowitz, M., Pines, A., and Mehring, M., *J. Chem. Phys.* **86**, 3172 (1987).
6. Zanardi, P., and Rasetti, M., *Phys. Rev. Lett.* **79**, 3306 (1997).
7. Lidar, D. A., Chuang, I. L., and Whaley, K. B., *Phys. Rev. Lett.* **81**, 2594 (1998).
8. Zanardi, P., and Rossi, F., *Phys. Rev. B* **59**, 8170 (1998).
9. Duan, L., and Guo, G., *Phys. Rev. A* **57**, 2399 (1998).
10. Duan, L. and Guo, G., *Phys. Lett. A* **243**, 265 (1998).
11. Zanardi, P., quant-ph/9809064 (1998).

Symmetrizing The Symmetrization Postulate

Michael York

975 S. Eliseo Dr. #9, Greenbrae, CA 94904, USA

Abstract. Reasonable requirements of (a) physical invariance under particle permutation and (b) physical completeness of state descriptions [1], enable us to deduce a Symmetric Permutation Rule(SPR): that by taking care with our state descriptions, it is always possible to construct state vectors (or wave functions) that are purely symmetric under pure permutation for all particles, regardless of type distinguishability or spin. The conventional exchange antisymmetry for two identical half-integer spin particles is shown to be due to a subtle interdependence in the individual state descriptions arising from an inherent geometrical asymmetry. For three or more such particles, however, antisymmetrization of the state vector for all pairs simultaneously is shown to be impossible and the SPR makes observably different predictions, although the usual pairwise exclusion rules are maintained. The usual caveat of fermion antisymmetrization – that composite integer spin particles (with fermionic constituents) behave only approximately like bosons – is no longer necessary.

I TERMINOLOGY

First let me express my deep gratitude to the organizers for the opportunity to present this paper at this conference.

The usual terminology of fermion/boson identifies the statistics of a particle with the symmetry or antisymmetry of the wave function for pairs of such particles. In this talk, I will show that this identification is, at best, insufficiently specific and, in some circumstances, erroneous. To avoid confusion, I will first define the terms I use.

Halfon : Particle with half-integer spin.

Fullon : Particle with integer spin.

Fermion : Particle which obeys Fermi-Dirac statistics. (Is *not* defined by antisymmetry of the wave function.)

Boson : Particle which obeys Bose-Einstein statistics. (Is *not* defined by symmetry of the wave function.)

Generalized Exclusion Rule : Pairs of identical particles, for which all other quantum numbers are the same, must have even composite spin. (Note: this is the same rule for both halfons and fullons.)

Spin-Statistics Theorem : Halfons are fermions. Fullons are bosons. (Consequence of the Generalized Exclusion Rule.)

The chief reason for this terminology, breaking the identification of fermion statistics with antisymmetry, is that it is possible to obtain the same observable fermionic behavior for halfons with wave functions that are pure permutation symmetric. However, to see this requires a more thorough analysis of state descriptions and uniqueness of state vectors than is usual.

We usually label state vectors by a set of variables which we shall call a *state description*. Typically, state descriptions are lists of quantum numbers. Physical transformations that change these quantum numbers therefore change one state vector into another. Conventionally we assume we can choose a unique state vector for a given set of quantum numbers from an infinite set mutually related by arbitrary phase factors.

However, there are some physically significant transformations that leave quantum numbers unchanged but nevertheless change state vectors, even if only by a phase. To distinguish such state vectors, and still choose them uniquely, we need state descriptions that carry more information than just the quantum numbers.

Physical Completeness (Of A State Description) : Requires the description of a state with sufficient precision to distinguish it from another state description related by any physically significant transformation – even if the quantum numbers are unchanged.

Uniqueness Principle : To choose a unique state vector for a given state description, the state description must be *physically complete*. (Follows from the definition above.)

The concept of *physical completeness* helps us to differentiate state vectors related by physically significant transformations that change only the phase. Since nature does not care about the order in which we describe individual particles in a multi-particle state, we can also write:

Permutation Invariance Principle : Permutation of individual particles in a multi-particle state description is not a physically significant transformation.

Symmetric Permutation Rule (SPR) : For any multi-particle state for which each particle has a physically complete state description, each of which is independent of all the other individual state descriptions or their order, it is always possible to choose a state vector that is unchanged by (i.e. symmetric under) *pure permutation*, regardless of particle spin or even identity. (Follows from *physical completeness* and *permutation invariance*).

Exchange Asymmetry : When order dependent state descriptions are used, permutation implies a reversal of the order dependence. This may introduce a physically significant "exchange" transformation in addition to pure permutation and may result in a change of sign of the state vector.

II PHYSICAL COMPLETENESS AND SPIN

We have argued that physical completeness is not a trivial matter of listing quantum numbers. We now need to explore what it means in the case of particles with spin. First, however, a few more terms need to be defined:

Spin Quantization Frame (SQF) : The frame of reference in which we measure the spin component.

Canonical Frame : The frame of reference in which we measure the position or momentum.

In general, the relative orientation of the SQF to the canonical frame need not always be a null rotation. Even when the axes of both frames coincide, for example, it could be given by a 2π rotation about some arbitrary axis.

We shall now consider those physically significant transformations that can change the phase of the state vector for a particle with spin.

A Single Particle SQFs

In the usual methodology [2]:

$$|Q, \boldsymbol{p}, s, m(\hat{\boldsymbol{n}})> = U(B(\boldsymbol{p})) |Q, \boldsymbol{0}, s, m(\hat{\boldsymbol{n}})> \qquad (1)$$

$|Q, \boldsymbol{0}, s, m(\hat{\boldsymbol{n}})>$ is a rest frame eigenstate of spin s and component m in the direction $\hat{\boldsymbol{n}}$. Q represents all other intrinsic quantum numbers. $U(B(\boldsymbol{p}))$ is boost operator for $\boldsymbol{0} \to \boldsymbol{p}$ (conventionally, but not necessarily, a *Lorentz* boost).

$$U(B(\boldsymbol{p})) = U(R(\hat{\boldsymbol{z}} \to \hat{\boldsymbol{p}}))U(B(p\hat{\boldsymbol{z}}))U(R^{-1}(\hat{\boldsymbol{z}} \to \hat{\boldsymbol{p}})) \qquad (2)$$

However, this is **ambiguous** because $\hat{\boldsymbol{n}}$ does not uniquely specify the rotation $S \to C$ (S = SQF, C = canonical frame).

Instead, we prefer:

$$|Q, \boldsymbol{p}, s, m>^S = U(B(\boldsymbol{p})) |Q, \boldsymbol{0}, s, m>^S \qquad (3)$$

and, in terms of a standard base frame B,

$$\begin{aligned}|Q, \boldsymbol{p}, s, m>^S &= |Q, \boldsymbol{p}, s, m(R_{BS})>_B \\ &= \sum_{m'} D^s_{m'm}(R_{BS})|Q, \boldsymbol{p}, s, m'>^B \end{aligned} \qquad (4)$$

where R_{BS} is the rotation $B \to S$. These latter two additions (B, R_{BS}) to the usual list of quantum numbers are clearly essential for a physically complete state description.

Here are some commonly used examples:

$$|Q, \boldsymbol{p}, s, \lambda >^H = |Q, \boldsymbol{p}, s, \lambda(N) >_H \tag{5}$$
$$|Q, \boldsymbol{p}, s, m >^C = |Q, \boldsymbol{p}, s, m(N) >_C \tag{6}$$
$$|Q, \boldsymbol{p}, s, m >^C = |Q, \boldsymbol{p}, s, m(R_{HC}) >_H$$
$$= \sum_\lambda D^s_{\lambda m}(R_{HC}) |Q, \boldsymbol{p}, s, \lambda >^H \tag{7}$$
$$|Q, \boldsymbol{p}, s, \lambda >^H = |Q, \boldsymbol{p}, s, \lambda(R_{CH}) >_C$$
$$= \sum_m D^s_{m\lambda}(R_{CH}) |Q, \boldsymbol{p}, s, m >^C \tag{8}$$

where C is the canonical frame, H is a helicity frame and N is a null rotation. For massive particles, it is reasonable to choose C for a base frame. To include massless particles, however, it is more reasonable to choose a frame H for which the spin is quantized along the direction of motion. Of course, this still leaves the $x-$ and $y-$ axes undefined, but specifying the rotation R_{CH} uniquely will resolve the matter.

B Two-Particle SQFs

For two or more particles we must define B for both particles. A common methodology is to choose $B = C$ for both particles and indeed it is quite possible to do so. However there is a very subtle complication that we must take care over: *an inherent geometrical asymmetry between any two vectors in a common frame of reference.*

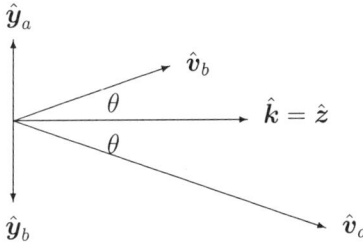

As shown in the figure, we can select a common z-axis symmetrically by choosing $\boldsymbol{k}$ which bisects the two vectors $\hat{\boldsymbol{v}}_a$ and $\hat{\boldsymbol{v}}_b$. But as soon as we get to the other axes we find an asymmetry. For example, each of the two choices of y-axis shown, and their accompanying x-axis, is asymmetric with respect to $\hat{\boldsymbol{v}}_a$ and $\hat{\boldsymbol{v}}_b$. A classic manifestation of this asymmetry lies in the relationship between the polar angles of two momentum vectors in their CM frame. Although θ_a and θ_b are symmetrically related by $\theta_b = \pi - \theta_a$, the relationship between ϕ_a and ϕ_b is asymmetric: e.g. $\phi_b = \pi + \phi_a$.

To restore and ensure symmetry, when defining two-particle SQFs, it is better to use a common method to define B *independently* for each particle. If we then rotate each separately to a common SQF when required, we can do so in a way that makes the asymmetry explicit. An helicity frame provides a good choice for this – as long as we specify all axes using a symmetric method.

For "current" particle c, where $c = a$ or b, and "other" particle o, we define the independent helicity frames by

$$\hat{z}_c = \hat{p}_c$$
$$\hat{y}_c = \hat{p}_c \times \hat{p}_o / |\hat{p}_c \times \hat{p}_o| \qquad (9)$$

To get to the canonical frame, we then have:

$$|Q_a, \boldsymbol{p}_a, s_a, m_a; Q_b, \boldsymbol{p}_b, s_b, m_b >^C$$
$$= |(Q_a, \boldsymbol{p}_a, s_a, m_a(R_a))_{H_a}; (Q_b, \boldsymbol{p}_b, s_b, m_b(R_b))_{H_b} >$$
$$= \sum_{\lambda_a \lambda_b} D^{s_a}_{\lambda_a m_a}(R_a) D^{s_b}_{\lambda_b m_b}(R_b) |Q_a, \boldsymbol{p}_a, s_a, \lambda_a; Q_b, \boldsymbol{p}_b, s_b, \lambda_b >^H \qquad (10)$$

Since the helicity state vector on the right uses order independent and physically complete state descriptions, the SPR tells us that it is permutation symmetric. As long as R_a, R_b are both uniquely specified and order independent then the canonical state vector on the left will also be permutation symmetric.

However, we can also write

$$R_b = R_a.R_{ba} = R_a.R_{\boldsymbol{k}}(\pm\pi) \qquad (11)$$

where R_{ba} takes the helicity frame of b into that of a and is a rotation by $\pm\pi$ about $\boldsymbol{k}$. The sign ambiguity encapsulates the problem with physical completeness for the two-particle state vector in the canonical frame, since, for a halfon, the difference between the two possible choices causes a sign difference between the two independent canonical state vectors that result from eqn. 10 for a given R_a.

If the state description does not specify unique choices of R_a, R_b or, equivalently, R_a, R_{ba}, then it will not be complete. We would then not be able to determine how the state vector transforms under permutation. This ambiguity persists even in the limit that the momenta coincide and is inherent in the conventional way we define state vectors by simple lists of quantum numbers.

One alternative to choosing a fixed value of R_{ba} is to fix the relative orientation between particle "1" and particle "2":

$$R_2 = R_1.R_{21} = R_1.R_{\boldsymbol{k}}(\pm\pi) \qquad (12)$$

This gives an order dependent method of specifying physical completeness:

$$|(Q_a, \boldsymbol{p}_a, s_a, m_a)^1; (Q_b, \boldsymbol{p}_b, s_b, m_b)^2 >^C$$
$$= |(Q_a, \boldsymbol{p}_a, s_a, m_a(R_a))_{H_a}; (Q_b, \boldsymbol{p}_b, s_b, m_b(R_a.R_{21}))_{H_b} >$$
$$|(Q_b, \boldsymbol{p}_b, s_b, m_b)^1; (Q_a, \boldsymbol{p}_a, s_a, m_a)^2 >^C$$
$$= |(Q_b, \boldsymbol{p}_b, s_b, m_b(R_b))_{H_b}; (Q_a, \boldsymbol{p}_a, s_a, m_a(R_b.R_{21}))_{H_a} > \qquad (13)$$

and we see that there are two cases which tells us how to compute the exchange phase. The first case is:

$$R_b = R_a.R_{21}$$
$$|(Q_b, \boldsymbol{p}_b, s_b, m_b)^1; (Q_a, \boldsymbol{p}_a, s_a, m_a)^2 >^C$$
$$= (-)^{2s_a}|(Q_a, \boldsymbol{p}_a, s_a, m_a)^1; (Q_b, \boldsymbol{p}_b, s_b, m_b)^2 >^C \qquad (14)$$

and the second case is:

$$R_a = R_b.R_{21}$$
$$|(Q_b, \boldsymbol{p}_b, s_b, m_b)^1; (Q_a, \boldsymbol{p}_a, s_a, m_a)^2 >^C$$
$$= (-)^{2s_b}|(Q_a, \boldsymbol{p}_a, s_a, m_a)^1; (Q_b, \boldsymbol{p}_b, s_b, m_b)^2 >^C \qquad (15)$$

In each case re-ordering has forced a rotation by 2π on one particle's SQF (the two cases differing by which particle's has been rotated and therefore the two versions of the exchanged state vector differ by a 2π rotation to both particles' SQFs). This is a clear example of the relationship: *Exchange = Permutation + Physically Significant Transformation*; the physically significant transformation being a rotation.

The conventional description corresponds to the order dependent case for a very simple reason; because the conventional description is not physically complete, the exchange phase is indeterminate, unless we associate the missing information (R_{ba} or R_{21}) with the only free variable available – the particle order.

III COMPOSITE SPIN AND OBSERVABLE EFFECTS

As we previously noted, however, the observable effects of permutation invariance for identical particle pairs are uniquely described not by wave function symmetry or antisymmetry but by the generalized exclusion rules relating to composite spin.

To compute eigenstates of composite spin, we need a common SQF for both particles. Again, we must either specifically include the information concerning R_{ba} or else assume an order dependence (fixed R_{21}). Clearly the conventional description corresponds to the latter case and we can prove the usual generalized exclusion rule for identical particles in the usual way. However, the "antisymmetry" usually associated with even composite spin for identical fermions is not real but is a *pseudo-antisymmetrization* due to a hidden order dependence in the *relative orientation* of the SQFs – even when their axes coincide. If we had used order independent SQFs, then the scalar coefficients would be symmetric for even composite spin, since the antisymmetric Clebsch-Gordon coefficients would be accompanied by phase factors differing by a sign, due to the difference between R_{ba} and $R_{ab}(= R_{ba}^{-1})$.

It should now be clear that the observed exclusion rules correspond to the effect of the asymmetric quality of eqn. 11 on eigenstates of composite spin. In particular, any given pair in a multi-particle state will obey the usual exclusion rules. In the conventional approach, which equates fermion exclusion with antisymmetry, this is usually described as requiring "complete" antisymmetrization, by which is meant the simultaneous antisymmetrization with respect to all possible pairwise exchanges. There are, of course, $N(N-1)/2$ such independent pairs for N identical halfons.

But the different statistics originate purely in the difference in the way fullons and halfons combine to give eigenstates of composite spin. Instead of notions of bosons and fermions being different "types" of particles defined by symmetry or antisymmetry, we must look at the allowed composite spin states. The maximum number of independent simultaneous composite spin eigenstates of N particles is only $N-1$. Contrary to the usual description, this permits at most $N-1$ simultaneous pseudo-antisymmetrizations of our N halfon state. (State vectors for different combinations of these $N-1$ pairs will be related to each other by possible sign changes [3].) The allowed composite eigenstates will differ from those predicted by complete antisymmetrization. Hence the difference between the SPR and the conventional complete antisymmetrization should be experimentally testable.

When composite fullons have constituent halfons then constituent antisymmetrization *between* these fullons forbids them from being true bosons. The usual explanation is that when the wave functions of the composite fullons overlap, their constituent halfons must be in excited states. The SPR, on the other hand, makes no distinction between composite or elementary fullons in terms of their bosonic behavior and constituent halfon exclusion between composite fullons is not dependent on its identification with antisymmetrization, but rather on interaction breaking up the composite fullon eigenstates [4].

REFERENCES

1. York, M. J., "Identity, Geometry, Permutation And The Spin-Statistics Theorem," http://xxx.lanl.gov/abs/quant-ph/9908078.
2. Wigner, E. P., "Relativistic Invariance and Quantum Phenomena," *Rev. Mod. Phys.* **29**, 255-268 (1957).
3. York, M. J., "Permutation Symmetry For Many Particles," http://xxx.lanl.gov/abs/quant-ph/9912057.
4. Ehrenfest, P. & Oppenheimer, J. R., "Note On The Statistics Of Nuclei," *Phys. Rev.* **37**, 333-338 (1931).

II. Theories of Violations of the Spin-Statistics Connection

Theories of Violation of Statistics

O.W. Greenberg

Center for Theoretical Physics
Department of Physics
University of Maryland
College Park, MD 20742-4111

Dedicated to George Abraham Snow, August 24, 1926-June 24, 2000.

Abstract. I discuss theories of violations of statistics, including intermediate statistics, parastatistics, parons, and quons. I emphasize quons, which allow small violations of statistics. I analyze the quon algebra and its representations, implications of the algebra including the observables allowed by the superselection rule separating inequivalent representations of the symmetric group, the conservation of statistics rules, and the rule for composite systems of quons. I conclude by raising the questions of possible origins of violations of statistics and of the level at which violations should be expected if they exist.

I INTRODUCTION

As far as I know this is the first international conference devoted entirely to the relation of spin and statistics and to the investigation of possible small violations of statistics. My purpose in this talk is to give an overview of theoretical issues connected with violations of statistics. I have divided the talk into five parts: (1) general theoretical remarks, (2) types of experiments to detect violations, (3) attempts to violate statistics, (4) quons, the best formalism so far to describe small violations, and (5) summary and open questions.

II GENERAL THEORETICAL REMARKS

The general principles of quantum theory do not require that all particles be either bosons or fermions. This restriction requires an additional postulate which A.M.L. Messiah [1] named the "symmetrization postulate," which I quote as Messiah defined it: "The states of a system containing N identical particles are necessarily either all symmetrical or all antisymmetrical with respect to permutations of the N particles." The symmetrization postulate can be restated as: "All states of identical particles are in one-dimensional representations of the symmetric group." With Messiah's definition the spin-statistics connection, that integer spin particles are bosons and odd-half-integer spin particles are fermions, is a separate statement.

Messiah and I gave a detailed discussion of quantum mechanics without the symmetrization postulate [2]. We emphasized that without the symmetrization postulate, a set of one-body measurements is never a maximal set; one needs additional measurements to fix the state of the system. Further there is a superselection rule separating states in inequivalent representations of the symmetric group. If identical particles can occur in states that violate the spin-statistics connection their transitions must occur in the same representation of the symmetric group. For example, in the experiment of Deilamian, et al [3] which looked for anomalous helium atoms in which the two electrons violated the exclusion principle and were in the symmetric state, the search was for transitions among the symmetric states rather than between symmetric and antisymmetric states. This point was also made by R. Amado and H. Primakoff [4]. In addition if one assumes that charged particles couple universally to the electromagnetic field, then the transitions among the anomalous states occur at the normal rate, so that isolated atoms will be in the lowest state of the anomalous system. Since the symmetrization postulate is not an intrinsic part of quantum theory, this postulate must be subjected both to theoretical study and to experimental tests. Quantitative tests require a theory in which the symmetrization postulate does not have to hold and in which the violation of the postulate is reflected in a parameter that departs from its standard value at which the symmetrization postulate and the spin-statistics connection do hold. It is certainly possible that violations of statistics are extremely small and require high-precision tests to be observed. This conference brings together leading workers in this search.

Because the notion that particles are identical requires that the Hamiltonian and, indeed, all observables must be symmetric in the dynamical variables associated with the identical particles, the observables can't change the permutation symmetry type of the wave function. In particular one can't introduce a small violation of statistics by assuming the Hamiltonian is the sum of a statistics-conserving and a small statistics-violating term,

$$H = H_S + \epsilon H_V \qquad (1)$$

as one can for violations of parity, charge conjugation, etc. Violation of statistics has to be introduced in a more subtle way.

If charged particles couple universally to the electromagnetic field, then there can't be two kinds of–say–electrons, "red" electrons and "blue" electrons, because then the lowest order pair production cross section,

$$\sigma(\gamma X \longrightarrow e^+ e^- X) \qquad (2)$$

would double. A high-precision measurement is not needed to rule this out.

A convenient way to parametrize violations or bounds on violations of statistics uses the two-particle density matrix. For fermions,

$$\rho_2 = (1 - v_F)\rho_a + v_F \rho_s; \qquad (3)$$

for bosons,

$$\rho_2 = (1 - v_B)\rho_s + v_B \rho_a; \qquad (4)$$

in each case the violation parameter varies between zero if the statistics is not violated and one if the statistics is completely violated.

III TYPES OF EXPERIMENTS

There are three basic types of experiments to detect violations of statistics: (1) transitions among anomalous states–these can occur in solids, liquids or gases, (2) accumulation of particles in anomalous states, and (3) deviations from the usual statistical properties of the identical particles. Since, as mentioned earlier, a superselection rule prevents transitions between normal and anomalous states, experiments searching for such transitions do not provide a valid test of violation of statistics.

Transitions among anomalous states can provide a very sensitive test, since in some cases a single such transition can be observed. The prototype of this kind of test is the experiment of Maurice and Trudy Goldhaber [5]. They asked the qualitative question, "Do the electrons from nuclear beta decay obey the same exclusion principle as electrons in atoms?" They knew that electrons from each source have the same charge, spin, and mass, etc., i.e. that the single-electron states in each case are identical, but there was no evidence that the many-electron states from each source are identical. They devised the following ingenious test: they let beta decay electrons from a nuclear source fall on a block of lead. They argued that if the many-electron states were not identical then the nuclear beta decay electrons would not obey the same exclusion principle as the electrons in the lead atoms. Then the beta decay electrons would not see the K shells in the lead atoms as filled and could fall into the K shells and would emit x-rays. A single such x-ray could be observed. They saw no such x-rays above background and thus answered their qualitative question in the affirmative. I estimate that their experiment gave the bound $v_F \leq 5 \times 10^{-2}$ for electrons.

E. Ramberg and G.A. Snow [6] developed this experiment into one which yields a high-precision bound on violations of the exclusion principle. Their idea was to replace the natural β source, which provides relatively few electrons, by an electric current, in which case Avogadro's number is on their side. The possible violation of the exclusion principle is that a given collection of electrons can, with different probabilities, be in different permutation symmetry states. The probability to be in the "normal" totally antisymmetric state presumably would be close to one, the next largest probability would occur for the state with its Young tableau having one row with two boxes, etc. The idea of the experiment is that each collection of electrons has a possibility of being in an "abnormal" permutation state. If the density matrix for a conduction electron together with the electrons in an atom has a projection onto such an "abnormal" state, then the conduction electron will not see the K shell of that atom as filled. Then a transition into the K shell with x-ray emission is allowed. Each conduction electron which comes sufficiently close to a given atom has an independent chance to make such an x-ray-emitting transition, and thus the probability of seeing such an x-ray is proportional to the number of conduction electrons which traverse the sample and the number of atoms which the electrons visit, as well as the probability that a collection of electrons can be in the anomalous state. Ramberg and Snow chose to run 30 amperes through a thin copper strip for about a month. They surrounded the experiment with veto scintillators to remove background x-rays. They estimated the energy of the modified x-rays which would be emitted due to the transition to the K shell. No excess of x-rays above background was found in this energy region. Ramberg and Snow set the limit

$$v_F \leq 1.7 \times 10^{-26} \tag{5}$$

for electrons. This is high precision indeed!

The Ramberg-Snow experiment may seem discouraging for the discovery of generalizations of bose and fermi statistics; however there are small numbers in physics which, if necessary, can occur in degree greater than one. For example the ratios

$$\frac{m_{proton}}{M_{Planck}} \sim 10^{-19}, \text{ and } \frac{G_N m_e^2}{e^2} \sim 10^{-43} \tag{6}$$

can provide numbers smaller than the Ramberg-Snow bound. In addition new physics effects such as violations of Lorentz invariance, spacetime discreteness, spacetime noncommutativity, etc. may provide small effects. Mohapatra and I gave an early survey of experimental bounds on violations of statistics [7].

Composite structure can mimic violations of statistics. This is not what I am considering here.

IV ATTEMPTS TO VIOLATE STATISTICS

A Gentile's "intermediate statistics"

The first attempt to go beyond bose and fermi statistics seems to have been made by G. Gentile [8] who suggested an "intermediate statistics" in which at most n identical particles could occupy a given quantum state. In intermediate statistics, fermi statistics is recovered for $n = 1$ and bose statistics is recovered for $n \to \infty$; thus intermediate statistics interpolates between fermi and bose statistics. However

Gentile's statistics is not a proper quantum statistics, because the condition of having at most n particles in a given quantum state is not invariant under change of basis [9]. For example, for intermediate statistics with $n = 2$, the state $|\psi\rangle = |k, k, k\rangle$ does not exist; however, the state $|\chi\rangle = \sum_{l_1,l_2,l_3} U_{k,l_1} U_{k,l_2} U_{k,l_3} |l_1, l_2, l_3\rangle$, obtained from $|\psi\rangle$ by the unitary change of single-particle basis, $|k\rangle' = \sum_l U_{k,l}|l\rangle$ does exist. By contrast, parafermi statistics of order n which I discuss just below is invariant under change of basis [10]. Parafermi statistics of order n not only allows at most n identical particles in the same state, but also allows at most n identical particles in a symmetric state. In the example just described, neither $|\psi\rangle$ nor $|\chi\rangle$ exist for parafermi statistics of order two.

B Parastatistics

H.S. Green [10] proposed the first proper quantum statistical generalization of bose and fermi statistics. Green noticed that the commutator of the number operator with the annihilation and creation operators is the same for both bosons and fermions

$$[n_k, a_l^\dagger]_- = \delta_{kl} a_l^\dagger. \tag{7}$$

The number operator can be written

$$n_k = (1/2)[a_k^\dagger, a_k]_\pm + \text{const.}, \tag{8}$$

where the anticommutator (commutator) is for the bose (fermi) case. If these expressions are inserted in the number operator-creation operator commutation relation, the resulting relation is *trilinear* in the annihilation and creation operators. Polarizing the number operator to get the transition operator n_{kl} which annihilates a free particle in state l and creates one in state k leads to Green's trilinear commutation relation for his parabose and parafermi statistics,

$$[[a_k^\dagger, a_l]_\pm, a_m^\dagger]_- = 2\delta_{lm} a_k^\dagger \tag{9}$$

Since these rules are trilinear, the usual vacuum condition,

$$a_k|0\rangle = 0, \tag{10}$$

does not suffice to allow calculation of matrix elements of the a's and $a^\dagger$'s; a condition on single-particle states must be added,

$$a_k a_l^\dagger |0\rangle = p \delta_{kl} |0\rangle. \tag{11}$$

Green found an infinite set of solutions of his commutation rules, one for each positive integer p, by giving an ansatz which he expressed in terms of bose and fermi operators. Let

$$a_k^\dagger = \sum_{\alpha=1}^{p} b_k^{(\alpha)\dagger}, \quad a_k = \sum_{\alpha=1}^{p} b_k^{(\alpha)}, \tag{12}$$

and let the $b_k^{(\alpha)}$ and $b_k^{(\beta)\dagger}$ be bose (fermi) operators for $\alpha = \beta$ but anticommute (commute) for $\alpha \neq \beta$ for the "parabose" ("parafermi") cases. This ansatz clearly satisfies Green's relation. The integer p is the order of the parastatistics. The physical interpretation of p is that, for parabosons, p is the maximum number of particles that can occupy an antisymmetric state, while for parafermions, p is the maximum number of particles that can occupy a symmetric state (in particular, the maximum number which can occupy the same state). The case $p = 1$ corresponds to the usual bose or fermi statistics. Later, Messiah and I [11] proved that Green's ansatz gives all Fock-like solutions of Green's commutation rules. Local observables have a form analogous to the usual ones; for example, the local current for a spin-1/2 theory is $j_\mu = (1/2)[\bar{\psi}(x), \psi(x)]_-$. From Green's ansatz, it is clear that the squares of all norms of states are positive, since sums of bose or fermi operators give positive norms. Thus parastatistics [12] gives a set of orthodox theories.

This is all well and good; however, the violations of statistics provided by parastatistics are gross. Parafermi statistics of order two has up to two particles in each quantum state. High-precision experiments are not necessary to rule this out for all particles we think are fermions.

C The Ignatiev-Kuzmin model

Interest in possible small violations of the exclusion principle was revived by a paper of Ignatiev and Kuzmin [13] in 1987. They constructed a model of one oscillator with three possible states: a vacuum state, a one-particle state and, with small amplitude β, a two-particle state. They gave trilinear commutation relations for their oscillator. Mohapatra and I noticed that the Ignatiev-Kuzmin oscillator could be represented by a modified form of the order-two Green ansatz. We suspected that a field theory generalization of this model having an infinite number of oscillators would not have local observables and set about trying to prove this. To our surprise, we found that we could construct local observables and gave trilinear relations which guarantee the locality of the current [14].

D Parons

Following Ignatiev and Kuzmin we introduced a parameter β that gives the deformation of the Green trilinear commutation relations. For $\beta \to 1$ the relations reduce to those of the $p = 2$ parafermi field; for $\beta \to 0$ the double occupancy is completely suppressed and the theory is equivalent to a fermi theory. A random state of two paronic electrons has the violation parameter $\beta^2/2$. Mohapatra and I checked that the norms are positive for states of up to three particles. At this stage, we were carried away with enthusiasm, named these particles "parons" since their algebra is a deformation of the parastatistics algebra, and thought we had found a local theory with small violation of the exclusion principle. Unknown to us Govorkov [15], using a detailed algebraic argument, already had shown in generality that any deformation of the Green commutation relations necessarily has states with negative squared norms in the Fock-like representation. For our model, the first such negative-probability state occurs for four particles in the representation of $\mathcal{S}_4$ with three boxes in the first row and one in the second. We were able to understand Govorkov's result qualitatively as follows: [16] Since parastatistics of order p is related by a Klein transformation to a model with exact $SO(p)$ or $SU(p)$ internal symmetry, a deformation of parastatistics which interpolates between Fermi and parafermi statistics of order two would be equivalent to interpolating between the trivial group whose only element is the identity and a theory with $SO(2)$ or $SU(2)$ internal symmetry. This is impossible, since there is no such interpolating group.

E The Doplicher-Haag-Roberts analysis

S. Doplicher, R. Haag and J. Roberts [17] made a general study of identical particle statistics using the algebraic field theory methods pioneered by Haag. They found parabose and parafermi statistics of positive integer orders which as mentioned above were introduced by Green. They also found another case which they called infinite statistics. Young patterns label the inequivalent irreducible representations of the symmetric group. In parabose (parafermi) statistics of order p the Young patterns have at most p rows (columns) corresponding to having at most p particles in an antisymmetric (a symmetric) state. In infinite statistics all irreducibles of the symmetric group occur. Doplicher, et al, did not give an operator realization of infinite statistics.

F Infinite statistics

In 1989 I gave an evening lecture at Wake Forest University. My talk was attended by physicists, philosophers, and among people in other disciplines, chemists. In my talk I mentioned the bose and fermi commutation relations. After the talk Roger Hegstrom, a chemist, asked "Why not average the bose and fermi commutation relations and consider the relation

$$a(k)a^\dagger(l) = \delta(k,l)?" \tag{13}$$

I was surprised to find that such a simple case had not been considered. Later I found out that it had, in the mathematical literature, by J. Cuntz [18]. With Hegstrom's permission I developed this case, which turned out to be the first operator example of infinite statistics [19]. In order to select the Fock-like representation, one must add the vacuum condition

$$a(k)|0\rangle = 0. \tag{14}$$

We can calculate all vacuum matrix elements of products of a's and $a^\dagger$'s using the commutation relation and the vacuum condition. There is no commutation relation involving two a's or two $a^\dagger$'s. There are $n!$ linearly independent n-particle states in Hilbert space if all quantum numbers are distinct; these states differ only by permutations of the order of the creation operators. (Later we will see that there are *not* that many independent density matrices or other observables.) The matrix of scalar products of these states is the identity matrix,

$$M_{P,Q}^n(q) = (Pa^\dagger(k_1)a^\dagger(k_2)\cdots a^\dagger(k_n)|0\rangle, Qa^\dagger(l_1)a^\dagger(l_2)\cdots a^\dagger(l_n)|0\rangle = \prod_{i=1}^n \delta(k_i, l_i)\delta(P, Q) \tag{15}$$

where P and Q are permutations from S_n; that is, the scalar product is zero unless there are the same number of creation operators on each side of the scalar product and they have the same quantum numbers in the same order. This algebra can be viewed as a deformation of either the bose or the fermi algebras. As is typical for deformed algebras, there is an element that is infinite degree in the generators of the algebra. In this case the number operator that obeys

$$[n(k), a^\dagger(l)]_- = \delta(k,l) a^\dagger(l) \tag{16}$$

is the operator of infinite degree; in terms of the a's and the $a^\dagger$'s,

$$n(k) = a^\dagger(k)a(k) + \sum_t a^\dagger(t)a^\dagger(k)a(k)a(t) + \sum_{t_1,t_2} a^\dagger(t_2)a^\dagger(t_1)a^\dagger(k)a(k)a(t_1)a(t_2) + \cdots. \tag{17}$$

There is an analogous formula for the transition operator, $n(k,l)$, that obeys

$$[n(k,l), a^\dagger(m)]_- = \delta(l,m) a^\dagger(k). \tag{18}$$

V QUONS

A The quon algebra

The quon algebra [20–22] is the best attempt so far to violate statistics by a small amount. The infinite statistics algebra just discussed is the average of the bose and fermi algebras. The quon algebra can be obtained as the convex sum of these two algebras,

$$\frac{1+q}{2}[a(k), a^\dagger(l)]_- + \frac{1-q}{2}[a(k), a^\dagger(l)]_+ = \delta(k,l), \tag{19}$$

or

$$a(k)a^\dagger(l) - qa^\dagger(l)a(k) = \delta(k,l). \tag{20}$$

As usual the Fock-like representation is selected by the vacuum condition

$$a(k)|0\rangle = 0. \tag{21}$$

Convexity requires $-1 \leq q \leq 1$; for this range the states have positive squared norms. Outside this range the squared norms become negative. Using the algebra (20) and the vacuum condition (21) all vacuum matrix elements of polynomials in the a's and $a^\dagger$'s can be calculated; for example,

$$(a^\dagger(k_1)a^\dagger(k_2)|0\rangle, a^\dagger(l_1)a^\dagger(l_2)|0\rangle) =$$

$$\delta(k_1,l_1)\delta(k_2,l_2) + q\delta(k_1,l_2)\delta(k_2,l_1) =$$

$$\frac{1+q}{2}[\delta((k_1,l_1)\delta(k_2,l_2) + \delta(k_1,l_2)\delta(k_2,l_1)] + \frac{1-q}{2}[\delta((k_1,l_1)\delta(k_2,l_2) - \delta(k_1,l_2)\delta(k_2,l_1)]. \tag{22}$$

The first proof of the positivity of the norms was given by D. Zagier [23], who gave a tour-de-force calculation of the determinant of the $n! \times n!$ matrix of scalar products (15) for arbitrary n,

$$det M^n_{P,Q}(q) = \prod_{k=1}^{n-1}(1 - q^{k(k+1)})^{\frac{(n-k)n!}{k(k+1)}}. \tag{23}$$

As shown above, at $q = 0$ the norms are positive and the determinant is one. In order for a norm to become negative the determinant has to change sign. From Zagier's formula this happens only when $q^{k(k+1)} = 1$, i.e., on the unit circle. This proves that the norms remain positive between negative one and one.

Speicher [24] gave an ingenious proof of the positivity of the norms using an ansatz for the Fock-like representation of quons analogous to Green's ansatz for parastatistics. Speicher represented the quon annihilation operator as the weak operator limit,

$$a_k = \lim_{N \to \infty} N^{-1/2} \sum_{\alpha=1}^{N} b_k^{(\alpha)}, \tag{24}$$

where the $b_k^{(\alpha)}$ are bose oscillators for each α, but with relative commutation relations given by

$$b_k^{(\alpha)} b_l^{(\beta)\dagger} = s^{(\alpha,\beta)} b_l^{(\beta)\dagger} b_k^{(\alpha)}, \alpha \neq \beta, \text{ where } s^{(\alpha,\beta)} = \pm 1. \tag{25}$$

This limit is taken as the limit, $N \to \infty$, in the vacuum expectation state of the Fock space representation of the $b_k^{(\alpha)}$. In this respect Speicher's ansatz differs from Green's, which is an operator identity. To get the Fock-like representation of the quon algebra, Speicher chose a probabilistic condition for the signs $s^{(\alpha,\beta)}$,

$$\text{prob}(s^{(\alpha,\beta)} = 1) = (1+q)/2, \tag{26}$$

$$\text{prob}(s^{(\alpha,\beta)} = -1) = (1-q)/2. \tag{27}$$

Speicher's rules reproduce the quon algebra. The norms are positive since the sums of bose or fermi operators have positive norms. The constraint on q follows because the probabilities have to lie between zero and one.

The number and transition operators for general q have infinite degree expansions analogous to but more complicated than those for the $q = 0$ case,

$$n(k,l) = a^\dagger(k)a(l) + \frac{1}{1-q^2}\sum_t (a^\dagger(t)a^\dagger(k) - qa^\dagger(k)a^\dagger(t))(a(l)a(t) - qa(t)a(l)) + \cdots. \tag{28}$$

The general formula for the number operator was given by S. Stanciu [25].

At $q = \pm 1$ only the symmetric (antisymmetric) representation of $\mathcal{S}_n$ occurs. The quon operators interpolate smoothly between fermi and bose statistics in the sense that as q departs from ± 1 the vectors formed by polynomials in the creation operators, which are superpositions of vectors in different irreducible representations of the symmetric group, have higher weights in the more symmetric (antisymmetric) representations, and as $q \to \mp 1$ the antisymmetric (symmetric) representations smoothly become more heavily weighted.

B   Observables in quon theory

It is important to note that although there are $n!$ linearly independent vectors in Fock space associated with a degree n monomial in creation operators that carry disjoint quantum numbers acting on the vacuum, there are fewer than $n!$ observables associated with such vectors. For example, for two identical quons 1 and 2 in orthogonal quantum states the two vectors $a^\dagger(1)a^\dagger(2)|0\rangle$ and $a^\dagger(2)a^\dagger(2)|0\rangle$ are orthogonal and each is normalized to one. Let

$$|\phi_{s,a}\rangle = N_{s,a}(a^\dagger(1)a^\dagger(2) \pm a^\dagger(2)a^\dagger(1))|0\rangle \tag{29}$$

be normalized states that are symmetric or antisymmetric under transposition of 1 and 2. The quon algebra gives

$$N_{s,a} = \frac{1}{\sqrt{2(1 \pm q)}}. \tag{30}$$

One can then calculate the expansion

$$a^\dagger(1)a^\dagger(2)|0\rangle = \alpha|\phi_s\rangle + \beta|\phi_a\rangle, \tag{31}$$

either using

$$a^\dagger(1)a^\dagger(2)|0\rangle = (1/2)[(a^\dagger(1)a^\dagger(2) + a^\dagger(2)a^\dagger(1)) + (a^\dagger(1)a^\dagger(2) - a^\dagger(2)a^\dagger(1))]|0\rangle \tag{32}$$

or using

$$a^\dagger(1)a^\dagger(2)|0\rangle = \langle\phi_s|a_1^\dagger a_2^\dagger|0\rangle|\phi_s\rangle + \langle\phi_a|a_1^\dagger a_2^\dagger|0\rangle|\phi_a\rangle. \tag{33}$$

Either way gives

$$\alpha = \sqrt{(1+q)/2}, \beta = \sqrt{(1-q)/2} \tag{34}$$

so that

$$a_1^\dagger a_2^\dagger |0\rangle = \sqrt{\frac{1+q}{2}}\phi_s + \sqrt{\frac{1-q}{2}}\phi_a \tag{35}$$

and

$$a_2^\dagger a_1^\dagger |0\rangle = \sqrt{\frac{1+q}{2}}\phi_s - \sqrt{\frac{1-q}{2}}\phi_a. \tag{36}$$

Then, dropping the cross terms that are excluded by the superselection rule separating symmetric and antisymmetric states of identical particles (and, indeed, states of identical particles in different representations of the symmetric group generally),

$$a_1^\dagger a_2^\dagger|0\rangle\langle 0|a_2 a_1 = a_2^\dagger a_1^\dagger|0\rangle\langle 0|a_1 a_2 = \frac{1+q}{2}|\phi_s\rangle\langle\phi_s| + \frac{1-q}{2}|\phi_a\rangle\langle\phi_a|. \tag{37}$$

This shows that since particles 1 and 2 are identical the same observable results follow when the labels 1 and 2 are transposed. That also shows that the relative phase in Eq.(35) and Eq.(36) is not observable. Equation (37) states that the density matrices for $a_1^\dagger a_2^\dagger|0\rangle$ and $a_2^\dagger a_1^\dagger|0\rangle$ are *identical*, which means that these two "states" correspond to exactly the same physical situation. We put quotation marks around the word "states" to indicate that these should really be represented by density matrices. Note that the sum of the coefficients of the two terms in the two-particle density matrix is one, as it should be. The general observable is a linear combination of projectors on the irreducibles of the symmetric group.

The parameters v_F and v_B that represent small violations of statistics can be written in terms of the q parameters; the result is,

$$q_F = 2v_F - 1 \text{ or } v_F = \frac{1}{2}(1+q_F); \quad q_B = 1 - 2v_B \text{ or } v_B = \frac{1}{2}(1-q_B). \tag{38}$$

C Properties of quon theory

Surprizingly several properties of relativistic theories that I would expect to fail for the quon theory (made relativistic kinematically) actually hold. These include Wick's theorem, cluster decomposition theorems and the CPT theorem. We are familiar with Wick's theorem for bosons which states that the vacuum matrix element of a product of free fields is the sum of all possible products of two-point functions, with each product occuring with factor one. For fermions the factors are plus or minus one, depending on the parity of the permutation between the order in the vacuum matrix element and the order in the product of two point functions. In Wick's theorem for quons the corresponding factors are q raised to the inversion number of the permutation between the order in the vacuum matrix element and the order in the product of two point functions. This result reduces to the usual Wick's theorem when $q \to \pm 1$. The inversion number can be found conveniently by drawing lines above the vacuum matrix element to indicate the pairs that are contracted into two-point functions. The minumum number of crossings of these lines is the inversion number. Since both cluster decomposition and the CPT theorem for vacuum matrix elements of free fields depend on the properties of two-point functions, these theorems hold for quon fields. Note that quon fields, which clearly violate the spin-statistics theorem, obey the CPT theorem; this emphasises the point made by R. Jost [26] that the CPT theorem requires only very weak assumptions.

If all the usual properties of relativistic field theory hold, then the spin-statistics theorem holds; thus some property must fail for quons. The property that does not hold is locality in the sense of the commutativity of observables at spacelike separation. Jost [27] showed that if locality holds in an open spacelike region then analyticity arguments prove that it holds everywhere outside the lightcone. This result does not hold if the violation of locality decreases–say–exponentially away from the lightcone. The experimental bounds on such a violation are not clear. Note that the nonrelativistic form of locality

$$[\rho(\mathbf{x}), \psi^\dagger(\mathbf{y})]_- = \delta(\mathbf{x} - \mathbf{y})\psi^\dagger(\mathbf{y}), \tag{39}$$

where ρ is the charge density, does hold.

D Conservation of statistics rules for quon theory

For the energies of systems that are widely (spacelike) separated to be additive, all terms in the Hamiltonian must be effective bose operators in the sense that

$$[H(x), \phi(y)]_- \to 0, \text{ as } x - y \to \infty \text{ spacelike} \tag{40}$$

for *all* fields ϕ. This condition imposes "conservation of statistics" rules, the simplest of which is that only an even number of fermi fields can appear in any term of the Hamiltonian. For parafields, which have local observables, Messiah and I [11] showed that parafields must occur in even degree, except that, for p odd, p parafields can occur. I defined paragrassmann and quongrassmann numbers which must be used in coupling para and quon operators to external sources and showed that for external parasources there are analogous restrictions [28]. Using these results, R.C. Hilborn and I [29] gave an heuristic argument relating the q parameter for electrons to that for photons. The result,

$$q_e^2 = q_\gamma, \qquad (41)$$

allows the very accurate bound from the Ramberg-Snow experiment to be carried over to photons with comparable accuracy. Similar arguments work for any particles that are coupled to electrons through any chain of reactions.

E Bound states of quons

The classical result about bound states of bosons and fermion due to E.P. Wigner [30] and to P. Ehrenfest and J.R. Oppenheimer [31] states that a bound state of bosons and fermions is a boson unless it has an odd number of fermions, in which case it is a fermion. Hilborn and I [32] showed that this result generalizes for quons. A bound state of n identical quons with parameter $q_{constituent}$ has parameter $q_{bound} = q_{constituent}^{n^2}$. This implies that if $q_{nucleus}$ is bounded within ϵ of ± 1 for a nucleus with A nucleons, then $q_{nucleon}$ is bounded within ϵ/A^2. Thus the bound on the nucleons is *stronger* than the bound on the nucleus. Analogously the bound on quarks is improved by $1/9$ over the bound on nucleons. Michael Berry [33] pointed out that in the context of the quon theory this result on q for bound states implies that either the layers of compositeness stop or all particles are bosons or fermions.

This result reduces to the usual one for bosons and fermions since q and q^2 are even or odd together.

VI SUMMARY AND OPEN QUESTIONS

Like any other physical property, statistics should be subjected to high-precision experimental tests. In order to interpret such tests we need a theory in which statistics can be violated and a parameter that gives a quantitative measure of the validity of statistics. A theory that allows violations of statistics cannot have all the properties we might like. So far quons are the best theory that allows small violations.

In summary the positive properties of quons as a field theory are (a) norms are positive, (b) a simple modification of Wick's theorem holds, (c) cluster decomposition theorems hold, (d) the CPT theorem holds, and (e) free fields can have relativistic kinematics. The negative properties are (a) spacelike commutativity of observables fails, and because of this (b) interacting relativistic field theory is in doubt. I do not have a concrete suggestion for the possible origin of small violations of the exclusion principle. One could turn this issue around and observe that the constraints of bose and fermi statistics are grafted onto the general structure of quantum theory in an ad hoc way and ask why these constraints are realized in nature. Study of the situation in which these constraints are violated may shed light on why they hold for the known particles. In any case a fundamental issue such as statistics should be subjected to experimental tests and to theoretical study, just as is being done for Lorentz and CPT invariance.

What we lack is an "external" motivation for violation of statistics—that is a connection of violations with some other physical property. We also don't have any insight into the level at which we can expect violations if they do occur. Possible external motivations for violation of statistics include (a) violation of CPT, (b) violation of locality, (c) violation of Lorentz invariance, (d) extra space dimensions, (e) discrete space and/or time and (f) noncommutative spacetime. Of these, (a) seems unlikely because the quon theory which obeys CPT allows violations, (b) seems likely because if locality is satisfied we can prove the spin-statistics connection and there will be no violations, (c), (d), (e) and (f) seem possible.

At the conference the question was raised whether the stability of matter which depends on the exclusion principle might set stringent bounds on possible violations of statistics. This certainly deserves careful study.

Hopefully either violations will be found experimentally or our theoretical efforts will lead to understanding of why only bose and fermi statistics occur in Nature.

REFERENCES

1. Messiah, A.M.L., *Quantum Mechanics, Vol. II*, North-Holland, Amsterdam, 1962, p595.
2. Messiah A.M.L., and Greenberg, O.W., *Phys. Rev.* **136**, B248 (1964).
3. Deilamian, K., Gillaspy, J.D., and Kelleher, D.E., *Phys. Rev. Lett.* **74**, 4787 (1995).
4. Amado, R., and Primakoff, H., *Phys. Rev.* **C22**, 1388 (1980).
5. Goldhaber, M., and Goldhaber, G.S., *Phys. Rev.* **73**, 1472 (1948).
6. Ramberg, E., and Snow, G.A.,*Phys. Lett. B* **238**, 438 (1990).
7. Greenberg, O.W., and Mohapatra, R.N., *Phys. Rev. D* **39**, 2032 (1989).
8. Gentile, G., *Nuovo Cimento* **17**, 493 (1940).
9. Greenberg, O.W., *Physica* **180**, 419 (1992).
10. Green, H.S., *Phys. Rev.* **90**, 270 (1952).
11. Greenberg, O.W., and Messiah, A.M.L., *Phys. Rev.* **138**, B1155 (1965).
12. Dell' Antonio, G.F., Greenberg, O.W., and Sudarshan, E.C.G., *Group Theoretical Concepts and Methods in Elementary Particle Physics*, ed. F. Gürsey, Gordon and Breach, New York, 1964, p403, introduced the word "parastatistics."
13. Ignatiev, A.U., and Kuzmin, V.A., *Yad. Fiz.* **46**, 786 (1987) [*Sov. J. Nucl. Phys.* **46**, 444 (1987)].
14. Greenberg, O.W., and Mohapatra, R.N., *Phys. Rev. Lett.* **59**, 2507 (1987).

15. Govorkov, A.B., *Teor. Mat. Fis.* **54**, 361 (1983) [*Sov. J. Theor. Math. Phys.* **54**, 234 (1983)]; *Phys. Lett. A* **137**, 7 (1989).
16. Greenberg, O.W., and Mohapatra, R.N.,*Phys. Rev. Lett.* **62**, 712 (1989).
17. Doplicher, S., Haag, R., and Roberts, J., *Commun. Math. Phys.* **23**, 199 (1971) and *ibid* **35**, 49 (1974).
18. Cuntz, J., *Commun. Math. Phys.* **57**, 173 (1977).
19. Greenberg, O.W., *Phys. Rev. Lett.* **64**, 705 (1990).
20. Arik, M., and Coon, D.D., and Lam, Y.-m., *J. Math. Phys.* **16**, 1776 (1975), introduced a finite set of quon operators in the context of the dual resonance model.
21. Polyakov, A.M., pointed out that the infinite statistics algebra is a special case of the quantum group relation $a(k)\tilde{a}(l) - q\tilde{a}(l)a(k) = \delta(k,l)$, where $\tilde{a}$ is $a^\dagger$ for q real and $\tilde{a}_q = a^\dagger_{q^*}$ for q complex, (private communication, 1990).
22. Greenberg, O.W., *Phys. Rev. D* **43**, 4111 (1991).
23. Zagier, D., *Commun. Math. Phys.* **57**, 173 (1997).
24. Speicher, R., *Lett. Math. Phys.* **27**, 97 (1993).
25. Stanciu, S., *Commun. Math. Phys.* **147**, 211 (1992).
26. Jost, R., *The Theory of Quantized Fields*, Am. Math Soc., Providence, 1965, pp 103-104.
27. *ibid.*, p85.
28. Greenberg, O.W., *Phys. Lett. A* **209**, 137 (1995).
29. Greenberg, O.W., and Hilborn, R.C., *Found. of Phys.* **29**, 397 (1999).
30. Wigner, E.P., *Math. und Naturwiss. Anzeiger der Ungar. Ak. der Wiss.* **46**, 576 (1929); translated in *Collected Works of E.P. Wigner*
31. Ehrenfest, P., and Oppenheimer, J.R., *Phys. Rev.* **37**, 333 (1931).
32. Greenberg, O.W., and Hilborn, R.C., *Phys. Rev. Lett.* **83**, 4460 (1999).
33. Berry, M., private communication, June 2000.

Connecting q-mutator Theory with Experimental Tests of the Spin-Statistics Connection

Robert C. Hilborn

Department of Physics
Amherst College
Amherst, MA 01002-5000

Abstract. The q-mutator theory is used to connect the value of $1-|q|$, the parameter measuring the "difference" between quons and ordinary bosons and fermions, to experiments that test the spin-statistics connection. Such calculations are best carried out using a density matrix formulation because a superselection rule prevents transitions between states associated with different representations of the permutation group. The interpretation of the experimental results, however, in terms of a quantitative limit on $1-|q|$ can be easily misled by the density matrix formulation. As a concrete example, the theory is applied to a spin-statistics test for photons. The formalism is then applied to spin-statistics tests for electrons in atomic helium and for ^{16}O nuclei in molecules. Finally, the analysis is used to extend experimental limits on composite systems such as ^{16}O nuclei to provide a test of the spin-statistics connection for the constituents of those composite systems (nucleons and quarks in the case of oxygen nuclei).

INTRODUCTION

At present, the q-mutator formalism [1, 2] is the only quantum field theoretic formalism that provides a description of "small" violations of the spin-statistics connection. By "small" we mean that the probability of finding two electrons, for example, in a permutation symmetric state (which, of course, violates the usual spin-statistics connection) is small compared to 1. The q-mutator is a generalization of the usual commutator and anticommutator relation between annihilation and creation operators

$$a_m a_n^\dagger - q a_n^\dagger a_m = \delta_{mn}. \tag{1}$$

Particles whose annihilation and creation operators satisfy Eq. (1) are called quons. In order to guarantee a Fock-space-type description, we must also require that $a_m |0\rangle = 0$. For spin ½ particles such as electrons, $\varepsilon_F = 1 + q$ is a measure of the deviation from the ordinary fermion anticommutator algebra for creation and annihilation operators (for which $q = -1$). For integer spin particles, $\varepsilon_B = 1 - q$ is the appropriate measure indicating a departure from the usual boson commutation relations for which $q = +1$. For electrons, nucleons, and photons, we know that these deviations are small because no experiment has yet found these particles in states that violate the usual spin-statistics connection.

The question addressed in this paper is how to express the results of experimental tests of the spin-statistics connection in terms of ε_F and ε_B. This procedure permits a comparison among different experiments designed to test the spin-statistics connection. The formalism also allows us to extend the limits for experiments on composite systems (such as nuclei) to provide limits on possible spin-statistics violations for their constituents [3]. We first provide some general discussion and then apply the formalism to three experimental tests of the spin-statistics connection.

DENSITY MATRIX FORMALISM

A superselection rule [4, 5] provides stringent restrictions on the kinds of experiments that can be used to detect a violation of the spin-statistics connection: for an identical particle system, transitions between states associated with different representations of the permutation group are absolutely forbidden within the framework of standard quantum mechanics. Since this restriction holds for all types of interactions, it is called a *superselection* rule. Quantum mechanics forbids transitions, for example, between a permutation antisymmetric state for two electrons and a permutation symmetric state. In essence, the Hilbert space for the identical particles breaks up into "non-interacting" sectors, each associated with a different representation of the permutation group. For a two-particle system, there are only two possibilities: symmetric states and antisymmetric states. For systems with three or more identical particles, there are—in addition to the symmetric and antisymmetric cases—other states associated with higher-dimensional representations of the permutation group.

Because of this superselection rule, it is often useful (and sometimes necessary) to describe the "state" of an identical particle system using a density matrix formalism with no "off-diagonal" terms linking the different sectors of the Hilbert space: the "state" is best described as a mixed state, in the sense commonly used with density matrices [6].

We will now give some general background showing how the density matrix formalism can be used to describe transition probabilities since the experiments to be discussed in detail look for spin-statistics violations by searching for atomic or molecular transitions between two states, both of which violate the spin-statistics connection. As usual, the probability for a (weak) transition from an initial state $|\psi_i\rangle$ to a final state $|\psi_f\rangle$ is given by

$$P_{if} = A|\langle \psi_f | H_{int} | \psi_i \rangle|^2 = A \langle \psi_i | H_{int}^\dagger | \psi_f \rangle \langle \psi_f | H_{int} | \psi_i \rangle, \tag{2}$$

where H_{int} is the appropriate interaction Hamiltonian and A contains all of the necessary numerical coefficients. We can write the transition probability as the expectation value of an effective operator H_{eff}, where

$$H_{eff} = H_{int}^{\dagger} |\psi_f\rangle\langle\psi_f| H_{int}. \qquad (3)$$

Using this effective operator, we can express the transition probability as

$$P_{if} = A\langle\psi_i | H_{eff} |\psi_i\rangle. \qquad (4)$$

From standard density matrix theory, we know that we can write the expectation value of an operator in terms of the trace of the product of that operator with the corresponding density *operator* ρ. The transition probability is then

$$P_{if} = A Tr(H_{eff} \rho_i), \qquad (5)$$

where ρ_i is the density operator for the initial state of the system [6]. Of course, to carry out calculations, the density operator must be expressed as a density matrix in some basis set. Let m and n label the set of quantum numbers needed to specify the states. The transition probability is then given by

$$P_{if} = A Tr(H_{eff} \rho_i) = A \sum_{m,n} \langle n|H_{eff}|m\rangle\langle m|\rho_i|n\rangle. \qquad (6)$$

For a "pure" state, the density operator can be written as

$$\rho = |\psi\rangle\langle\psi| = \sum_{m,n} c_m c_n^* |m\rangle\langle n|. \qquad (7)$$

For a mixed state, the density operator is given by

$$\rho = \sum_a W_a |\psi^{(a)}\rangle\langle\psi^{(a)}|, \qquad (8)$$

where W_a is the weight given to each of the contributing states. The weights, of course, must sum to unity.

DENSITY MATRIX FOR QUONS

We now apply these ideas to an initial state consisting of two identical particles whose creation and annihilation operators satisfy the q-mutator relation, Eq. (1). We shall then describe how to extend these results to more general states.

The first goal is to construct the density matrix for a two-quon system and to demonstrate how it separates into permutation symmetric and antisymmetric parts. Consider two identical quons in orthogonal quantum states with "mode" labels 1 and 2. The state $a_1^{\dagger} a_2^{\dagger} |0\rangle$ has norm 1 since states $a_1^{\dagger}|0\rangle$ and $a_2^{\dagger}|0\rangle$ are orthonormal. The normalization is most easily established using the general result

$$\langle 0|a_i a_j a_k^\dagger a_l^\dagger|0\rangle = \frac{1}{2}(1+q)(\delta_{jk}\delta_{il}+\delta_{ik}\delta_{jl})$$
$$+\frac{1}{2}(1-q)(\delta_{jk}\delta_{il}-\delta_{ik}\delta_{jl}), \tag{9}$$

which can easily be derived using the basic q-mutator relation, Eq. (1).

The states that are symmetric or antisymmetric under the interchange of the mode labels 1 and 2 can be expressed as

$$|\phi_{s,a}\rangle = N_{s,a}\left(a_1^\dagger a_2^\dagger \pm a_2^\dagger a_1^\dagger\right)|0\rangle. \tag{10}$$

The normalization factors are

$$N_{s,a} = \frac{1}{\sqrt{2(1\pm q)}}. \tag{11}$$

The original two-quon state can be written as a linear combination of the symmetric and antisymmetric states:

$$a_1^\dagger a_2^\dagger|0\rangle = \alpha|\phi_s\rangle + \beta|\phi_a\rangle. \tag{12}$$

To find the coefficients α and β, we write

$$a_1^\dagger a_2^\dagger|0\rangle = \tfrac{1}{2}\left(a_1^\dagger a_2^\dagger + a_2^\dagger a_1^\dagger\right)|0\rangle + \tfrac{1}{2}\left(a_1^\dagger a_2^\dagger - a_2^\dagger a_1^\dagger\right)|0\rangle$$
$$= \frac{1}{2N_s}|\phi_s\rangle + \frac{1}{2N_a}|\phi_a\rangle \tag{13}$$

or alternatively

$$a_1^\dagger a_2^\dagger|0\rangle = \langle\phi_s|a_1^\dagger a_2^\dagger|0\rangle|\phi_s\rangle + \langle\phi_a|a_1^\dagger a_2^\dagger|0\rangle|\phi_a\rangle. \tag{14}$$

We then find that

$$\alpha = \sqrt{(1+q)/2} \text{ and } \beta = \sqrt{(1-q)/2}. \tag{15}$$

Thus, the original two-quon state in Eq. (12) can be written as

$$a_1^\dagger a_2^\dagger|0\rangle = \sqrt{(1+q)/2}\,|\phi_s\rangle + \sqrt{(1-q)/2}\,|\phi_a\rangle \tag{16}$$

and

$$a_2^\dagger a_1^\dagger|0\rangle = \sqrt{(1+q)/2}\,|\phi_s\rangle - \sqrt{(1-q)/2}\,|\phi_a\rangle. \tag{17}$$

We now construct the density operator for the two-particle quon state. First, writing the density operator using a specific ordering of the creation and annihilation operators gives

$$\rho^{(2)} = a_1^\dagger a_2^\dagger|0\rangle\langle 0|a_2 a_1$$
$$= \frac{1+q}{2}|\phi_s\rangle\langle\phi_s| + \frac{1-q}{2}|\phi_a\rangle\langle\phi_a| \tag{18}$$
$$+ \frac{\sqrt{1-q^2}}{2}|\phi_a\rangle\langle\phi_s| + \frac{\sqrt{1-q^2}}{2}|\phi_s\rangle\langle\phi_a|.$$

The last two terms will not make a contribution when we compute the expectation value of operators that are permutation symmetric. In other words, operators that commute with the permutation operator satisfy

$$\langle \phi_s | \hat{B} | \phi_a \rangle = 0. \tag{19}$$

(This is the formal statement of the superselection rule.)

It is instructive to see explicitly how the last terms of Eq. (18) drop out: Using the standard density operator procedure we get

$$\langle \hat{B} \rangle = Tr\left(\hat{B} \rho^{(2)}\right) = \sum_{m,n} \langle m | \hat{B} | n \rangle \langle n | \rho^{(2)} | m \rangle . \tag{20}$$

We will do the *Tr* operation two ways. First, we use the symmetric and antisymmetric states for $|m\rangle$ and $|n\rangle$. Using the notation $|s,a\rangle = |\phi_s, \phi_a\rangle$, we find that

$$\langle \hat{B} \rangle = \langle s | \hat{B} | s \rangle \langle s | \rho^{(2)} | s \rangle + \langle s | \hat{B} | a \rangle \langle a | \rho^{(2)} | s \rangle$$
$$+ \langle a | \hat{B} | s \rangle \langle s | \rho^{(2)} | a \rangle + \langle a | \hat{B} | a \rangle \langle a | \rho^{(2)} | a \rangle. \tag{21}$$

The second and third terms on the right side of Eq. (21) vanish because of the superselection rule leaving

$$\langle \hat{B} \rangle = \langle s | \hat{B} | s \rangle \frac{1+q}{2} + \langle a | \hat{B} | a \rangle \frac{1-q}{2} . \tag{22}$$

The two-quon state acts like an "incoherent" mixture of a symmetric part with weight $(1+q)/2$ and an antisymmetric part with weight $(1-q)/2$.

For the second approach, we use the states $|m\rangle = |n\rangle = a_1^\dagger a_2^\dagger |0\rangle$, the basic two-quon state. In that case we find that

$$\langle \hat{B} \rangle = Tr(\hat{B} \rho^{(2)}) = \langle n | \hat{B} | n \rangle \langle n | \rho^{(2)} | n \rangle$$
$$= \left\{ \alpha^2 \langle s | \hat{B} | s \rangle + \alpha \beta \langle s | \hat{B} | a \rangle + \beta \alpha \langle a | \hat{B} | s \rangle + \beta^2 \langle a | \hat{B} | a \rangle \right\} \tag{23}$$
$$\times \left\{ \begin{array}{l} \dfrac{1+q}{2} \langle n | s \rangle \langle s | n \rangle + \dfrac{1-q}{2} \langle n | a \rangle \langle a | n \rangle \\ + \dfrac{\sqrt{1-q^2}}{2} \langle n | a \rangle \langle s | n \rangle + \dfrac{\sqrt{1-q^2}}{2} \langle n | s \rangle \langle a | n \rangle \end{array} \right\}.$$

Recognizing that $\langle s | n \rangle = \alpha$ and $\langle a | n \rangle = \beta$, we find that Eq. (23) reproduces the result given in Eq. (22).

Dropping the density operator "cross terms" that give no contribution to matrix elements (due to the superselection rule), we may write the density operator for a two-quon system as

$$a_1^\dagger a_2^\dagger |0\rangle\langle 0| a_2 a_1 = \frac{1+q}{2} |\phi_s\rangle\langle\phi_s| + \frac{1-q}{2} |\phi_a\rangle\langle\phi_a| . \tag{24}$$

Alternatively, this result can be expressed as

$$\rho^{(2)} = \frac{1+q}{2}\rho_s^{(2)} + \frac{1-q}{2}\rho_a^{(2)} . \tag{25}$$

We see that the two-quon density operator can be written as a symmetric part with weight $(1+q)/2$ and an antisymmetric part with weight $(1-q)/2$. Note that the sum of the coefficients of the two terms in the two-particle density operator is equal to unity, as it should for a set of "mixed" states.

The analogous calculation for the case with the mode labels 1 and 2 transposed gives

$$a_2^\dagger a_1^\dagger |0\rangle\langle 0| a_1 a_2 = \frac{1+q}{2}|\phi_s\rangle\langle\phi_s| + \frac{1-q}{2}|\phi_a\rangle\langle\phi_a| . \tag{26}$$

This result shows that we have the *same* density operator and hence the same expectation values for observed quantities for states generated by the creation operators acting in either order. In other words, the relative phase in Eq. (16) and Eq. (17) does not lead to any observable consequences. Equations (24) and (26) show that the density operators for $a_1^\dagger a_2^\dagger |0\rangle$ and $a_2^\dagger a_1^\dagger |0\rangle$ are *identical*, which means that these two "states" correspond to exactly the same physical situation. We put quotation marks around the word "states" to indicate that we should actually use density operators. Since the two Fock states $a_1^\dagger a_2^\dagger |0\rangle$ or $a_2^\dagger a_1^\dagger |0\rangle$ have the same density operator, we can, without loss of generality, do computations with either of them.

For states with three or more identical particles, we argue that the same kind of decomposition of the density operator obtains; that is, the density operator can be written as a sum of terms each associated with one of the representations of the permutation group and weighted by a polynomial in q. For a three-particle state, there are three inequivalent permutation symmetry representations (the completely symmetric, the completely antisymmetric, and the mixed symmetry representations) with weights proportional to $(1+q)(1+q+q^2)/6$ for the (3) (symmetric) representation, $(1-q)(1-q+q^2)/6$ for the (1,1,1) (antisymmetric) representation, $(1+q)^2(1-q)/3$ and $(1+q)(1-q)^2/3$ for the two (2,1) representations [2].

THE TWO-PHOTON ABSORPTION EXPERIMENT

We now proceed to apply this formalism to calculate the transition probabilities for the two-photon absorption experiment carried out by DeMille, Budker, Derr, and Devaney [7]. We shall show that the q-mutator formalism indeed predicts a non-zero transition probability for the two-photon absorption if the spin-statistics connection is violated for photons. If the photons are always in symmetric states ($q = +1$), then the transition probability vanishes.

The formalism developed in the previous section seems to indicate that we can think of the two-quon Hilbert space as a mixture of a symmetric sector and an antisymmetric sector. Thus, it is tempting to say that the transition probability for the two-photon experiment ought to be directly proportional to $(1-q)/2$, the weight factor for the antisymmetric sector. We shall show that in fact the transition **probability** is directly proportional to $(1-q)^2$.

The details of the two-photon experiment are given in the paper by D. DeMille and colleagues in this volume. Here we give just a brief description. The experiment investigates the two-photon transition between an atomic ground state with angular momentum $J = 0$ and an excited state with the same parity and $J = 1$. Angular momentum selection rules require that the two photons have orthogonal linear polarizations or opposite circular polarizations. As we shall see, this transition is forbidden if the two photons have the same frequency and are ordinary bosons. If photons are quons, the transition is allowed with a probability proportional to $(1-q)^2$.

For the two-photon absorption experiment, the Hamiltonian for an atom interacting with two photon modes, one labeled by s and the other by t, can be written as

$$H_{int} = F \sum_{s,t,\sigma,\nu,j} \frac{\hat{d} a_{s\nu} |j\rangle\langle j| a_{t\sigma} \hat{d}}{E_g - E_j + \hbar\omega_t} + \frac{\hat{d} a_{t\sigma} |j\rangle\langle j| a_{s\nu} \hat{d}}{E_g - E_j + \hbar\omega_s}, \qquad (27)$$

where j labels the intermediate atomic states that are connected to the initial state and the final state by the interaction. ν and σ label the polarization modes of the photons. E_j is the energy of the jth atomic intermediate state. $\hat{d}$ is the atomic dipole operator, assumed to be a Hermitian operator, as usual. F contains all the necessary numerical coefficients. This expression is just the result of ordinary second-order perturbation theory in the rotating-wave and dipole approximations [8]. Note that in this form H_{int} is not a Hermitian operator because the terms involving creation operators have been left out as part of the rotating-wave approximation. The atomic dipole operators are assumed to behave in the normal fashion; only the photons are described using q-mutators. Note that the photon annihilation operators appear in opposite order in the two parts of the operator corresponding to the two alternative routes from initial state to final state. Thus, we see that this operator is permutation symmetric and the superselection rule applies.

Greenberg [9] has argued that the interaction Hamiltonian must be constructed carefully when quons are involved in order to maintain the additivity of energies for well-separated systems. The implications of that issue for the two-photon experiment are discussed elsewhere [10].

The effective transition operator whose expectation value we calculate to find the transition probability is given by

$$H_{eff} = H_{int}^\dagger |f\rangle\langle f| H_{int}. \qquad (28)$$

First, let us focus on a simple two-photon initial state $|1,1\rangle = a^\dagger_{u\beta} a^\dagger_{v\gamma} |0\rangle$, which can be written as a sum of a permutation symmetric state and an antisymmetric state. We then use Eq. (5) to find the transition probability:

$$P_{if} = Tr(H_{eff} \rho^{(2)})$$
$$= \frac{1+q}{2} \langle s|H_{eff}|s\rangle + \frac{1-q}{2} \langle a|H_{eff}|a\rangle. \quad (29)$$

For notational simplicity we have suppressed the atomic part of the state labeling.

Since H_{eff} is expressed in terms of annihilation operators, we re-express the antisymmetric and symmetric states in terms of the original two-photon states (temporarily dropping the polarization mode labels) as

$$a^\dagger_v a^\dagger_u |0\rangle \text{ and } a^\dagger_u a^\dagger_v |0\rangle. \quad (30)$$

The result is

$$P_{if} = \tfrac{1}{2}\langle 0|a_u a_v H_{eff} a^\dagger_v a^\dagger_u |0\rangle + \tfrac{1}{2}\langle 0|a_v a_u H_{eff} a^\dagger_u a^\dagger_v |0\rangle$$
$$= \langle 0|a_v a_u H_{eff} a^\dagger_u a^\dagger_v |0\rangle \equiv |M_{if}|^2. \quad (31)$$

The last equality defines the transition amplitude M_{if}.

It is instructive to calculate the relevant matrix elements separately for the antisymmetric and symmetric parts. Using Eq. (29) and Eqs. (10) and (11), we see that the symmetric part is given by

$$P^{sym}_{if} = \frac{1+q}{2} \frac{1}{2(1+q)} \langle 0|(a_v a_u + a_u a_v) H_{eff} (a^\dagger_u a^\dagger_v + a^\dagger_v a^\dagger_u)|0\rangle, \quad (32)$$

and the antisymmetric part is

$$P^{anti}_{if} = \frac{1-q}{2} \frac{1}{2(1-q)} \langle 0|(a_v a_u - a_u a_v) H_{eff} (a^\dagger_u a^\dagger_v - a^\dagger_v a^\dagger_u)|0\rangle. \quad (33)$$

For a two-photon initial state, the photon part of the matrix elements connecting the initial state to the final state (with no photons) are all of the form

$$\langle 0|a_i a_j a^\dagger_k a^\dagger_l |0\rangle = \frac{1}{2}(1+q)(\delta_{jk}\delta_{il} + \delta_{ik}\delta_{jl})$$
$$+ \frac{1}{2}(1-q)(\delta_{jk}\delta_{il} - \delta_{ik}\delta_{jl}). \quad (34)$$

Using $D_{\gamma\beta}$ to label the Cartesian components of the products of atomic dipole matrix elements

$$D_{\gamma\beta} = (d_{fj})_\gamma (d_{jg})_\beta, \quad (35)$$

we find that the antisymmetric part of the transition *amplitude* is given by

$$M^{anti}_{if} = \frac{F}{2} \sum_j \left\{ \frac{D_{\gamma\beta} - q D_{\gamma\beta}}{E_i - E_j + \hbar\omega_u} + \frac{q D_{\beta\gamma} - D_{\beta\gamma}}{E_i - E_j + \hbar\omega_v} \right\}. \quad (36)$$

In the experimental case of interest, the two frequencies are the same, but the polarization directions are different. In that case the antisymmetric part of the transition amplitude is given by

$$M_{if}^{anti} = \frac{F}{2}(1-q)(D_{\gamma\beta} - D_{\beta\gamma})\sum_j \frac{1}{E_i - E_j + \hbar\omega} \ . \tag{37}$$

A straightforward exercise in Clebsch-Gordan algebra shows that $D_{\gamma\beta} = -D_{\beta\gamma}$ for a $J = 0$ to $J = 1$ atomic transition.

We see from Eq. (37) that the transition amplitude vanishes if the photons are "pure" bosons with $q = 1$. It is also easy to show that the symmetric state part of the amplitude is proportional to $(1+q)(D_{\gamma\beta} + D_{\beta\gamma})$. Thus that part of the amplitude vanishes for any value of q (if the two photon frequencies are the same). The conclusion is that the transition probability is proportional to $(1-q)^2$. We might (naively) have expected the transition probability to go as $(1-q)$ because that is the weight assigned to the antisymmetric (spin-statistics violating) part of the density matrix.

To generalize to more complex initial states, we next treat the case of general number states. The q-deformed initial (number) state is constructed by applying quon creation operators to the photon vacuum [11]

$$|n_u, n_v\rangle = \frac{(a_u^\dagger)^{n_u}(a_v^\dagger)^{n_v}|0\rangle}{\sqrt{[n_u]_q! [n_v]_q!}} \ , \tag{38}$$

where $[n]_q$ (a so-called q-deformed number) is defined to be

$$[n]_q = \frac{1-q^n}{1-q} \ , \tag{39}$$

and the q-factorial is $[n]_q! = [n]_q [n-1]_q \ldots 1$.

The matrix elements needed are all polynomials in q and can be written in the following form:

$$\langle f|a_v a_u|i\rangle = g_{n_u n_v}(q)$$
$$\langle f|a_u a_v|i\rangle = q g_{n_u n_v}(q) \tag{40}$$

Although the general form of $g_{n_u n_v}(q)$ is not known, the results of this calculation for several number states are listed in Table 1.

We can see some general features by examining Table 1: Except for the (1,1) initial state, $g_{n_u n_v}(q)$ is zero when $q = -1$. This result should be expected because $q = -1$ corresponds to anticommutators for the operators and hence leads to the Exclusion Principle: all matrix elements that involve states with more than one fermion per mode must equal zero. For the (1,1) case the two matrix elements have the opposite sign when $q = -1$ as required for amplitudes that differ by the interchange of two identical fermions. On the other hand, for $q = +1$, we expect the amplitude to be

Table 1. The polynomial $g_{n_u n_v}(q)$ for several photon number states.

initial state n_u, n_v	$g_{n_u n_v}(q)$
1, 1	1
2, 1	$q + q^2$
1, 2	$1 + q$
2, 2	$q + 2q^2 + q^3$

proportional to $(n_u n_v)^{1/2}$ in order to have the probability be proportional to the product of the numbers of photons in each of the two modes. The normalization in Eq. (38) requires $g_{n_u n_v}(q) = (n_u(n_u!)n_v(n_v!))^{1/2}$ for $q = +1$, which result agrees with the listings in Table 1.

Even without knowing the general form of $g_{n_u n_v}(q)$, we can draw the following conclusions: The general transition amplitude can be written in the form

$$M_{if} = F g_{n_u n_v}(q) \sum_j \frac{D_{\gamma\beta}}{E_g - E_j + \hbar\omega_u} + q \frac{D_{\beta\gamma}}{E_g - E_j + \hbar\omega_v}. \tag{41}$$

If we consider the case when the two modes have the same frequency (but not the same polarization), the transition amplitude takes the form

$$M_{if} = F g_{n_u n_v}(q) \sum_j \frac{[D_{\gamma\beta} + q D_{\beta\gamma}]}{E_g - E_j + \hbar\omega}$$

$$= F g_{n_u n_v} \sum_j \frac{1}{E_g - E_j + \hbar\omega} \left\{ \left(\frac{1+q}{2}\right)(D_{\gamma\beta} + D_{\beta\gamma}) + \left(\frac{1-q}{2}\right)(D_{\gamma\beta} - D_{\beta\gamma}) \right\}. \tag{42}$$

As we saw previously, for a $J = 0$ to $J = 1$ two-photon transition, the sum of the two D terms in Eq. (42) vanishes (in the electric dipole approximation). For ordinary boson behavior ($q = +1$), the second term vanishes also. However, for $q < 1$, the second term can be nonzero, thus signaling a violation of the spin-statistics connection for photons with an *amplitude* proportional to $(1 - q)$. The transition *probability* is proportional to $(1 - q)^2$. In a forthcoming paper Greenberg and Hilborn [10] show that the same result obtains when the initial photon states are coherent states.

The experiment of DeMille et al [7] shows that the relative probability of the forbidden two-photon transition is less than 1.2×10^{-7}. This provides a limit of $\varepsilon_B = 1 - q < 3 \times 10^{-4}$ for photons.

A spin-statistics violating probability proportional to $(1 - q)^2$ was also obtained in the analysis [12] of experiments involving the so-called dispersive interaction of Rydberg atoms with microwave photons in a cavity. Gerry and Hilborn showed that in principle those experiments can provide a test of the spin-statistics connection for photons. Unfortunately in the Rydberg atom experiments, various systematic effects limit the possible sensitivity for detecting a spin-statistics violation.

TESTS FOR ATOMIC ELECTRONS - ATOMIC HELIUM

We now apply a similar formalism to the atomic helium test of the spin-statistics connection for electrons [13]. The experiment looks for atomic transitions between the spin-statistics violating metastable permutation symmetric state 1s2s ^{1}S and the permutation symmetric 1s3p ^{1}P state, which has an energy slightly different from the normal (antisymmetric) 1s3p ^{3}P state [14]. We use the following shorthand notation to denote the relevant single-particle states (1 = 1s, 2 = 2s, 3 = 3p). Then the initial atomic state can be represented by

$$|i\rangle = c_1^\dagger c_2^\dagger |0\rangle, \qquad (43)$$

where $c_j^\dagger$ is the creation operator for an electron in the single-particle state labeled by j. By interacting with a laser field, the atom makes a transition to the final state

$$|f\rangle = c_1^\dagger c_3^\dagger |0\rangle. \qquad (44)$$

The appropriate interaction operator can be expressed (in the rotating wave approximation) as

$$H_{int} = V a_k c_3^\dagger c_2. \qquad (45)$$

This operator removes an electron from the 2s state and creates one in the 3p state while annihilating a single photon with wavevector k. The symbol V incorporates all the detailed numerical factors for the transition.

We assume that the cs are described by the q-mutator relation, Eq. (1). The initial atomic state can be written in terms of the permutation symmetric and antisymmetric states:

$$c_2^\dagger c_1^\dagger |0\rangle = \sqrt{\frac{1+q}{2}} |s\rangle + \sqrt{\frac{1-q}{2}} |a\rangle. \qquad (46)$$

Writing things the other way around yields

$$|s\rangle = \frac{1}{\sqrt{2(1+q)}} \left(c_1^\dagger c_2^\dagger + c_2^\dagger c_1^\dagger \right)|0\rangle$$

$$|a\rangle = \frac{1}{\sqrt{2(1-q)}} \left(c_1^\dagger c_2^\dagger - c_2^\dagger c_1^\dagger \right)|0\rangle. \qquad (47)$$

Because of the superselection rule, the transition rate involves the permutation symmetric and permutation antisymmetric sectors independently. We also note that the atomic energies will be different for the symmetric and antisymmetric states [14]. That means that in practice we can tune our excitation laser to look for transitions between symmetric states or for transitions between antisymmetric states.

Let us now calculate the atomic matrix elements needed to see how the transition probability depends on the parameter q. The matrix elements we need to evaluate are given by

$$\frac{1}{2(1\pm q)}\langle 0|(c_3 c_1 \pm c_1 c_3)c_3^\dagger c_2 \left(c_2^\dagger c_1^\dagger \pm c_1^\dagger c_2^\dagger\right)|0\rangle , \qquad (48)$$

where the first factor comes from the normalization of the s and a states. The + signs are for the symmetric states while the − signs are for the antisymmetric states. Straightforward application of the quon algebra shows that the matrix element of the operators is given by $(1 \pm q)^2$; so that taking normalization into account, the transition **amplitude** goes as $(1 \pm q)$. The transition probabilities are then proportional to $(1 \pm q)^2$. Again we see that the simple argument based on the weight given to the symmetric or antisymmetric part of the density operator gives the wrong dependence on $1 - |q|$.

The experiment of Dailamian, Gillaspy, and Kelleher [13] shows that the relative probability for finding two electrons in a permutation symmetric state in atomic helium is less than 5×10^{-6}. From that result we conclude that the experiment sets a limit $\varepsilon_F = 1 - |q| < 2.5 \times 10^{-3}$ for electrons.

^{16}O NUCLEI IN MOLECULES

We now turn our attention to tests of the spin-statistics connection for nuclei. Several experiments [15-18] using molecular spectroscopy have set rather stringent limits on finding molecules containing identical nuclei (in this case ^{16}O nuclei, which have spin 0) in states that violate the spin-statistics connection. To put these tests into the q-mutator formalism, we need to use field operators that (at least symbolically) create a nucleus at a particular spatial location. As usual, these field operators can be expressed in terms of the basic creation operators:

$$\psi^\dagger(x) = \sum_k u_k(x) a_k^\dagger , \qquad (49)$$

where the us are some complete orthonormal set of states. In a nonrelativistic theory, we can write a creation operator for a molecular state labeled by a set of quantum numbers indicated by α. For example, for a diatomic molecule, the creation operator can be written as

$$A_\alpha^\dagger = \int dx_1 dx_2 f_\alpha(x_1, x_2) \psi^\dagger(x_1) \psi^\dagger(x_2) , \qquad (50)$$

where $f_\alpha(x_1, x_2)$ is the appropriate wave function. (We have suppressed all dependence on electron coordinates and electron operators.) It is easy to show that the field operators satisfy q-mutator relations of the following form:

$$\psi(x)\psi^\dagger(x') - q\psi^\dagger(x')\psi(x) = \delta(x - x') . \qquad (51)$$

As before, it is helpful to write these operators in terms of symmetric and antisymmetric combinations:

$$A'_\alpha(s,a) = N_{s,a} \int dx_1 dx_2 f_\alpha(x_1,x_2) \left[\psi^\dagger(x_1)\psi^\dagger(x_2) \pm \psi^\dagger(x_2)\psi^\dagger(x_1) \right], \quad (52)$$

where N is the normalization factor.

The amplitude for a transition from one molecular state to another is proportional to

$$\langle 0 | A_{\alpha_f}(s,a) A^\dagger_{\alpha_f} A_{\alpha_i} A^\dagger_{\alpha_i}(s,a) | 0 \rangle. \quad (53)$$

Straightforward application of the quon algebra shows that the transition probability for the symmetric part is proportional to $(1+q)^2$ while the transition probability for the antisymmetric (spin-statistics violating) part is proportional to $(1-q)^2$. The experiment of Modugno, et al [18] shows that this probability for the SSC violation transition is less than 1×10^{-11} of that for the "allowed" transition in molecular CO_2. Thus we find that the experiment sets a limit of $\varepsilon_B = 1 - q < 3 \times 10^{-6}$ for ^{16}O nuclei.

QUON STATISTICS FOR COMPOSITE SYSTEMS

Greenberg and Hilborn [3] have shown that the q parameter for a composite system (such as the ^{16}O nucleus) can be related to the q parameter of the particles that make up the composite system. That result is the quon generalization of the arguments first put forward by Wigner [19] and by Ehrenfest and Oppenheimer [20] in the early days of quantum mechanics. The usual result can be stated as follows: if the composite systems contains an odd number of fermions, then the composite is a fermion. Otherwise, the composite is a boson. For the quon case, a composite consisting of N quons is described by the following relation [3]

$$q_{composite} = q_{constituent}^{N^2}. \quad (54)$$

Note that the quon relationship reduces to the usual rule in the limit when $q = \pm 1$ since N and N^2 are even or odd together. In terms of deviations from $q = \pm 1$, we find that

$$\varepsilon_{composite} = N^2 \varepsilon_{constituent} \quad (55)$$

for small deviations from "pure" boson or fermion behavior. Hence, we see that experimental limits on the spin-statistics violation for a composite system such as ^{16}O can yield even more stringent limits on a possible violation for the constituents, say the nucleons. Using the results from Mudgno, et al [18], we find that

$$\varepsilon_{nucleon} < 10^{-8} \text{ and } \varepsilon_{quark} < 10^{-9}. \quad (56)$$

DISCUSSION

We have shown how the algebra of q-mutators can be linked to experimental results to set limits on the deviation of q from the usual $q = \pm 1$ for bosons and fermions, respectively. Since the superselection rule forbids transitions between states associated with different representations of the permutation group, the calculations are best framed in terms of density operators with different weights for the various sectors associated with different representations dependent on q. However, the straightforward use of Fock states includes all the possible representations and thereby removes the necessity of finding the weight factors. Applying these results to recent experiments for photons, electrons and ^{16}O nuclei shows that these experiments are still not highly sensitive to deviations of $|q|$ from 1.

ACKNOWLEDGMENTS

This work was supported in part by the National Science Foundation, the Howard Hughes Medical Institute, Amherst College, and the University of Nebraska-Lincoln. I thank O. W. Greenberg, D. DeMille, and D. Budker for many stimulating conversations.

REFERENCES

1. Greenberg, O.W., *Phys. Rev. Lett.* **64**, 705–708 (1990).
2. Greenberg, O.W., *Phys. Rev. D* **43**, 4111-4120 (1991).
3. Greenberg, O.W., and Hilborn, R.C., *Phys. Rev. Lett.* **83**, 4460-4463 (1999).
4. Messiah, A.M.L., and Greenberg, O.W., *Phys. Rev.* **136**, B248-B267 (1964).
5. Amado, R.D., and Primakoff, H., *Phys. Rev. C* **22**, 1338-1340 (1980).
6. Blum, K., *Density Matrix Theory and Applications*, Plenum Press, New York, 1996.
7. DeMille, D., Budker, D., Derr, N., and Deveney, E., *Phys. Rev. Lett.* **83**, 3978-3981 (1999).
8. Weissbluth, M., *Atoms and Molecules*, Academic Press, New York, 1978.
9. Greenberg, O.W., *Phys. Lett. A* **209**, 137-142 (1995).
10. Greenberg, O.W., and Hilborn, R.C., In preparation. (2000).
11. Agarwal, G.S., and Chaturvedi, S., *Mod. Phys. Lett. A* **7**, 2407–2413 (1992).
12. Gerry, C.G., and Hilborn, R.C., *Phys. Rev. A* **55**, 4126-4130 (1997).
13. Deilamian, K., Gillaspy, J.D., and Kelleher, D.E., *Phys. Rev. Lett.* **74**, 4787-4790 (1995).
14. Drake, G.W.F., *Phys. Rev. A* **39**, 897-899 (1989).
15. de Angelis, M., Gagliardi, G., Gianfrani, L., and Tino, M., *Phys. Rev. Lett.* **76**, 2840-2843 (1996).
16. Hilborn, R.C., and Yuca, C.L., *Phys. Rev. Lett.* **76**, 2844-2847 (1996).
17. Modugno, G., Inguscio, M., and Tino, G.M., *Phys. Rev. Lett.* **81**, 4790-4793 (1998).
18. Modugno, G., et al., in this volume.
19. Wigner, E.P., *Anz. Ung. Ak. Wiss.* **46**, 576 (1929).
20. Ehrenfest, P., and Oppenheimer, J.R., *Phys. Rev.* **37**, 333-338 (1931).

Quon Theories in Quantum Optics

Allan I. Solomon

Quantum Processes Group,
The Open University
Milton Keynes MK7 6AA, United Kingdom

Abstract. We start from the viewpoint of quantum canonical transformations, analogues of the usual classical canonical transformations. We show that in the quantum case, such transformations lead naturally to the definition of some important quantum optics states, including the vacuum, coherent states, squeezed vacuum and squeezed states. We then go on to consider *quons*, operators satisfying deformed versions of the canonical commutation relations. Extending the idea of a canonical transformation to *deformed* quantum commutators gives quon analogues of these states, quon states. We finally show that this approach provides us with defining relations for some quantum groups.

INTRODUCTION

The objectives of this note are threefold:

- Firstly, we wish to present a unified view of certain important states in standard quantum optics, namely; the vacuum, coherent states, squeezed states and Kerr states. Our point of view is very simple. Essentially, these states are just vacuum states for a transformed annihilation operator. And each transformation is simply a canonical transformation of the usual commutation relations. The set of transformations corresponds to the *automorphism* group of the canonical commutation relation algebra; this algebra is the Heisenberg-Weyl algebra and, in general, the automorphisms are not *inner*.

- Secondly, we introduce *quons*, which are a special type of deformation of the usual bosons, and adopt a modification of this approach; namely, we define canonical transformations of the quon commutation relations, and the analogues of the usual quantum optics states are simply the new vacua of the transformed quon annihilation operator. We shall find that it is necessary to introduce non-commuting base fields in order to define these canonical transformations, and this leads us to the third part of our article.

- Thirdly, we shall see that the automorphisms of the quonic commutation relations correspond to *quantum groups*. Thus, in the case where we define the analogue of the displacement operator, we obtain a q-analogue of the coherent state, and the quantum displacement group. In the cases of the q-squeezed states, we obtain some defining relations for the quantum groups $SU_q(1,1)$ and (a subgroup of) $ISU_q(1,1)$.

CANONICAL TRANSFORMATIONS

Classical canonical transformations are defined in terms of the Poisson Bracket, which for two functions such as $F(q,p)$, $G(q,p)$ takes the form

$$\{F,G\} = \frac{\partial F}{\partial q}\frac{\partial G}{\partial p} - \frac{\partial F}{\partial p}\frac{\partial G}{\partial q}. \tag{1}$$

The coordinate q and momentum p satisfy $\{q,p\} = 1$. The transformation

$$q \to Q(q,p) \quad p \to P(q,p) \tag{2}$$

is canonical if $\{Q,P\} = 1$. In quantum mechanics, where q and p are now operators, we use the commutator bracket

$$[q,p] \equiv qp - pq \tag{3}$$

which for conjugate Hermitian variables has the value i (times the unit operator). Generally we do not avail ourselves in quantum mechanics of the wealth of canonical transformations possible in the classical case, usually restricting to those described by unitary transformations U. Thus, expressing the q, p Hermitian conjugate variables in the more common form of the annihilation and creation operators $a, a^\dagger$ of quantum optics,

$$a \equiv \frac{q+ip}{\sqrt{2}} \quad a^\dagger \equiv \frac{q-ip}{\sqrt{2}} \tag{4}$$

we have $[a, a^\dagger] = I$. We shall consider canonical transformations of the form

$$a \to A = UaU^\dagger \quad a^\dagger \to A^\dagger = Ua^\dagger U^\dagger. \tag{5}$$

The approach of this note is to consider the relevant quantum optics states as vacuum states for the transformed operator A. The *vacuum state* is defined as the normalized state satisfying

$$a|0\rangle = 0. \tag{6}$$

We note that the vacuum state corresponding to A, defined by

$$A|0'\rangle = 0 \tag{7}$$

is given by
$$|0'\rangle = U|0\rangle. \tag{8}$$

In the spirit of our approach, the ordinary vacuum state corresponds to the trivial unitary transformation $U = I$,
$$a \to A = IaI^\dagger \qquad a^\dagger \to A^\dagger = Ia^\dagger I^\dagger \tag{9}$$
with
$$|0'\rangle = I|0\rangle. \tag{10}$$

Less trivially, we shall now consider some important states which may be considered in the same light.

Coherent States

These states correspond to the (next) most simple canonical transformation, that corresponding to
$$\begin{aligned} a &\to A = a - \alpha I \\ a^\dagger &\to A^\dagger = a^\dagger - \alpha^* I \end{aligned} \tag{11}$$

where I is the unit operator. This canonical transformation is generated by the unitary displacement operator
$$U \equiv D(\alpha) = \exp(\alpha a^\dagger - \alpha^* a)$$
which operates on a by
$$a \to A = UaU^\dagger \equiv D(\alpha)aD^\dagger(\alpha) = a - \alpha I. \tag{12}$$

The vacuum state of A is the coherent state
$$|0'\rangle = |\alpha\rangle = D(\alpha)|0\rangle$$

And the vacuum equation for A is $A|0'\rangle = 0$ which is equivalent to $a|\alpha\rangle = \alpha|\alpha\rangle$. This is the usual coherent state [1]
$$|\alpha\rangle = \mathcal{N}(\alpha) \sum_{n=0}^\infty \frac{\alpha^n}{\sqrt{n!}} |n\rangle \tag{13}$$

where $\mathcal{N}(\alpha) = \exp(-|\alpha|^2)$ is the normalization, and has all the well-known properties, such as minimizing the Heisenberg Uncertainty Principle.

Squeezed Vacuum

This state corresponds to the canonical transformation $U = S(\xi)$,

$$a \to A = S(\xi)aS^\dagger(\xi) = \lambda a + \mu a^\dagger$$
$$a^\dagger \to A^\dagger = S(\xi)a^\dagger S^\dagger(\xi) = \lambda^* a^\dagger + \mu^* a \qquad (14)$$

where $\lambda(\xi)$ and $\mu(\xi)$ satisfy $|\lambda|^2 - |\mu|^2 = 1$. Here $S(\xi)$ is the squeezing operator, given by

$$S(\xi) = \exp\frac{1}{2}(\xi^* a^2 - \xi a^{\dagger 2}), \quad (\xi = r\exp(i\phi)) \qquad (15)$$

and where we have put

$$\lambda = \cosh r, \quad \mu = \exp(i\phi)\sinh r.$$

The squeezed vacuum satisfying $A|0'\rangle = 0$ is given by $|\xi\rangle = S(\xi)|0\rangle$; that is, it satisfies

$$(\lambda a + \mu a^\dagger)|\xi\rangle = 0. \qquad (16)$$

The 2-parameter operator $S(\xi)$ is not the most general element of the 3-parameter vacuum-squeezing group $SU(1,1)$. An additional operator $P(\theta) = \exp(i\theta\hat{n})$, where $\hat{n} \equiv a^\dagger a$, supplies the third parameter and completes the generators of the group.

Squeezed States

Standard squeezed states correspond to the canonical transformation $U(\eta) = S(\xi)D(\alpha)$ $(\eta = \eta(\xi, \alpha))$ where

$$a \to U(\eta)aU^\dagger(\eta) = \lambda a + \mu a^\dagger - \alpha I$$
$$a^\dagger \to U(\eta)a^\dagger U^\dagger(\eta) = \lambda^* a^\dagger + \mu^* a - \alpha^* I \quad |\lambda|^2 - |\mu|^2 = 1. \qquad (17)$$

The squeezed state $|\eta\rangle = U(\eta)|0\rangle$ thus defined may be written explicitly as

$$|\eta\rangle = U(\eta)|0\rangle$$
$$= S(\xi)D(\alpha)|0\rangle$$
$$= \exp\frac{1}{2}(\xi^* a^2 - \xi a^{\dagger 2})\exp(\alpha a^\dagger - \alpha^* a)|0\rangle \qquad (18)$$

and satisfies [2]

$$(\lambda a + \mu a^\dagger)|\eta\rangle = \alpha|\eta\rangle. \qquad (19)$$

One may refer to the Automorphism Group of the Canonical Commutation relations generated by $U(\eta)$ as the squeezing group $\mathcal{G}$. This is a 5-parameter group corresponding to the action of the semi-direct sum of the 3-dimensional Heisenberg-Weyl algebra (the central element of this algebra acts trivially) and the 3-dimensional algebra $su(1,1)$ on $\{a, a^\dagger\}$, and is a 5-parameter subgroup of the inhomogeneous pseudounitary group $ISU(1,1)$ with fundamental representation

$$G \ni g = \begin{bmatrix} \lambda & \mu & \alpha \\ \mu^* & \lambda^* & \alpha^* \\ 0 & 0 & 1 \end{bmatrix} \quad (\lambda, \mu, \alpha \in \mathbb{C};\ |\lambda|^2 - |\mu|^2 = 1). \tag{20}$$

Kerr State

Consider the canonical transformation:

$$a \to A = \exp(iF(\hat{n}))a \exp(-iF(\hat{n})) = \exp iG(\hat{n})a, \quad G(\hat{n}) = F(\hat{n}) - F(\hat{n}+1). \tag{21}$$

This map is not, strictly speaking, an automorphism of the Heisenberg-Weyl algebra (except when $F(\hat{n})$ is linear, when we obtain a (trivial) phase transformation of the operators $\{a, a^\dagger\}$), since it maps to an isomorphic algebra rather than the algebra itself[1]. However, it is of particular interest in optical physics, since the case $F(\hat{n}) = -\frac{1}{2}\gamma\hat{n}(\hat{n}-1)$ with corresponding transformation $K(\gamma) \equiv \exp(-\frac{1}{2}i\gamma\hat{n}(\hat{n}-1))$

$$a \to A = K(\gamma)aK^\dagger(\gamma) = e^{i\gamma\hat{n}}a \tag{22}$$

leads to the *Kerr* state [3]. The Kerr State is defined by the canonical transformation $U = K(\gamma)D(\alpha)$ leading to the usual form for this state

$$U|0\rangle = \mathcal{N}(\alpha) \sum_{n=0}^{\infty} \frac{\exp(-\frac{1}{2}i\gamma n(n-1))\alpha^n}{\sqrt{n!}}|n\rangle \quad \mathcal{N}(\alpha) = \exp(-|\alpha|^2). \tag{23}$$

This completes the first part of our strategy; namely, to describe each of the foregoing quantum optics states as a unitary transformation of the vacuum state, where the unitary operator provides a canonical transformation of the canonical commutation relations. The corresponding quonic states will now be described in a similar fashion, but with the canonical commutation relations replaced by the appropriate quonic commutation relation.

QUONIC STATES

We now introduce quonic analogues of the optics states described above, based on the *quon* [4], whose annihilation operator a satisfies

---

[1] I am indebted to Dr. Hongchen Fu for this remark.

$$[a, a^\dagger]_q \equiv aa^\dagger - qa^\dagger a = I. \tag{24}$$

where q is a real parameter (usually taken to lie between -1 and 1 and to be non-zero). Note that in this Section we use the same symbol a for the quon annihilation operator as for the usual boson, this latter simply corresponding to one value of q, namely, $q = 1$. It is our objective to define states analogous to those discussed in the previous Section for the ($q = 1$) case. Of course, there is no unique way to extend from $q = 1$. But we shall use as our guiding principle the strategy adopted above; namely, we shall consider as the q-analogues those states which are vacuum states under canonical transformations of the quonic operators, canonical in the sense that they preserve the quonic commutation relation Eq.(24).

Quon coherent states

One direct (and generally adopted) approach is to define the q-coherent states as normalized eigenstates of the quon annihilation operator [4]. It is straightforward to show that $a|z\rangle = z|z\rangle$ where

$$|z\rangle = \mathcal{N}(z) E_q(z a^\dagger)|0\rangle. \tag{25}$$

Here we have used the Jackson q-exponential function [5]

$$E_q(x) = \sum_{n=0}^{\infty} \frac{x^n}{[n]!} \tag{26}$$

where $[n] = \frac{1-q^n}{1-q}$ is the *basic number* of classical q-analysis [6] and

$$[n]! \equiv [n][n-1]\ldots[1] \qquad [0]! \equiv 1. \tag{27}$$

A remark about the number operator $\hat{n}$ associated with the quons. This is an operator which counts the number of quons produced by the repetitive action of $a^\dagger$ on the vacuum. Number states are normalized eigenstates of $\hat{n}$. The number operator $\hat{n}$ satisfies $[\hat{n}, a] = -a$ as in the usual ($q = 1$) case; however, unlike this latter case we may not write $a^\dagger a = \hat{n}$. The appropriate relation for quons is $a^\dagger a = [\hat{n}]$.

It is possible and consistent to take this counting operator equal to the usual number operator $\hat{n}$, that is, $\hat{n} = a^\dagger_{q=1} a_{q=1}$. However, in the current context of fixed $q \neq 1$ there is no *usual* case, and here it is more appropriate to adopt, as our defining relation between a and $\hat{n}$, $[a, q^{\hat{n}}]_q = 0$. Under a canonical transformation $a \mapsto A$ the new number operator $\hat{N}$ satisfies $[A, q^{\hat{N}}]_q = 0$.

We now express the q-coherent state Eq.(25) in terms of number states; we have

$$|z\rangle = \mathcal{N}(z) \sum_{n=0}^{\infty} \frac{z^n}{\sqrt{[n]!}} |n\rangle \tag{28}$$

where the normalization $\mathcal{N}(z) = E_q(zz*)^{-\frac{1}{2}}$. Clearly Eq.(28) has the correct limit for $q = 1$, namely Eq.(13) with $\alpha = z$.

However, in line with the overall approach outlined here, we demand that the displacement

$$a \to A = a + \alpha \tag{29}$$

be "canonical" in the sense that

$$I = [a, a^\dagger]_q = [a + \alpha, a^\dagger + \alpha^*]_q \tag{30}$$

The constraints

$$[\alpha, \alpha^*]_q = 0 \quad [a, \alpha^*]_q = 0 \tag{31}$$

are satisfied by

$$\alpha = zq^{\hat{n}} \quad [z, z^*]_q = 0. \tag{32}$$

It is noteworthy that we can construct a unitary displacement operator U to implement this canonical transformation. Consider Eq.(12) in the equivalent form

$$a \to A = U^\dagger a U \equiv D^\dagger(\alpha) a D(\alpha) = a + \alpha I. \tag{33}$$

We therefore need a q-analogue of the operator U which produces the displacement

$$a \to A = a + zq^{\hat{n}}. \tag{34}$$

It may be shown [7] that if instead of using $\mathbb{C}$ as the base field, one uses a deformation of the complex plane, $\mathbb{C}_q$ [8] [9,10], then such an operator may be constructed. Formally $\mathbb{C}_q$ is the algebra $\mathbb{C}[z, z^*]/B_q$ i.e. the quotient of the involutive algebra freely generated by z and z^*, and B_q, the bi–ideal determined by the re–ordering rule

$$zz^* = qz^*z. \tag{35}$$

Explicitly, the unitary displacement operator $U \equiv D_q(z, z^*)$ is given by

$$D_q(z, z^*) = E_q(-z^*z)^{\frac{1}{2}} E_q(za^\dagger) E_{q^{-1}}(-z^*a). \tag{36}$$

We now define the coherent (displaced vacuum) states by

$$|z, z^*\rangle = D_q(z, z^*)|0\rangle \tag{37}$$

whence it can be shown that $|z, z^*\rangle$ is a " right eigenstate" of a with eigenvalue z,

$$a|z, z^*\rangle = |z, z^*\rangle z. \tag{38}$$

We may generalize this approach to find the quon analogues of squeezed states.

Quon squeezed vacuum states

As in the conventional case, we define the q-squeezed vacuum by $S_q(\xi)|0\rangle$, where $S_q(\xi)$ is a unitary transformation implementing the q-canonical transformation

$$a \to A = S_q(\xi)aS_q^\dagger(\xi) = \lambda a + \mu a^\dagger$$
$$a^\dagger \to A^\dagger = S_q(\xi)a^\dagger S_q^\dagger(\xi) = \lambda^* a^\dagger + \mu^* a, \qquad (39)$$

which is "q-canonical" in the sense that

$$I = \left[a, a^\dagger\right]_q = \left[\lambda a + \mu a^\dagger, \lambda^* a^\dagger + \mu^* a\right]_q. \qquad (40)$$

This is satisfied by

$$[\lambda, a] = 0 = [\mu, a] \qquad (41)$$

together with the relations

$$[\lambda, \mu^*]_q = 0$$
$$[\mu, \lambda^*]_q = 0$$
$$\lambda\lambda^* - \lambda^*\lambda = q\mu^*\mu - q^{-1}\mu\mu^*$$
$$\lambda\lambda^* - q\mu^*\mu = 1. \qquad (42)$$

Although we do not have an explicit form of $S_q(\xi)$, we give immediately below a non-unitary form in the more general q-squeezed state case.

Quon squeezed states

In the same way, we demand that the transformation

$$a \to U_q(\eta)aU_q^\dagger(\eta) = \lambda a + \mu a^\dagger - \alpha$$
$$a^\dagger \to U_q(\eta)a^\dagger U_q^\dagger(\eta) = \lambda^* a^\dagger + \mu^* a - \alpha^*. \qquad (43)$$

be q-canonical; the conditions imposed result in the previous q-displacement $\alpha = zq^{\hat{n}}$ and squeezed vacuum λ, μ relations, together with the additional constraints

$$[\lambda, z^*] = 0 \quad [\mu, z^*]_{q^2} = 0. \qquad (44)$$

These conditions on the parameters give a quantum analogue of the inhomogeneous group $\mathcal{G}$ in the conventional case, as we shall see in the next section.

The appearance of the q^2-commutator had already been noted some while ago in the context of q-squeezing [12] where it was shown that a suitable candidate $|\xi, \alpha\rangle$ for a q-analogue of the squeezed state of the previous section may be obtained (up to normalization) as

$$|\xi,\alpha\rangle = E_{q^2}(-\frac{1}{[2]_q}\xi a^{\dagger 2})E_q(\alpha a^\dagger)|0\rangle . \tag{45}$$

The state $|\xi,\alpha\rangle$, which is a normalizable state for $|q| < 1$, satisfies

$$(a + \xi a^\dagger)|\xi,\alpha\rangle = \alpha|\xi,\alpha\rangle$$

provided $\xi\alpha = q^{-2}\alpha\xi$. This was an early indication of the necessity for a non-commutative base field for these quonic states.

Quon Kerr States

The transformation defined by Eq.(21) is still a canonical transformation in the quon case. We may therefore define a quonic Kerr State by the canonical transformation $U = K(\gamma)D_q(z,z^*)$ leading to a form for this state

$$U|0\rangle = \mathcal{N}\sum_{n=0}^{\infty}\frac{\exp(-\frac{1}{2}i\gamma n(n-1))\alpha^n}{\sqrt{[n]!}}|n\rangle \tag{46}$$

which differs little from the conventional case.

QUANTUM GROUPS

We have seen that the canonical transformations which define the quantum optics states in the $q = 1$ case are generated by various transformation groups, subgroups of the the automorphism group of the canonical commutation relations. When we extend this approach to quons, we have seen that these transformations must be expressed in terms of noncommuting variables. The structures that emerge are *quantum groups* and what we have done in the foregoing is simply to give defining relations for some specific quantum groups.

A prototypical quantum group is $GL_q(2)$ as defined by Manin [13]. We may consider this as the set of matrices

$$GL_q(2) = \left\{\begin{bmatrix} a & b \\ c & d \end{bmatrix} \middle| \begin{array}{lll} ab = qba & cd = qdc & ac = qca \\ bd = qdb & bc = cb & ad = da + (q - 1/q)bc \end{array}\right\} \tag{47}$$

with $det_q = ad - qbc$. The quantum group $GL_q(2)$ is essentially the (Hopf) algebra generated by products and sums of the elements a, b, c, d; these may be interpreted as operators acting on ordinary 2×2 matrices. For example, the entry c produces the lower left-hand element of the 2×2 matrix on which it operates. Note that in general these structures are not *groups* in spite of the name. For example, although the product of two members of $GL_q(2)$ as defined above, whose entries are assumed to commute mutually, does have entries which satisfy the same defining relations, this is not true of powers of members.

We now use the quon canonical transformations to define some quantum groups.

Quantum Displacement Group

The transformation defined by Eq.(12) may be written as

$$\begin{bmatrix} a \\ a^\dagger \\ q^{\hat{n}} \end{bmatrix} \mapsto \begin{bmatrix} 1 & 0 & z \\ 0 & 1 & z^* \\ 0 & 0 & 1 \end{bmatrix} \begin{bmatrix} a \\ a^\dagger \\ q^{\hat{n}} \end{bmatrix} \qquad (48)$$

which, as we have shown above, conserves the quon commutation relation.

In the spirit of our introduction to this section, we define the associated quantum group by giving a matrix of operators with appropriate commutation relations. This leads to the definition of the quantum displacement group $\mathcal{D}_q$

$$\mathcal{D}_q = \left\{ \begin{bmatrix} \lambda & 0 & \zeta \\ 0 & \lambda^* & \zeta^* \\ 0 & 0 & \chi \end{bmatrix} \middle| \begin{array}{lll} [\lambda, \lambda^*] = 0 & [\lambda, \zeta^*] = 0 & [\lambda, \zeta]_q = 0 \\ [\zeta, \zeta^*]_q = 0 & [\lambda, \chi] = 0 & [\lambda^*, \chi] = 0 \\ [\zeta, \chi]_q = 0 & [\chi, \zeta^*]_q = 0 & \lambda \lambda^* = 1 \end{array} \right\} \qquad (49)$$

with the remaining relations obtained by formal conjugation.

$SU_q(1,1)$

The transformation defined by Eq.(39)

$$\begin{bmatrix} a \\ a^\dagger \end{bmatrix} \mapsto \begin{bmatrix} \lambda & \mu \\ \mu^* & \lambda^* \end{bmatrix} \begin{bmatrix} a \\ a^\dagger \end{bmatrix} \qquad (50)$$

leads to a (partial) definition of the quantum group analogue of $SU(1,1)$

$$SU_q(1,1) = \left\{ \begin{bmatrix} \lambda & \mu \\ \mu^* & \lambda^* \end{bmatrix} \middle| \begin{array}{ll} [\lambda, \mu^*]_q = 0 & [\mu, \lambda^*]_q = 0 \\ \lambda\lambda^* - \lambda^*\lambda = q\mu^*\mu - q^{-1}\mu\mu^* & \lambda\lambda^* - q\mu^*\mu = 1 \end{array} \right\}. \qquad (51)$$

By a slight abuse of notation we have used the same symbols $\lambda, \mu, \ldots$ for the operators in $SU_q(1,1)$ as the matrix elements on which they act. Identification of the elements in Eq.(51) with those of the prototype Eq.(47) shows that not all the defining relations are obtained this way. Further relations would depend on making assumptions about the commutation relations between more than one mode.

Quantum Squeezing Group

We may finally define the quantum analogue of the squeezing group $\mathcal{G}$ defined by Eq.(17). The q-canonical transformation may be written

$$\begin{bmatrix} a \\ a^\dagger \\ q^{\hat{n}} \end{bmatrix} \mapsto \begin{bmatrix} \lambda & \mu & z \\ \mu^* & \lambda^* & z* \\ 0 & 0 & 1 \end{bmatrix} \begin{bmatrix} a \\ a^\dagger \\ q^{\hat{n}} \end{bmatrix} \qquad (52)$$

leading to a definition of the quantum squeezing group $\mathcal{G}_q$

$$\mathcal{G}_q = \left\{ \begin{bmatrix} \lambda & \mu & \zeta \\ \mu^* & \lambda^* & \zeta^* \\ 0 & 0 & \chi \end{bmatrix} \middle| \begin{array}{ll} [\lambda, \mu^*]_q = 0 & [\lambda, \zeta^*] = 0 \\ \lambda\lambda^* - \lambda^*\lambda = q\mu^*\mu - q^{-1}\mu\mu^* & \lambda\lambda^* - q\mu^*\mu = 1 \\ [\mu, \chi]_{q^2} = 0 & [\mu, \zeta^*]_{q^2} = 0 \end{array} \right\} \quad (53)$$

with the relations being supplemented by formal conjugation and those of Eq.(49). The interested reader should compare the quantum group relations obtained here for $\mathcal{G}_q$ from physical analogy with those for the (containing) quantum group $IGL_q(2)$ given in, for example, references [14,15].

CONCLUSIONS

In this essentially pedagogical article, our aim has been to describe some basic quantum optics states from a unified viewpoint. This consists in treating these states as being generated from the vacuum by a unitary representation of a canonical transformation of the usual bosonic commutation relation

$$[a, a^\dagger] \equiv aa^\dagger - a^\dagger a = 1.$$

In the conventional case, this leads quite naturally to a concise description of coherent, squeezed and Kerr states. One aim here has been to generalize this approach to a special q-*deformation* of the usual ($q = 1$) boson. In general, there is an unlimited variety of q deformations available; however, the most popular deformations are of two types (although there are links between them, see [16,17]). The type of q-boson on which we have exclusively focussed in this note, and to which we have referred as *quons*, were first discovered by Arik and Coon [4]; they satisfy

$$[a, a^\dagger]_q \equiv aa^\dagger - qa^\dagger a = 1.$$

This type leads naturally to the *basic numbers* of classical q-analysis and so have also been termed *maths*-type q-bosons.

The second q-boson seems to have been first discussed in connection with the representation theory of quantum groups [18–20]. This type, variously termed q-*oscillators* or *physics*-type q-bosons, satisfies

$$aa^\dagger - qa^\dagger a = q^{-\hat{n}}.$$

The coherent states of both types of q-boson have been extensively studied [21–23].

In the case of quons, relating the quantum optics states to the canonical transformation induced by a unitary operator gives a standpoint from which we can attempt to describe the analogous quonic states; however, if we wish to give an *explicit* description of such states it may be difficult to find the appropriate unitary transformation. We have explicitly given the unitary operator in the simplest case, that of quonic *coherent* states (and the trivial extension to quonic Kerr States). We have also exhibited a non-unitary form for the quon squeezed state case Eq.(45). One challenge is to produce the corresponding unitary operator for, *inter alia*, these quon squeezed states.

This approach also gives a motivation for the introduction of *Quantum Groups* which appear quite naturally here as associated with Canonical Transformations of the Quonic commutation relations. By this means we were able to describe the quantum group analogues of the Displacement and Squeezing groups. It should be emphasized that this physics-based approach is an essentially naive one; we are able to describe these special quantum groups in an algebraic, purely formal manner; however, no attempt is made to construct an analytic context in which the relations are actually implemented. Further, not all the defining relations for the Quantum Groups are obtained in the method outlined here; a full discussion would have to treat consistently the problem of more than one mode.

The perspicacious reader will have noticed that the approach here mirrors the introduction of *noncommutative geometry* in the definition of quantum groups. For example, the description of the squeezing group $\mathcal{G}_q$ by Eq.(53) is, apart from conjugation, similar to demanding that the transformation

$$\begin{bmatrix} x \\ y \\ z \end{bmatrix} \mapsto \begin{bmatrix} X \\ Y \\ Z \end{bmatrix} = \begin{bmatrix} a & b & g \\ c & d & h \\ 0 & 0 & k \end{bmatrix} \begin{bmatrix} x \\ y \\ z \end{bmatrix} \quad (54)$$

on the co-ordinates $\{x, y, z\}$ maintains the relations

$$[x, y]_q = 0, [x, z]_q = 0, [z, y]_q = 0.$$

In the present context the 'co-ordinates' $\{a, a^\dagger, q^{\hat{n}}\}$ and their transformed versions satisfy (essentially) these constraints. (The full set of defining relations for $IGL_q(2)$ requires also transformation of the differentials $\{dx, dy, dz\}$.) Note that in our case, the role of the third co-ordinate is played by $q^{\hat{n}}$. This provides us with our definition of the transformed number operator, using $\{a, a^\dagger, q^{\hat{n}}\} \mapsto \{A, A^\dagger, q^{\hat{N}}\}$.

ACKNOWLEDGEMENTS

I am pleased to acknowledge many interesting discussions with my colleagues in the Quantum Processes Group at the Open University, especially Dr.Hongchen Fu, as well as with Professor P. Kulish, Dr. Roger McDermott and Mr. Deepak Parashar. Naturally, they are not responsible for any shortcomings present in the article.

REFERENCES

1. Glauber, R.J., *Phys. Rev.* **130**, 2529 (1963).
2. Yuen, H.P., *Physics Letters* **56A**,105(1976).

3. Kitagawa, M., and Yamamoto,Y., *Phys. Rev.* **A 34**,3974 (1986).
4. Arik, M., and Coon, D.D., *J. Math. Phys.* **17**, 524 (1976).
5. Jackson, F.H., *Mess. Math.* **38**, 62 (1909).
6. Exton, H., *q-Hypergeometric Functions and Applications*, Ellis Horwood, Chichester, 1983.
7. McDermott, R.J., and Solomon, A.I., *J.Phys.* A **27**, 2037(1994).
8. Kowalski, K., and Rembielinski, J., *J. Math. Phys.* **34**, 2155 (1993).
9. Brzezinski, T., Dabrowski, H., and Rembielinski, J., *J. Math. Phys.* **33**, 19 (1992).
10. Brzezinski, T., and Rembielinski, J., *J. Phys. A.* **25**, 1945 (1992).
11. Katriel. J., and Solomon, A. I., *Phys. Rev.A.* **49**, 5149(1994).
12. Solomon, A.I., "Quantum Groups in Quantum Optics," in *XXI International Conference on Differential Geometric Methods in Physics-1992* edited by C.N. Yang et al., World Scientific, Singapore, 1993, p.435.
13. Manin, Yu. I., *Montreal Univ. Preprint CRM - 1561* (1988).
14. Schlieker, M., Weich, W., and Weixler, R., *Lett.Math.Phys.***27**, 335 (1993).
15. Castellani, L., *Phys.Lett* **298**, 217 (1993).
16. Kulish, P.P., and Damaskinsky, E.V., *J. Phys. A.* **23**, 415 (1990).
17. Chakrabarti, R., and Jagannathan, R., *J. Phys. A.* **24**, L711 (1991).
18. Macfarlane, A.J., *J. Phys. A.* **22**, 4581 (1989).
19. Biedenharn, L.C., *J. Phys. A.* **22**, L873 (1989).
20. Sun, C.P., and Fu, H., *J. Phys. A.* **22**, L983 (1989)
21. Chaichian. M., Ellinas, D., and Kulish, P.P., *Phys. Rev. Lett.* **65**, 980(1990).
22. Katriel, J., and. Solomon, A.I., *J. Phys. A.* **24**, 2093 (1991).
23. Buzek, V., *J. Mod. Opt.* **38**, 801 (1991).

Deformed Quantum Statistics of Quons

Stefan Kirchner[a] and Akira Inomata[b]

[a] *Institut für Theorie der Kondensierten Materie, Universität Karlsruhe, 76128 Karlsruhe, Germany*
[b] *Department of Physics, State University of New York at Albany, Albany, NY 12222, USA*

Abstract. Thermodynamical properties of a model quon gas are reported. The quons considered are the objects obeying the deformed commutator $a\,a^\dagger - q\,a^\dagger a = 1$ with $-1 \leq q \leq 1$. Although $q = 1$ and $q = -1$ correspond respectively to boson and fermion statistics, it is not trivial to generate a quon gas which interpolates between the boson gas and the fermion gas without Pauli's exclusion principle. The proposed model, interpolating between the two limits, can realize the Maxwell-Boltzmann limit at $q = 0$ and simulate a free anyon gas. Under the restriction $-1 \leq q \leq 1$, the gas is deformed but bosonic in nature. It shows the Bose-Einstein condensation even in two dimensions.

I INTRODUCTION

Quons, so named by Greenberg [1], are the objects whose creation and annihilation operators obey the minimally deformed commutation relations,

$$a_i^\dagger a_j - q\, a_j\, a_i^\dagger = \delta_{ij} \qquad (1)$$

where $-1 \leq q \leq 1$. The q-deformed algebra of this type was first discussed by Arik and Coon [2]. Evidently, the quon commutators (1) reduce to the usual commutators for bosons when $q = 1$ and the anticommutators for fermions in the limit $q \to -1$. Furthermore, the special case with $q = 0$ could correspond to the Maxwell-Boltzmann statistics. There are models of quons that have the bosonic limit [3]. Here we shall present a class of quon systems which unify the bosonic and fermionic limits, and show that the model can simulate a classical gas and an anyon gas. Some of the special examples have been reported previously [4,5], but what we are presenting here is a more general version.

II THE QUON GAS MODEL

The quon gas we consider is a system characterized by the Hamiltonian,

$$H = \sum_i \epsilon_i \left\{ \left(\frac{2}{[2]}\right)^{s(\hat{N}-1)} a_i^\dagger a_i + s\gamma(\hat{N}_i) \right\}, \tag{2}$$

where ϵ_i is the energy of the ith mode which is assumed to be non-negative, s is a positive parameter, $\hat{N}$ is the undeformed number operator and $\gamma(\hat{N}_i)$ is a q-dependent function which vanishes in both the boson and the fermion limits. In the above we have also used the $[2] = (1-q^2)/(1-q)$ which is a special case of deformed number $[n] = (1-q^n)/(1-q)$. Note that the eigenvalues of $a^\dagger a = [\hat{N}]$ are indeed $[n]$ with $n = 0, 1, 2, \ldots$. In the following we choose the $\gamma(n_i)$ as

$$\gamma(n) = \gamma_n = c_n[2](2-[2]) + d_n[2]^2(2-[2])^2, \tag{3}$$

where c_n and d_n are constants to be specified later. If $s=0$, the Hamiltonian (2) reduces to the standard Hamiltonian,

$$H = \sum_i \epsilon_i a_i^\dagger a_i, \tag{4}$$

for q-deformed bosons, considered by many authors [6]. Even if $s \neq 0$, since $[2] = 1+q$ equals 2 if $q=1$, the Hamiltonian (2) becomes that of the free bosons and takes the form identical to (4). However, in the limit $q \to -1$, the Hamiltonian (2) does not assume the standard fermion form. Nevertheless, as we shall see later, the fermion distribution factor as well as the boson factor can be obtained from (2).

III THERMODYNAMICS OF THE QUON GAS

In order to study statistical properties of the quon gas, we work with the grand canonical partition function

$$\Xi = e^{-\beta\Omega} = \text{Tr } \exp\{-\beta(\hat{H} - \mu\hat{N})\} \tag{5}$$

where Ω is the grand canonical potential, μ is the chemical potential, and $\hat{N}$ is the usual number operator. The effect of q-deformation enters into our calculation only through the Hamiltonian $\hat{H}$. Note that (5) is well-defined for all values of $q \in [-1, 1]$ and can be used to calculate in principle all physical quantities of interest. In the present case, the grand canonical partition function can be factorized as a product of single mode partition functions Z_i,

$$\Xi = \prod_i Z_i = \prod_i \left(\sum_{n=0}^\infty \xi_n(\epsilon_i) z^n \right) \tag{6}$$

Here $\xi_{n_i}(\epsilon_i) = e^{-\beta\epsilon_i \mathcal{E}(n_i)}$, and $z = e^{\beta\mu}$ which is the fugacity. $\mathcal{E}(n_i) = (2/[2])^{s(n-1)} + \gamma(n_i)$ are the eigenvalues of $\hat{H}$. From (5) and (6) follows the grand canonical potential:

$$\Omega = -\frac{1}{\beta} \sum_i \ln Z_i.$$

As usual, the above summation over all modes may be replaced by an integration over the momentum space. Considering the D-dimensional background, we make the following replacement:

$$\sum_i \to \frac{V_D}{(2\pi)^D} \int d^D\mathbf{k},$$

where V_D is the D-dimensional volume occupied by the gas. We assume a dispersion relation of the form $\epsilon(\mathbf{k}) = A|\mathbf{k}|^p$ independent of q. To carry out the energy integration we first expand $\ln Z$ by using the fugacity power series,

$$\ln Z = \sum_{n=1}^{\infty} \chi_n(\epsilon) z^n$$

where

$$\chi_n(\epsilon) = \sum_{r=1}^{n} \sum_{r_1, r_2, \cdots r_n} (-1)^{r-1} \frac{(r-1)!}{r_1! r_2! \cdots r_n!} \xi_1^{r_1} \xi_2^{r_2} \cdots \xi_n^{r_n},$$

the above summations being subjected to the constraints:

$$\sum_j^n r_j = r, \qquad \sum_j^n j\, r_j = n.$$

The grand canonical potential is then given by

$$\Omega = -\frac{\Lambda_D V_D}{\beta^{\nu+1}} \sum_{n=1}^{\infty} b_n z^n \tag{7}$$

with the expansion coefficients

$$b_n = \int_0^\infty d\epsilon\, \epsilon^{\nu-1} \chi_n(\epsilon) = \sum_{r=1}^n \sum_{r_1, r_2, \cdots r_n} (-1)^{r-1} \frac{(r-1)!}{r_1! r_2! \cdots r_n!} \left[\sum_k^n \mathcal{E}_k r_k\right]^{-\nu} \tag{8}$$

where $\Lambda_D = [2^{D-1} p A^{1-p(1-\nu)} \sqrt{\pi}^D \Gamma(D/2)]^{-1}$ and $\nu = D/p$. The average number of quons may be computed by

$$N = -\frac{\partial \Omega}{\partial \mu} = \Lambda_D V_D \int_0^\infty d\epsilon\, \epsilon^{\nu-1} \bar{n}(\epsilon) \tag{9}$$

where $\bar{n}(\epsilon) = z \partial \ln Z / \partial z$ is the distribution function. Notice that this occupation number is the mean value of $\hat{N}$ rather than $a^\dagger a = [\hat{N}]$, which differs from that of ref. [6].

IV THE BOSON, FERMION AND CLASSICAL LIMITS

Now we examine the bosonic, fermionic and classical limits of the present model by considering the fact that $[n] = n$ if $q = 1$, $[2] = 0$ if $q = -1$, and $[2] = 1$ if $q = 0$.

Bosonic Limit: When $q = 1$, $\mathcal{E}(n)$ reduces to $\mathcal{E}(n) = n$, leading to $\xi_n = e^{-\beta\epsilon n}$. Hence

$$\lim_{q \to 1} \ln Z = \sum_{n=1}^{\infty} \frac{1}{n} e^{-n\beta\epsilon} z^n.$$

¿From this follows the Bose-Einstein distribution:

$$\lim_{q \to 1} \bar{n}(\epsilon) = \frac{1}{e^{\beta(\epsilon-\mu)} - 1}.$$

Fermion Limit: In the limit $q \to -1$, $\mathcal{E}(n)$ blows up when $n \geq 2$ so that $\xi_n \to 0$ for $n \geq 2$. This gives rise effectively to the exclusion principle by terminating the sum of (6) at $n = 1$.[1] Since $\xi_1 = e^{-\beta\epsilon}$,

$$\lim_{q \to -1} \ln Z = \sum_{n=1}^{\infty} \frac{(-1)^n}{n} e^{-n\beta\epsilon} z^n,$$

which results in the Ferm-Dirac distribution:

$$\lim_{q \to 1} \bar{n}(\epsilon) = \frac{1}{e^{\beta(\epsilon-\mu)} + 1}.$$

Classical Limit: If $e^{\beta(\mu-\epsilon)} \ll 1$, of course, the above two limiting distributions reduce into the Maxwell-Boltzmann distribution

$$\bar{n}(\epsilon) \approx e^{\beta(\mu-\epsilon)}$$

However, without taking the usual classical approximation, we can also anticipate that the case of $q = 0$ corresponds to the classical limit. For $q = 0$, $[\hat{N}] = 1$ for all n. Hence the creation and annihilation operators commute, i.e., $\hat{a}\,\hat{a}^\dagger = \hat{a}^\dagger\,\hat{a}$, and behave like c-numbers. Indeed the quon gas characterized by the Hamiltonian (2) is able to produce the classical limit as q vanishes when $s = 0$ and $c_n + d_n + 1 = 0$ for $n > 0$. Although the Maxwell-Boltzmann limit is known as facing the Gibbs paradox [1,7], the present model is free from the paradox [8].

V THE ANYONIC BEHAVIOR

In order to examine the anyonic behavior of the quon gas, we calculate the virial coefficients following refs. [4,9,10]. The fugacity expansion of the pressure is

$$P = -\left(\frac{\partial \Omega}{\partial V_D}\right)_{\mu,T} = \Lambda_D \beta^{-(\nu+1)} \sum_{n=1}^{\infty} b_n z^n.$$

---

[1] This may be compared to the exclusion of doubly occupied sites in the infinite-U Anderson model due to strong Coulomb repulsion $\propto U$

The coefficients of the virial expansion,

$$P = \beta^{-1} \sum_{l=1}^{\infty} B_l \rho^l,$$

with the density $\rho = N/V_D$ are given by

$$B_l = \frac{1}{l!} \left[\frac{d^{l-1}}{dz^{l-1}} \left(\frac{V_D z}{N(z)} \right) \right]_{z=0}. \tag{10}$$

In the limits $q = 1$ and $q = -1$, the above virial coefficient B_2 reduces to that for the boson gas and for the fermion gas, respectively,

$$\lim_{q=\to\pm 1} B_2 = \mp \frac{\beta^\nu}{4\Lambda_D}.$$

It is known that the second virial coefficient for the anyon gas is [11-13]

$$B_2^{anyon} = \frac{1}{4}\lambda_T^2 \left\{ 1 - 2(1-\alpha)^2 \right\},$$

where $\lambda_T = (2\pi\beta\hbar^2/m)^{1/2}$ and α is the statistical parameter of an anyon gas. The same can be realized from B_2 of (10) by choosing constants appropriately [4]. In this case, if the γ-terms is absent, then α can be related to q by

$$\alpha = 1 - \{[2]/2\}^{(s-1)/2}.$$

However, if we wish to simulate the third virial coefficient of an anyon gas, we have to assume a non-vanishing γ-term [10]. Again, making an appropriate choice of c_n and d_n, we can realize for arbitrary s the same behavior of B_3 around the limiting points $\alpha = 0$ and $\alpha = 1$ or $q = \pm 1$ as that obtained by other methods [14]:

$$B_3^{anyon} \approx \lambda_T^4 \left\{ \frac{1}{36} + \frac{1}{12}\delta^2 \right\}$$

where δ is a small derivation from the two limiting points $q = \pm 1$, that is, $q = 1 - \delta/g$, $(\alpha = \delta)$ near the boson limit and $q = -1 + \delta/g$, $(\alpha = 1 - \delta)$ near the fermion limit. Here $g = 8(385)^{-1/2}$.

VI BOSE-EINSTEIN CONDENSATION

Here we wish to study the Bose-Einstein condensation (BEC) of our quon gas. We restrict ourselves to a two-dimensional nonrelativistic gas ($D = 2, p = 2$) and choose $c_n = 0$ and $d_n = 0$ in (2) for all n. In the vicinity of the boson limit $q = 1 - \eta \approx 1$, the Hamiltonian (2) can be approximated by

$$\mathcal{H} \approx \sum_i \epsilon_i \left\{ \left(1 - \frac{1}{2}(s-1)\eta\right) b_i^\dagger b_i + \frac{1}{2}(s-1)\eta \left(b_i^\dagger b_i\right)^2 \right\},$$

where b_i's are the ordinary boson operators. Apparently, the q-deformation (for $\eta \ll 1$) acts as a self-interaction, holding the quons together. Therefore one might expect Bose-Einstein condensation (BEC) for $q \neq 1$ even in two dimensions [5]. To show this we rewrite the total number of quons (9) as the sum of the ground state occupation and the remainder N^{re}:

$$N^{re}(z) = \frac{Amk_bT}{2\pi\hbar^2} \sum_{n=1}^{\infty} b_n(q)z^n, \qquad (11)$$

where b_n are given by (8) with $D = 1$ and $\nu = 1$. Since $q^{(l-1)/2} l \leq [l] \leq l$ holds for $q \in [0,1]$, it is straight forward to establish an upper and lower limits for the expansion coefficients b_n:

$$0 < b_n < \frac{1}{n}\left(\frac{1+q}{2}\right)^{s(n-1)/2} q^{-(n-1)/4}.$$

The first inequality assures that the expansion coefficients are all positive and that $N^{re}(z)$ is a monotonically increasing function of z (or a monotonically decreasing function of $|\mu|/k_BT$). The second inequality leads us to

$$N^{re}(z) \leq -\varphi(q,s)^{-1} \ln[1 - \varphi(q,s)z] \quad \text{iff} \quad \varphi(q,s) < 1,$$

where $\varphi(q,s) = [(1+q)/2]^{s/2} q^{-1/4}$. Now the condition for BEC to occur becomes $(1+q)^{2s}/q < 2^{2s}$. Evidently, in the pure bosonic case $q = 1$, no condensation can take place irrespective of the value of s. For $s > 1$, however, there is always a finite range of q, i.e., $q \in [a,b]$ ($a > 0$, $b < 1$), where BEC will occur. The transition temperature as a function of q is shown in Fig. 1 for three different values of s.

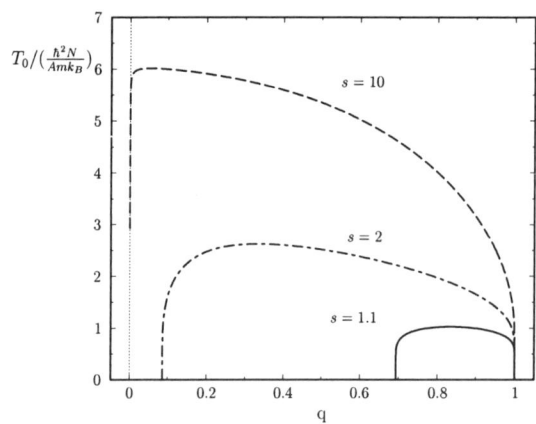

FIGURE 1. Transition temperature T_0 for $s = 1.1$, $s = 2$ and $s = 10$

VII CONCLUDING REMARKS

In the above, we have presented a simple quon model for deformed quantum statistics, based on the minimally q-deformed commutators (1). The model interpolates between the bosonic and fermionic limits and exhibits the anyonic behavior up to the third virial coefficient. The classical gas limit can also be realized when $q = 0$, without facing the Gibbs paradox. As is well-known, in two dimensions, the free boson gas does not show the Bose condensation. However, for a certain range of q, the quon gas can undergo a phase transition and the major portion of quons will be condensed in the ground state. It is uncertain if a toy model of this kind will be meaningful in the real world. Yet it is significant that a model which describes the bosonic, fermionic and classical gases in a unified manner can be constructed out of the minimally q-deformed commutators. It is also important to recognize that the present model supplies a simple scheme that explicitly reveals possible thermodynamical behaviors of a system of q-deformed quantum objects.

ACKNOWLEDGEMENTS

We thank A. Mildenberger, A. Mishra and G. Sellier for stimulating discussions.

REFERENCES

1. Greenberg O.W., *Phys. Rev. Lett.* **64**, 705-708 (1990).
2. Arik M., Coon D.D., *J. Math. Phys.* **17**, 524-527 (1976).
3. Chaichian M., Gonzalez Felipe R., Montonen C., *J. Phys. A* **26**, L1117-24 (1993).
4. Inomata A., *Phys. Rev. A* **52**, 932-935 (1995).
5. Inomata A., Kirchner S., *Phys. Lett. A* **231** 311-314 (1997).
6. Daoud M., Kibler M., *Phys.Lett. A* **206**, 13-17 (1995).
7. Goodison J.W., Toms D.J., *Phys. Lett. A* **195**, 38-42 (1994).
8. Inomata A., Kirchner S., *Phys. Lett. A* **222** 213-219 (1996).
9. Dalton S., Inomata A., *Phys. Lett. A* **199**, 315-319 (1995).
10. Inomata A., Dalton S., Kirchner S., in *Proceedings of VII Int. Conf. on Symmetry Methods in Physics*, JINR, Dubna, 1996, pp.260-265.
11. Arovas D., Schrieffer J.R., Wilczek F., A. Zee, *Nucl. Phys.* **B251** 117-126 (1985).
    D. Arovas, in *Geometric Phases in Physics*, World Scientific, Singapore, 1989, pp.284-322.
12. Comtet A., Georgelin Y., Ouvry S., *J. Phys. A* **22**, 3917-3925 (1989).
13. A. Inomata, P.C. Zhu, in *Path Integrals from meV to MeV: Tutzing '92*, World Scientific, Singapore, 1993, pp. 136-140.
14. S. Ouvry, in *ISATQP-Shanxi'92*, Science Press, Beijin, 1993, pp. 124;
    Myrheim J., Olaussen K., *Phys. Lett. B* **299** 267-272 (1993).

Generalized Fock Spaces and New Forms of Quantum Statistics

A. K. Mishra [*†] and G. Rajasekaran [†]

[*] Max-Planck Institute for Physics of Complex Systems, Nothnitzer Str. 38, D-01187 Dresden, Germany, and
[†] Institute of Mathematical Sciences C.I.T. Campus, Madras - 600 113, India
e-mail: mishra@imsc.ernet.in; graj@imsc.ernet.in

Abstract. The recent discoveries of new forms of quantum statistics require a close look at the under-lying Fock space structure. This exercise becomes all the more important in order to provide a general classification scheme for various forms of statistics, and establish interconnections among them whenever it is possible. We formulate a theory of generalized Fock spaces, which has a three tired structure consisting of Fock space, statistics and algebra. This general formalism unifies various forms of statistics and algebras, which were earlier considered to describe different systems. Besides, the formalism allows us to construct many new kinds of quantum statistics and the associated algebras of creation and destruction operators. Some of these are: orthostatistics, null statistics or statistics of frozen order, quantum group based statistics and its many *avatars*, and 'doubly-infinite' statistics. The emergence of new forms of quantum statistics for particles interacting with singular potential is also highlighted.

1. INTRODUCTION

In recent years, many new kinds of quantum statistics have been postulated. In spite of large literature which now exists, a unified picture for various statistics and their associated algebra has not emerged. The aim of the present work is to provide such a general formalism. This is achieved by introducing the concept of generalized Fock spaces. Starting with this basic notion, it is possible to show that more than one statistics can be postulated in a given Fock space, and many different algebraic realizations can be constructed for any particular statistics.

We introduce new forms of quantum statistics, *viz.* null, orthofermi and Hubbard statistics, and doubly-infinite statistics in the next section. The null statistics corresponds to a situation wherein no permutation is allowed and particles are frozen in their initial order. Orthofermi and Hubbard statistics satisfy an exclusion principle which is more exclusive than the Pauli's exclusion principle: an orbital state shall not contain more than one particle irrespective of their spin directions. Such a situation arises when the Coulomb repulsion U between two electrons occupying a same orbital state becomes infinity. The U infinity model has been extensively used in the context of strongly correlated electron systems [1]. A deformation of the orthofermi algebra subsequently leads to doubly-infinite statistics.

The theory of generalized Fock spaces is formulated in Sec.3. The key element is the notion of independence of the permutation ordered states. The largest linear vector space constructed in this way is the super Fock space. The subsequent specification of a subset of states in this space as null states leads to many reduced Fock spaces. All these spaces are collectively called as generalized Fock spaces. We construct creation $(c^\dagger)$, annihilation (c) and number (N) operators in the generalized Fock spaces. The creation and annihilation operators, even for a particular Fock space, are not unique. Consequently, many statistics and algebras can exist in a given Fock space. On the other hand, a universal representation for the number operator valid for all forms of statistics and algebra exists.

In quantum mechanical calculations, algebraic relations involving only c and $c^\dagger$ are required. One does not explicitly need a $c\,c$ relation. However, given a $c\,c^\dagger$ expression, corresponding $c\,c$ relation can be obtained in an elegant manner. This is demonstrated in Sec.4, and based on this approach, fractional statistics in one dimension is constructed. Sec.5 is devoted to summary and conclusions.

2. NEW FORMS OF QUANTUM STATISTICS

Newer forms of quantum statistics have been constructed by deforming the canonical commutation relations. For example, the deformation

$$[c_j, c_k^\dagger]_+ = \delta_{jk} \quad \rightarrow \quad [c_j, c_k^\dagger]_q = c_j c_k^\dagger + q c_k^\dagger c_j = \delta_{jk} \quad ; \quad -1 < q < 1 \qquad (1)$$

gives rise to infinite statistics [2,3]. Here no $c\,c$ relation exists, and all permuted states are linearly independent.

As a counterpoint to infinite statistics, null statistics can also be constructed [4]. As mentioned earlier, no permutation is allowed in the null statistics. The defining algebra for this statistics is :

$$c_k c_j^\dagger = 0 \text{ for } k \neq j \quad ; \quad c_j c_j^\dagger = 1 - \sum_{k<j} c_k^\dagger c_k \quad ; \quad c_i c_j = 0 \text{ for } i < j \qquad (2)$$

Singular interparticle interactions also lead to new kinds of statistics. When in addition to Pauli's exclusion principle, an infinite repulsion exists between two particles occupying the same orbital state (k or m) but having different spin directions (α and β), we have

$$c_{k\alpha} c_{k\beta} = 0 \qquad (3)$$

For usual fermions, the above relation is valid only when $\alpha = \beta$. Consistent with the above exclusion principle, two different statistics, $viz.$ orthofermi statistics

$$c_{k\alpha} c_{m\beta}^\dagger + \delta_{\alpha\beta} \sum_\gamma c_{m\gamma}^\dagger c_{k\gamma} = \delta_{km} \delta_{\alpha\beta} \quad ; \quad c_{k\alpha} c_{m\beta} + c_{m\alpha} c_{k\beta} = 0. \qquad (4)$$

and Hubbard statistics

$$\left.\begin{array}{r}c_{k\alpha}c_{m\beta}^{\dagger}+(1-\delta_{km})c_{m\beta}^{\dagger}c_{k\alpha} = \delta_{km}\delta_{\alpha\beta}\left(1-\sum_{\gamma}c_{k\gamma}^{\dagger}c_{k\gamma}\right) \\ \\ c_{k\alpha}c_{m\beta}+(1-\delta_{km})c_{m\beta}c_{k\alpha} = 0\end{array}\right\} \quad (5)$$

can be postulated [5]. The orthofermi statistics is formulated in a representation invariant manner. The Hubbard statistics is not invariant under unitary transformation, and it depends on the representation.

The algebra for the orthobose statistics is obtained by replacing the positive signs by negative signs in Eq.(4).

Usually states of a system are characterized by a set of individual indices describing position, spin, internal degrees of freedom etc.. These are then mapped to a set of single indices. The symmetry properties are then postulated with respect to these composite indices. As a result, symmetries with respect to the exchange of individual indices get correlated. For the ortho and Hubbard statistics, spatial and spin indices can not be mapped to a single composite index. But for Hubbard statistics, exchange between $k\alpha$ and $m\beta$ is still permissible. In orthostatistics, exchanges between k, m and between α, β are uncorrelated. The former exchange leads to fermi or bose statistics, where as the later satisfies infinite statistics.

A deformation of $c\ c^{\dagger}$ algebra for the orthostatistics, that is,

$$c_{k\alpha}c_{m\beta}^{\dagger}+q\delta_{\alpha\beta}\sum_{\gamma}c_{m\gamma}^{\dagger}c_{k\gamma} = \delta_{km}\delta_{\alpha\beta} \quad = \quad -1 < q < 1 \quad (6)$$

gives rise to doubly-infinite statistics; one with respect to the pair of indices (k,m) and the other for the pair (α, β).

3. GENERALIZED FOCK SPACES

Given a set of quantum numbers $g, h, i...$ with the respective occupancy being $n_g, n_h, n_i...$, all possible multiparticle state vectors are

$$|n_g, n_h \ldots n_m; \mu\rangle \quad , \quad \mu = 1, 2 \ldots s \quad (7)$$

where s is the total number of distinct permutations and μ labels each of these permuted states. we assume the existence of a unique vacuum state

$$|0\rangle \equiv |0, 0, 0 \ldots 0\rangle \quad (8)$$

All these states are linearly independent, but need not be orthogonal or normalized

$$\langle n_g', n_h' \ldots n_m'; \mu | n_g, n_h \ldots n_m; \nu \rangle = \delta_{n_g' n_g} \delta_{n_h' n_h} \ldots \delta_{n_m' n_m} M_{\mu\nu} \quad (9)$$

with M being a $s \times s$ hermitian matrix. We choose it to be positive definite.

From the set of linearly independent state vectors, an orthonormal set of vectors $\{\| n_g \ldots n_m; \mu \gg\}$ can be obtained

$$\| n_g \ldots n_m; \mu \gg = \sum_\nu X_{\nu\mu} | n_g \ldots n_m; \nu \rangle \tag{10}$$

Alternatively, starting with the orthonormal vectors $\{\| n_g \ldots n_m; \mu \gg\}$, the vectors $\{| n_g, n_h \ldots n_m; \mu \rangle\}$ can be constructed by taking the inverse of relation (10). X is a nonsingular matrix. Although X is not unique and depends on the particular orthogonalization procedure, we have

$$M^{-1} = XX^\dagger \tag{11}$$

Choosing a nonsingular matrix X and determining the inner product matrix M as above will ensure the positivity of the matrix M.

The set of state vectors considered here constitute super Fock space. Infinite statistics resides in this Fock space. Using the projection operator

$$P(n_g \ldots n_k \ldots n_m) = \sum_{\lambda,\nu} | n_g \ldots n_m; \nu \rangle (M^{-1})_{\nu\lambda} \langle n_g \ldots n_m; \lambda | \tag{12}$$

the number operator can be written as

$$N_k = \sum_{n_g \ldots n_k \ldots n_m} n_k P(n_g \ldots n_k \ldots n_m) \tag{13}$$

which satisfies the following properties

$$N_k | n_g \ldots n_k \ldots n_m; \mu \rangle = n_k | n_g \ldots n_k \ldots n_m; \mu \rangle \quad ; \quad [N_k, N_j]_- = 0 \tag{14}$$

The creation operator is defined as

$$c_j^\dagger = \sum_{n_g \ldots n_j \ldots n_m} \sum_{\mu\nu} A_{\mu\nu} | 1_j n_g \ldots n_j \ldots n_m; \mu \rangle \langle n_g \ldots n_j \ldots n_m; \nu | \tag{15}$$

and c_j as the hermitian conjugate of $c_j^\dagger$. $A_{\mu\nu}$ are a set of arbitrary (complex) numbers. Even at this stage, it is possible to verify that

$$[c_j^\dagger, N_k]_- = -c_j^\dagger \delta_{jk} \tag{16}$$

The ordered state vectors can be constructed using a string of $c^\dagger$ acting on the vacuum state. Consequently we also have

$$\sum_\nu A_{\mu\nu} M_{\nu\lambda} = \delta_{\mu\lambda} \quad ; \quad A = M^{-1} \tag{17}$$

We have provided here a unique representation of the number operator. But many different representations of creation and annihilation operators are possible through different choices of matrices A, X and M. The number operator N can be expressed in terms of $c^\dagger$ and c by solving Eqs.(12,13) and (15). Since $c^\dagger$ and c are not uniquely defined many different expressions for N in terms of $c^\dagger$, c can be obtained.

Next we consider reduced Fock spaces. All known forms of statistics other than the infinite statistics reside in reduced Fock spaces, which are obtained by postulating relations like

$$\sum_\mu B_\mu^p |n_g, n_h \ldots; \mu> = 0 \; ; \; p = 1, 2, \ldots r \qquad (18)$$

where $r < s$ and B_μ^p are constants. The vector space dimension in the sector $\{n_g, n_h \ldots\}$ is now reduced to $d = s - r$. The formalism developed for the super Fock space is also valid for reduced Fock spaces. But μ and ν now ranges from 1...d, and X, M and A are $d \times d$ matrices.

Arbitrariness in the matrix A appearing in the creation operator expression (15) can be exploited to generate many relations involving $c^\dagger c^\dagger$. Therefore, many different forms of statistics specified by different $c^\dagger c^\dagger$ relations can be constructed in a given Fock space. All these statistics are interconnected. No connection exists between the statistics and the associated algebras residing in different Fock spaces. It may also be mentioned here that multiplicity of statistics are not possible in the super Fock space and in the Fock space of frozen order. Only infinite and null statistics respectively reside in these two Fock spaces. But even here, many algebras involving c and $c^\dagger$ are possible. For example, depending on the different choice of the the inner product matrix M in the super Fock space, new $c\,c^\dagger$ relations in addition to the one given in Eq.(1) are possible. Some of these are ($p > 1$ or $p < -1$):

$$c_i c_j^\dagger - q\delta_{ij} \sum_k c_k^\dagger c_k = \delta_{ij} \; ; \; -1 < q < \infty \qquad (19)$$

$$c_i c_j^\dagger - c_j^\dagger c_i = \delta_{ij} p^{2\sum_{k<i} N_k} p^{N_i} \qquad (20)$$

$$c_i c_j^\dagger - p^{-1} c_j^\dagger c_i = 0 \text{ for } i \neq j \; ; \; c_i c_i^\dagger - c_i^\dagger c_i = p^{N_i} \qquad (21)$$

The q in Eq.(1) can be made a complex number, provided the indices are ordered.

$$c_i c_j^\dagger - q c_j^\dagger c_i = 0 \text{ for } i < j \qquad (22)$$

For completeness this relation has to be supplemented with (p real)

$$c_i c_i^\dagger - p c_i^\dagger c_i = 1 \qquad (23)$$

$p = |q|$ corresponds to infinite statistics when $|q| < 1$. $p = -1$ and $|q| < 1$ leads to infinite statistics with an exclusion principle.

Similarly, many algebraic relations can be obtained in the bosonic Fock space ($d = 1$). Taking q and p to be complex numbers and ϕ as any arbitrary function of number operator, a general $c_i\, c_i^\dagger$ relation in this space is written as

$$c_j c_j^\dagger - p c_j^\dagger c_j = |q|^{2\sum_{i<j} N_i} f(N_j) \; ; \; f(N_j) = \left|\frac{\phi(N_j)}{\phi(N_j+1)}\right|^2 - p\left|\frac{\phi(N_j-1)}{\phi(N_j)}\right|^2 \qquad (24)$$

Most interestingly, the corresponding $c_i\, c_j^\dagger$ relation for i < j is still given by Eq.(22). Thus replacing Eq.(23) by Eq.(24) takes us from super Fock space and infinite statistics to bosonic Fock space and deformed bose statistics satisfying the symmetry relation

$$c_i^\dagger c_j^\dagger - q c_j^\dagger c_i^\dagger = 0 \quad \text{for} \quad i > j \tag{25}$$

Various limiting cases, like ϕ being a constant, or p = 0, or $f = 1$ and p real are possible for Eq.(24). A particular interesting case is $p = |q|^2$. The complete algebra (Eqs.(22,24,25)) now becomes covariant under $SU_q(n)$ or quantum group transformation [6-8]. This shows that the algebra covariant under quantum group is a particular case of the more general algebra that can be derived from the formalism of generalized Fock spaces.

The canonical Bose statistics as well as the q-bose statistics given by relation (25) reside in the bosonic Fock space and are interconnected [8]. The underlying configuration space for the q-statistics is non-commutative. It has been shown that a complete Fock space realization of the differential calculus on a non-commutative space leads to a new concept of simultaneous transmutation between quanta satisfying different quantum statistics [9].

Restricting state occupancy to zero and one in the bosonic Fock space leads to the fermionic Fock space. Note that no restriction is placed on the symmetry properties of the state vectors. Consequently, it is possible to construct anticommuting bosons in the bosonic Fock space and commuting fermions in the fermionic Fock space. In addition, many distinct algebras can be obtained in the fermionic Fock space too.

A detailed list of statistics and algebras in various Fock spaces corresponding to single, and two-indexed systems (e.g., orthofermi, and particles obeying 'doubly-infinite' statistics) are provided in reference [8].

4. CC RELATIONS FROM $CC^\dagger$ ALGEBRA

A general form of $c\, c^\dagger$ algebra which allows us to to calculate vacuum matrix element of any polynomial in c and $c^\dagger$ is given as

$$c_i c_j^\dagger = A_{ij} + \sum_{k,m} B_{ijkm} c_k^\dagger c_m \tag{26}$$

where A_{ij} and B_{ijkm} are constants or functions of number operators.

Symmetry of particles under exchange is obtained by making the operator $Q_{ij} \equiv c_i c_j - q' c_j c_i$ a null operator ($Q_{ij} = 0$). This can be achieved if it can be shown that

$$Q_{ij} c_k^\dagger = \sum_{i'j'k'} F_{ijk;i'j'k'}\, c_{k'}^\dagger\, Q_{i'j'} \tag{27}$$

for all i,j and k where $F_{ijk;i'j'k'}$ may be a c-number or operator. The successive applications of the above equation over any string of creation operators, and then allowing both side of the resulting expression to act on the vacuum state $|0\rangle$ finally leads to the operator identity $Q_{ij} = c_i c_j - q' c_j c_i = 0$, which is the $c\,c$ relation sought after.

Employing the above methodology with the $c\,c^\dagger$ relations given in Eqs.(22,23), we can show that fractional statistics with (without) exclusion principle occurs in one dimension when $q = e^{i\theta}$ and $p = -1$ ($\neq -1$) [10]. For other values of q, no $c\,c$ relation exists. This provides an analytical method to prove the absence of a $c\,c$ relation for infinite statistics.

5. SUMMARY AND CONCLUSIONS

By decoupling the notion of the underlying Fock space from c and $c^\dagger$, we are able to define different forms of statistics in a representation independent manner. Subsequently, one can construct creation, annihilation operators and their algebra in any desired representation.

The general formalism not only unifies and classifies various forms of quantum statistics, but also enables us to construct many new kinds of statistics and algebras for single and two-indexed systems in a systematic manner. Some of these are: (i) null statistics, (ii) orthostatistics, (iii) doubly-infinite statistics, (iv) complex q or fractional statistics in one dimension. Many $c\,c^\dagger$ algebras representing these statistics are also constructed. Besides, the notion of generalized Fock space leads to the concept of statistical transmutation in a quantum plane.

REFERENCES

1. Fulde P., *Electron Correlations in Molecules and Solids: Springer Series in Solid-State Sciences* **vol. 100**, Springer, Berlin-Hiedelberg, 1995, pp. 281-345.
2. Greenberg O. W., *Phys. Rev. Lett.* **64**, 705-708 (1990).
3. Greenberg O. W., *Phys. Rev.* **D43**, 4111-4120 (1991).
4. Mishra A. K., and Rajasekaran G., *Phys. Lett. A* **203**, 153-156 (1995).
5. Mishra A. K., and Rajasekaran G., *Pramana - J. Phys.* **36**, 537-555 (1991); **37**, 455 (E) (1991).
6. Pusz W., and Woronowicz S. L., *Rep. Math. Phys.* **27**, 231-257 (1989).
7. Wess J., and Zumino B., *Nucl. Phys. (Proc. Suppl.)* **B 18**, 302-312 (1990).
8. Mishra A. K., and Rajasekaran G., *Pramana - J. Phys.* **45**, 91-139 (1995); hep-th/9605204.
9. Mishra A. K., and Rajasekaran G., *J. Math. Phys.* **38**, 466-475 (1997).
10. Mishra A. K., and Rajasekaran G., *Mod. Phys. Lett. A* **9**, 419-426 (1994).

Quantum Field Theory for Orthofermions and Orthobosons

A. K. Mishra [*†] and G. Rajasekaran [†]

[*] Max-Planck Institute for Physics of Complex Systems, Nothnitzer Str. 38, D-01187 Dresden, Germany, and
[†] Institute of Mathematical Sciences C.I.T. Campus, Madras - 600 113, India
e-mail: mishra@imsc.ernet.in; graj@imsc.ernet.in

Abstract. Orthofermi statistics is characterized by an exclusion principle which is more "exclusive" than Pauli's exclusion principle: an orbital state shall not contain more than one particle, no matter what the spin direction is. The wavefunction is antisymmetric in spatial indices alone with arbitrary symmetry in the spin indices. Orthobose statistics is corresponding Bose analog: the wavefunction is symmetric in spatial indices, with arbitrary symmetry in spin indices. We construct the quantum field theory of particles obeying these new kinds of quantum statistics. Non-relativistic as well as relativistic quantum field theories with interactions are considered.

1. INTRODUCTION

The constraint that an orbital state shall not contain more than one particle irrespective of their spin directions has led us to the formulation of a new family of quantum statistics, namely, orthostatistics [1]. These statistics are described through the algebra

$$c_{k\alpha} c_{p\beta}^\dagger \pm \delta_{\alpha\beta} \sum_\gamma c_{p\gamma}^\dagger c_{k\gamma} = \delta_{kp} \delta_{\alpha\beta} \quad (1)$$

$$c_{k\alpha} c_{p\beta} \pm c_{p\alpha} c_{k\beta} = 0 \quad (2)$$

where $c_{k\alpha}$ and $c_{p\beta}^\dagger$ are the annihilation and creation operators of particles with momenta k and p and spins α and β respectively. The positive sign corresponds to the orthofermi and negative sign is for the orthobose statistics.

Greenberg has earlier constructed infinite statistics wherein wavefunctions with arbitrary symmetries are allowed [2,3]. Infinite statistics also follows from Eq.(1) if the spatial indices k and p are suppressed [4]. On the other hand, if spin indices are suppressed, canonical commutation relations for boson and anticommutaors for fermion are obtained. Thus orthostatistics describe a generalized class of statistics wherein different indices exhibit different and uncorrelated symmetry properties.

One of the important property of orthofermions is that only one particle can be accommodated among the set of states $\{k\alpha\}$, irrespective of the range of α. This contrast with the occupancy status of parafermions. In parafermi statistics of order n, utmost n parafermions can occupy the same state [5].

Until now, it has been possible to construct a local relativistic quantum field theory (LRQFT) only for parafermions and parabosons, of which fermion and boson are specific examples [6]. No such formulation is possible for infinite statistics [2,3]. Since the orthofermions (orthobosons) have properties common with fermions (bosons) as well as with the particles obeying infinite statistics, we are motivated to examine whether a LRQFT for orthoparticles can be constructed? We show here that if the second index α in $c_{k\alpha}$ is reinterpreted as a new degree of freedom which can be excited at some higher energy scale yet to be probed experimentally, it is possible to construct LRQFT for orthofermions [7]. However, the relativistic quantum field theory for the orhobosons remains nonlocal.

The nonrelativistic quantum field theory for the both kinds of particles can be constructed even when α denotes the usual kinematic spin variable. This is considered in Sec.2. Formulations of Dirac field as an orthofermi field and Klein-Gordon field as an orthobose field, provided α characterizes the new degree of freedom, are given in sections 3 and 4, respectively. Interactions are considered in Sec.5, followed by a section on summary and conclusions.

2. NONRELATIVISTIC QUANTUM FIELD THEORY

The orthoparticles obey quantization rules different from the canonical commutators and anticommutators. The Poisson brackets involving canonical coordinate ψ and canonical momentum π, that is, $[\psi, \pi]_{PB}$ and $[\psi, \psi]_{PB}$ are now replaced by the new quantum brackets corresponding to Eqs.(1) and (2), or equivalently

$$\psi_a(\mathbf{x}, 0)\psi_b^\dagger(\mathbf{y}, 0) \pm \delta_{ab} \sum_c \psi_c^\dagger(\mathbf{y}, 0)\psi_c(\mathbf{x}, 0) = \delta_{ab}\delta^3(\mathbf{x} - \mathbf{y}) \tag{3}$$

$$\psi_a(\mathbf{x}, 0)\psi_b(\mathbf{y}, 0) \pm \psi_a(\mathbf{y}, 0)\psi_b(\mathbf{x}, 0) = 0 \tag{4}$$

In spite of different quantization procedure, the corresponding nonrelativistic field theory can be developed in a consistent manner. Consider the nonrelativistic field $\psi_a(\mathbf{x}, t)$ satisfying the Schrodinger equation

$$\left(i\frac{\partial}{\partial t} + \frac{1}{2m}\nabla^2\right)\psi_a(\mathbf{x}, t) = 0 \tag{5}$$

The fourier expansion for ψ_a is

$$\psi_a(\mathbf{x}, t) = \sum_{\mathbf{k},\alpha} c_{\mathbf{k}\alpha}\xi_a^{(\alpha)} e^{i\mathbf{k}\cdot\mathbf{x} - i\omega_k t}, \tag{6}$$

where $\xi_a^{(\alpha)}$ is the spin wavefunction for the spin-component α, and $\omega_k = k^2/(2m)$ is the nonrelativistic energy.

The Hamiltonian of the system is

$$H = \sum_{\mathbf{k},\alpha} \omega_k c_{\mathbf{k}\alpha}^\dagger c_{\mathbf{k}\alpha} \tag{7}$$

From Eqs.(1) and (2) (with either sign), the following commutation relation can be derived

$$\left[c_{\mathbf{k}\alpha}, \sum_\gamma c_{\mathbf{p}\gamma}^\dagger c_{\mathbf{p}\gamma} \right] = \delta_{\mathbf{k}\mathbf{p}} c_{\mathbf{k}\alpha} \tag{8}$$

Now it can be shown that the Schrodinger Eq.(5) follows from the Heisenberg equation of motion for ψ_a and the relation (8).

The Schrodinger equation and the Hamiltonian can also be obtained from the canonical formalism. In fact the Schrodinger Eq.(5) is the Euler-Lagrange equation of motion for the usual Lagrange density

$$\mathcal{L} = i \sum_a \psi_a^* \frac{\partial \psi_a}{\partial t} - \frac{1}{2m} \sum_a \nabla \psi_a^* \cdot \nabla \psi_a \tag{9}$$

The Legendre transformation gives the corresponding total Hamiltonian as

$$H = \int d^3x \mathcal{H} = \frac{1}{2m} \int d^3x \sum_a \psi_a^* \nabla^2 \psi_a \tag{10}$$

Substitution of Eq.(6) into Eq.(10) leads to the Hamiltonian given by Eq.(7).

Thus the analysis given in this section shows that our quantization is consistent, and a nonrelativistic quantum field theory based on a canonical Lagrangian formalism can be constructed for the orthoparticles.

3. RELATIVISTIC QUANTUM FIELD THEORY FOR ORTHOFERMIONS

We consider the Dirac Hamiltonin

$$H = \int \psi^\dagger (\boldsymbol{\alpha} \cdot \mathbf{p} + \beta m) \psi d^3x \tag{11}$$

and the usual four component Dirac field $\psi_a(\mathbf{x},t)$ ($a = 1,2,3,4$). Using the expansion

$$\psi_a(\mathbf{x},t) = \sum_{\mathbf{k}} \sum_{\alpha=1}^{4} \left(\frac{m}{E_k}\right)^{1/2} c_{\mathbf{k}\alpha} u_a^{(\alpha)}(\mathbf{k}) e^{i\mathbf{k}\cdot\mathbf{x} - iE_k^{(\alpha)}t}, \tag{12}$$

and the relation

$$(\boldsymbol{\alpha} \cdot \mathbf{k} + \beta m) u^{(\alpha)}(\mathbf{k}) = E_k^{(\alpha)} u^{(\alpha)}(\mathbf{k}) \tag{13}$$

in Eq.(11), the Hamiltonian can be written as

$$H = \sum_k E_k \left[\sum_{\alpha=1,2} c^\dagger_{k\alpha} c_{k\alpha} - \sum_{\alpha=3,4} c^\dagger_{k\alpha} c_{k\alpha} \right] \quad (14)$$

$u_a^{(\alpha)}(\mathbf{k})$ in expansion (12) are Dirac spinors, $\alpha = 1$ and 2 being the positive energy spinors ($E_k^{(\alpha)} \equiv E_K = +(m^2 + k^2)^2)$, and $\alpha = 3$ and 4 being the negative energy spinors ($E_k^{(\alpha)} \equiv -E_k$). α, β are the Dirac matrices and $\mathbf{p} = -i\nabla$.

If $c_{k\alpha}$ and $c^\dagger_{k\alpha}$ satisfy the orthofermionic algebra given in Eqs.(1) and (2), and the second index α for an orthfermion is identified as the Dirac index α going from 1 to 4, then neither $\sum_{\alpha=1,2} c^\dagger_{k\alpha} c_{k\alpha}$ nor $\sum_{\alpha=3,4} c^\dagger_{k\alpha} c_{k\alpha}$ occurring in the Hamiltonian (Eq.(14)) are number operators. It can be verified from Eq.(8) that only the sum $\sum_{\alpha=1-4} c^\dagger_{k\alpha} c_{k\alpha}$ is a number operator for particles of momenta $\mathbf{k}$. Therefore as such, the Hamiltonian in Eq.(14) can not be reexpressed as the sum of energies of particles, and our quantization procedure seems to fail. Note that if $c^\dagger$ and c in Eq.(14) denote creation and annihilation operators for the usual fermion, the Hamiltonian can be reexpressed as the sum of energies of particles and antiparticles by introducing the Dirac vacuum state.

The above problem can be circumvented by introducing a new degree of freedom indexed by A, B, C, D.... The orthofermi algebra with the new indices is written as

$$c_{k\alpha A} c^\dagger_{p\beta B} + \delta_{AB} \sum_D c^\dagger_{p\beta D} c_{k\alpha D} = \delta_{\mathbf{kp}} \delta_{\alpha\beta} \delta_{AB} \quad (15)$$

$$c_{k\alpha A} c_{p\beta B} + c_{p\beta A} c_{k\alpha B} = 0 \quad (16)$$

The Eq.(14) for the Hamiltonian now gets replaced as

$$H = \sum_k E_k \left[\sum_{\alpha=1,2} \sum_A c^\dagger_{k\alpha A} c_{k\alpha A} - \sum_{\alpha=3,4} \sum_A c^\dagger_{k\alpha A} c_{k\alpha A} \right] \quad (17)$$

It can be verified that

$$n_{k\alpha} = \sum_A c^\dagger_{k\alpha A} c_{k\alpha A} \quad (18)$$

is the number operator for particles of momentum $\mathbf{k}$ and Dirac index α. Note that in orthostatistics, number operators $n_{k\alpha A}$ with all the indices specified, do not exist [8]. The Dirac vacuum is introduced as the filled negative energy sea. The vacuum state is infinitely degenerate since orthofermions of negative (as well as positive) energy can have arbitrary index A. One way of lifting this degeneracy is to choose the vacuum state as the normalized sum of all the states with the new index taking the full range of values.

The creation and annihilation operators for antiparticles are defined as

$$d_{k\alpha A} = c^\dagger_{k\alpha+2,A} \; ; \; d^\dagger_{k\alpha A} = c_{k\alpha+2,A} \; ; \; \alpha = 1,2 \quad (19)$$

The number operator for antiparticles is

$$\bar{n}_{\mathbf{k}\alpha} = 1 - n_{\mathbf{k}\alpha+2} = 1 - \sum_A c^\dagger_{\mathbf{k}\alpha+2,A} c_{\mathbf{k}\alpha+2,A} \quad \text{for } \alpha = 1,2 \tag{20}$$

which can be rewritten as $\bar{n}_{\mathbf{k}\alpha} = d^\dagger_{\mathbf{k}\alpha A} d_{\mathbf{k}\alpha A}$, for any A. The A-independence of the product $d^\dagger_{\mathbf{k}\alpha A} d_{\mathbf{k}\alpha A}$ follows from the Eqs.(15) and (19). Using Eqs.(18) and (20), the Eq.(17) can be rewritten as

$$H = \sum_k E_k \sum_{\alpha=1,2} (n_{\mathbf{k}\alpha} + \bar{n}_{\mathbf{k}\alpha}) - \sum_k E_k \sum_{\alpha=1,2} 1 \tag{21}$$

Note that no summation over A appears in the last term describing the vacuum energy. In orthofermi statistics, for each $\mathbf{k}$ and α, there is only one particle irrespective of the value of A. Subtracting the vacuum energy, we finally get the modified Hamiltonian

$$\widetilde{H} = H - <0|H|0> = \sum_k E_k \sum_{\alpha=1,2} (n_{\mathbf{k}\alpha} + \bar{n}_{\mathbf{k}\alpha}) \tag{22}$$

which is positive definite and is expressed as the sum of energies of particles and antiparticles. That the Heisenberg equation of motion for $\psi_{aA}(\mathbf{x},t)$ is consistent with the Dirac equation

$$i\frac{\partial \psi_{aA}}{\partial t} = (\boldsymbol{\alpha} \cdot \mathbf{p} + \beta m) \psi_{aA} \tag{23}$$

and the algebra of c and $c^\dagger$ (Eq.(15), can be easily verified. The Hamiltonian (Eq.(17)) can also be derived from the Lagrangian density

$$\mathcal{L} = \sum_{aA} i\psi^\dagger_{aA} \frac{\partial \psi_{aA}}{\partial t} - \sum_{aA} \psi^\dagger_{aA}(-i\boldsymbol{\alpha} \cdot \nabla + \beta m)\psi_{aA} \tag{24}$$

Thus the consistency of quantization rule as given by Eqs.(15,16), or equivalently by the relations

$$\psi_{aA}(\mathbf{x},0)\psi^\dagger_{bB}(\mathbf{y},0) + \delta_{AB}\sum_D \psi^\dagger_{bD}(\mathbf{y},0)\psi_{aD}(\mathbf{x},0) = \delta_{AB}\delta_{ab}\delta^3(\mathbf{x}-\mathbf{y}) \tag{25}$$

$$\psi_{aA}(\mathbf{x},0)\psi_{bB}(\mathbf{y},0) + \psi_{bA}(\mathbf{y},0)\psi_{aB}(\mathbf{x},0) = 0. \tag{26}$$

is established.

In spite of the modified structure of the basic commutation relations in Eqs.(25) and (26), bilinear observables such as the current density $j_\mu(\mathbf{x},t)$ at two different points commute for space like separation, thus satisfying the microcausality condition. We define the current density four-vector $j_\mu = \sum_A \psi^\dagger_A \gamma_0 \gamma_\mu \psi_A$ where $\gamma_0 = \beta$ and $\boldsymbol{\gamma} = \gamma_0 \boldsymbol{\alpha}$. It is straightforward to show that

$$[j_\mu(\mathbf{x},0), j_\nu(\mathbf{y},0)] = 0 \text{ for } x \neq y \tag{27}$$

Because of the relativistic invariance, commutativity of $j_\mu(\mathbf{x},t)$ and $j_\nu(\mathbf{y},t')$ for arbitrary space like separations follows from the above relation.

4. SCALAR FIELD AS AN ORTHOBOSE FIELD

The Eqs.(15) and (16) with the positive signs being replaced by negative signs define the algebra for orthobosons. We consider a real (Hermitian) scalar field

$$\phi_A(\mathbf{x},t) = \phi_A^\dagger(\mathbf{x},t) = \sum_k \frac{1}{(2\omega_k)^{1/2}}(c_{\mathbf{k}A}e^{i\mathbf{k}\cdot\mathbf{x}-i\omega_k t} + c_{\mathbf{k}A}^\dagger e^{-i\mathbf{k}\cdot\mathbf{x}+i\omega_k t}) \qquad (28)$$

$\omega_k = +(k+m^2)^{1/2}$. The Hamiltonian is taken as

$$H = \frac{1}{2}N\int \sum_A (\nabla\phi_A \cdot \nabla\phi_A + m^2\phi_A\phi_A + \dot{\phi}_A\dot{\phi}_A)d^3x = \sum_\mathbf{k}\omega_k n_\mathbf{k} \qquad (29)$$

where N denotes the normal ordering operator. The number operator for definite momentum $\mathbf{k}$ is $n_{\mathbf{k}A} = \sum_A c_{\mathbf{k}A}^\dagger c_{\mathbf{k}A}$. If the Hamiltonian had been defined without normal ordering, ω_k in Eq.(29) would have been scaled by a term $(1+n)/2$, where n is the range of the A. The operator N removes this unacceptable scaling term along with the zero point energy.

The consistency of the quantization can be established by showing that Klein-Gordon equation for ϕ follows from the Heisenberg equation of motion and orthobose algebra. The Hamiltonian density $\mathcal{H}(x)$ in Eq.(29) can also be derived by starting with the Lagrangian density

$$\mathcal{L} = \sum_A \frac{1}{2}\left\{\frac{\partial\phi_A}{\partial t}\frac{\partial\phi_A}{\partial t} - \nabla\phi_A\cdot\nabla\phi_A - m^2\phi_A\phi_A\right\} \qquad (30)$$

and employing the Legendre transformation.

However, if we take the Hamiltonian density as a local operator, it can be shown that $\mathcal{H}(\mathbf{x},0)$ and $\mathcal{H}(\mathbf{y},0)$ do not commute for $\mathbf{x}\neq\mathbf{y}$. The commutation relations among ϕ^S needed for this purpose can be derived using the Eq.(28) and orthobose algebra. Thus the relativistic field theory for orthobose does not satisfy the microcausality.

5. INTERACTIONS

The interactions can be introduced in the relativistic theory in usual way. A brief comment about the nature of new degree of freedom, which has been left unspecified so far, will help us in formulating appropriate interactions. We may consider that the new degree of freedom is also described by a compact Lie group symmetry such as $SU(n)$ or $SO(n)$ etc., so that the field ψ_A and ϕ_A form representations of Lie algebra with index A labelling the components of multiplet as in the usual quantum field theory. Of course, the physical consequences in the present case will be different.

The interaction terms can be now constructed which are invariant under the Lie group transformations. For example, if ψ_A is an orthofermi field which is a multiplet under $SU(n)$, and ϕ is an ordinary Bose field which is singlet under $SU(n)$, an interaction term is given as $\sum_A \psi_A^\dagger \beta \psi_A \phi$.

6. SUMMARY AND CONCLUSIONS

In contrast to infinite statistics, we have been able to construct local relativistic quantum field theory for orthofermions. The two factors responsible for this are: (i) the number operators are bilinear in $c^\dagger$ and c, and (ii) in orthofield theory, quadratic relation in c exists.

However, our success has been achieved by introducing a new degree of freedom. In fact we have circumvented the problem faced by infinite statistics by allowing the conventional degrees of freedom to be associated with the fermionic (antisymmetry) or bosonic (symmetric) behaviour in the wave function and assigning the new property of infinite statistics (arbitrary symmetry) to the new degree of freedom.

It has been pointed out that the infinite statistics can not be based on a canonical Lagrangian formalism [9]. In contrast, orthofield theory is based on a canonical Lagrangian formalism for the Schrodinger and Dirac fields.

It is most remarkable that microcausality is satisfied for the orthofermi field. Thus, we have shown that even with a modified quantization procedure, it is possible to obtain a consistent quantum field theory, and thus the framework of quantum field theory has been enlarged.

REFERENCES

1. Mishra A. K., and Rajasekaran G., *Pramana - J. Phys.* **36**, 537-555 (1991); **37**, 455 (E) (1991).
2. Greenberg O. W., *Phys. Rev. Lett.* **64**, 705-708 (1990).
3. Greenberg O. W., *Phys. Rev.* **D43**, 4111-4120 (1991).
4. Mishra A. K., and Rajasekaran G., *Phys. Lett. A* **188**, 210-214 (1994).
5. Green H. S., *Phys. Rev.* **90**, 270-273 (1953).
6. Fredenhagen K., *Commun. Math. Phys.* **79**, 141-151 (1981).
7. Mishra A. K., and Rajasekaran G., *Mod. Phys. Lett. A* **7**, 3425-3437 (1992); **11**, 1031 (E) (1996).
8. Mishra A. K., and Rajasekaran G., *Pramana - J. Phys.* **45**, 91-139 (1995); hep-th/9605204.
9. Chaturvedi S., Kapoor A. K., Sandhya R., Srinivasan V., and Simon R., *Phys. Rev.* **A43**, 4555-4557 (1991).

III. Supersymmetry

Spin-Statistics Connection and Supersymmetry

F. Iachello

Center for Theoretical Physics, Sloane Physics Laboratory,
Yale University, New Haven, CT 06520-8120
E-mail: francesco.iachello@yale.edu

Abstract. The recent confirmation of the occurrence of supersymmetry in nuclei is reviewed in light of the spin-statistics connection.

I SUPERSYMMETRY

Supersymmetry is a complex type of symmetry that deals simultaneously with two different types of particles: bosons and fermions. In contrast with "normal" symmetries in which symmetry operations transform bosons into bosons *or* fermions into fermions, in supersymmetry the symmetry operations transform bosons into bosons, fermions into fermions *and* bosons into fermions (and viceversa). For a supersymmetry to occur, the following conditions should be met: (i) To each bosonic quantum state there should be an associated fermionic state with related energy. What states are there and their energy relation is given by the irreducible representations of the supergroup, $\mathcal{G}^*$, and by the supersymmetric Hamiltonian H (for non-relativistic situations), or Lagrangian $\mathcal{L}$ (for quantum fields) that describe the system. (ii) All observables in the bosonic and fermionic system are likewise related by supersymmetry transformations.

The early history of supersymmetry is rather confused. In 1966, Miyazawa [1] introduced a construction of supersymmetry by generalizing the flavor symmetry of Gell-Mann and Ne'eman and Gürsey and Radicati, but this paper went unnoticed. Ramond [2] and Neveu and Schwarz [3] reintroduced it in the context of dual models. However, supersymmetry did not acquire prominence in physics until 1973-74, following the work of Volkov and Akulov [4] and Wess and Zumino [5]. An example of supersymmetric field theory is provided by the Wess-Zumino Lagrangian

$$\mathcal{L}(\S) = -\frac{1}{2}\left(\partial_\mu A(x)\right)^2 - \frac{1}{2}\left(\partial_\mu B(x)\right)^2$$
$$-\frac{1}{2}i\bar{\psi}(x)\gamma^\mu \partial_\mu \psi(x)$$

$$-\frac{1}{2}m^2 A^2(x) - \frac{1}{2}m^2 B^2(x) - \frac{1}{2}im\bar{\psi}(x)\psi(x)$$
$$-gmA(x)\left[A^2(x) + B^2(x)\right]$$
$$-\frac{1}{2}g^2\left[A^2(x) + B^2(x)\right]$$
$$-ig\bar{\psi}(x)\left[A(x) - \gamma_5 B(x)\right]\psi(x) \qquad (1)$$

The fundamental ingredients in this Lagrangian are two scalar fields $A(x), B(x)$ (bosonic) and a spinor field $\psi(x)$ (fermionic). The Lagrangian is invariant under supersymmetry transformations between the bosonic and fermionic fields. As a consequence, the coefficients in front of the various terms are not independent, but related by supersymmetry. There is thus only one mass m and one coupling constant g.

II SPIN-STATISTICS CONNECTION

The spin-statistics connection states that:
(a) For a purely bosonic system, the total wave function must be totally symmetric. Thus, if the single boson wave functions form a basis for the representations of $U(n)$, the wave functions for N bosons must be totally symmetric irreducible representations of $U(n)$ with Young tableau

$$[N] = \Box\Box...\Box \qquad (2)$$

(b) For a purely fermionic system, the total wave function must be totally antisymmetric. Thus, if the single fermion wave functions form a basis for the representations of $U(m)$, the wave functions of M fermions must be totally antisymmetric irreducible representations of $U(m)$ with Young tableau

$$\{M\} = \begin{array}{c} \Box \\ \Box \\ ... \\ \Box \end{array} \qquad (3)$$

These statements are generalized to mixed systems of bosons and fermions:
(c) If the single boson and fermion form a basis for the representations of the superalgebra $\mathcal{G}^* \equiv \mathcal{U}(\backslash/\updownarrow)$, the wave functions for N bosons and M fermions must be totally supersymmetric irreducible representations of $U(n/m)$ with Young supertableau

$$[\mathcal{N}\} = \Diamond\Diamond...\Diamond$$
$$\mathcal{N} = N + M \qquad (4)$$

A different symbol has been used here in order to indicate that the supertableau (4) has a different meaning than an ordinary tableau in that the bosonic indices are symmetrized while the fermionic indices are antisymmetrized. Representations of $U(n/m)$ are discussed by Bars [6], who uses a box with a slash instead of a diamond to distinguish representations of supergroups from ordinary groups.

If the spin-statistics connection holds, then only the representation mentioned above with $\mathcal{N} = N + M$ are allowed.

III SUPERSYMMETRY IN NUCLEI

Although supersymmetry was originally introduced for applications in particle physics, soon afterwards it was applied to other fields of physics. In 1980, the author of this contribution suggested that the spectra of some heavy nuclei could be linked by a supersymmetry transformation and showed evidence for it [7].

Supersymmetry in nuclei is based on a model of nuclear structure, the Interacting Boson Model, introduced earlier [8], [9], in which nucleons pair together into angular momentum $J = 0$ and $J = 2$ (s- and d-wave pairing). Pairs are treated as bosons. The building blocks of the model are six boson operators $b_\alpha (\alpha = 1, ..., 6) \equiv s, d_\mu(\mu = 0, \pm 1, \pm 2)$. These boson operators span a six-dimensional space with algebraic structure $U(6)$. The elements of $U(6)$ are the bilinear products of boson creation and annihilation operators $G_{\alpha\beta} = b_\alpha^\dagger b_\beta$. The Hamiltonian and transition operators are

$$H = E_0 + \sum_{\alpha\beta} \varepsilon_{\alpha\beta} b_\alpha^\dagger b_\beta + \sum_{\alpha\alpha'\beta\beta'} v_{\alpha\alpha'\beta\beta'} b_\alpha^\dagger b_{\alpha'}^\dagger b_\beta b_{\beta'} + ...,$$

$$T = t_0 + \sum_{\alpha\beta} t_{\alpha\beta} b_\alpha^\dagger b_\beta + ... \quad . \tag{5}$$

These operators can be rewritten in terms of elements of $U(6)$ by replacing the bilinear products of creation and annihilation operators by $G_{\alpha\beta}$.

The Interacting Boson Model has beeen used extensively to describe properties of nuclei with an even number of protons and neutrons (even-even nuclei). In this description, a major role is played by dynamic symmetries of the Hamiltonian. These are situations in which the Hamiltonian contains only certain operators (called invariant or Casimir operators, C_i) of a chain of subalgebras originating from $U(6)$ and terminating with the angular momentum algebra $O(3)$ (the problem is rotational invariant). In other words, H is a functional of invariants only,

$$H = f(C_i). \tag{6}$$

For the Interacting Boson Model, there are three and only three such situations corresponding to the breakings:

$$U(6) \supset U(5) \supset O(5) \supset O(3) \supset O(2) \quad (I),$$
$$U(6) \supset SU(3) \supset O(3) \supset O(2) \quad\quad\quad (II),$$
$$U(6) \supset O(6) \supset O(5) \supset O(3) \supset O(2) \quad (III). \tag{7}$$

When a dynamic symmetry occurs, all properties can be calculated in explicit analytic form. In particular, the energies of the states are given by

$$E^{(I)}(N, n_d, v, n_\Delta, L, M_L) = E_0 + \varepsilon n_d + \alpha n_d(n_d + 4)$$
$$+ \beta v(v + 3) + \gamma L(L + 1)$$
$$E^{(II)}(N, \lambda, \mu, K, L, M_L) = E_0 + \kappa(\lambda^2 + \mu^2 + \lambda\mu +$$
$$+ 3\lambda + 3\mu) + \kappa' L(L + 1)$$
$$E^{(III)}(N, \sigma, \tau, \nu_\Delta, L, M_L) = E_0 + A\sigma(\sigma + 4)$$
$$+ B\tau(\tau + 3) + CL(L + 1) \quad (8)$$

Here N is the total number of bosons, L their angular momentum and M_L its projection on a fixed axis. The other quantum numbers, $n_d, v, n_\Delta, ...$, label the representations of the groups in the chains (I),(II),(III).

[These formulas are similar to Bohr's formula for the hydrogen atom

$$E(n, \ell, m_\ell) = -\frac{A}{n^2} \quad (9)$$

(with dynamic symmetry $O(4)$), and to mass formulas of particle physics.]

Since their introduction in the years 1976-79, several examples of dynamic symmetries in nuclei have been found [9].

The Interacting Boson Model and its symmetries provide a classification of spectra of even-even nuclei. However, one has also odd-even and even-odd nuclei. For these nuclei, at least one particle must be unpaired. Furthermore, in even-even nuclei, at higher energies, some of the pairs may break and one must consider explicitly states with two unpaired fermions. A more general model is thus one in which there are at the same time pairs (bosons) and individual particles (fermions). This model, called Interacting Boson-Fermion Model, is used to describe more general properties of nuclei [10], [11].

The building blocks of the Interacting Boson-Fermion Model are boson, $b_\alpha(\alpha = 1, ..., 6) \equiv s(J = 0); d_\mu(J = 2; \mu = 0, \pm 1, \pm 2)$, and fermion, $a_i(i = 1, ..., \Omega) \equiv a_{jm}(j; m = \pm j, \pm(j-1), ..., \pm\frac{1}{2})$, operators. The model Hamiltonian is

$$H = H_B + H_F + V_{BF} \quad (10)$$

with

$$H_B = E_0 + \sum_{\alpha\beta} \varepsilon_{\alpha\beta} b_\alpha^\dagger b_\beta + \sum_{\alpha\alpha'\beta\beta'} v_{\alpha\alpha'\beta\beta'} b_\alpha^\dagger b_{\alpha'}^\dagger b_\beta b_{\beta'}$$
$$H_F = E_0' + \sum_{ik} \eta_{ik} a_i^\dagger a_k + \sum_{ii'kk'} u_{ii'kk'} a_i^\dagger a_{i'}^\dagger a_k a_{k'}$$
$$V_{BF} = \sum_{\alpha\beta ik} w_{\alpha\beta ik} b_\alpha^\dagger b_\beta a_i^\dagger a_k \quad (11)$$

Here H_B and H_F denote the boson and fermion Hamiltonian and V_{BF} their interaction.

Within the context of the Interacting Boson-Fermion Model, a supersymmetry occurs whenever: (i) the boson and fermion single particle states are related by a supersymmetry transformation; and (ii) the coefficients in the Hamiltonian H describing the bosonic and fermionic degrees of freedom are also related by supersymmetry transformations. Supersymmetries require a new type of mathematical construction for their description: Graded Lie Algebras (sometimes called superalgebras). The basic properties of these algebraic constructions are: there are two types of generators, called bosonic, G_α, and fermionic, F_i, respectively, satisfying the following commutation and anticommutation relations

$$[G_\alpha, G_\beta] = \sum_\gamma c^\gamma_{\alpha\beta} G_\gamma$$
$$[G_\alpha, F_i] = \sum_j d^j_{\alpha i} F_j$$
$$\{F_i, F_j\} = \sum_\alpha g^\alpha_{ij} G_\alpha \qquad (12)$$

The presence of anticommutators makes the algebra not a Lie algebra.

The mathematical background needed to describe supersymmetries in nuclei was developed in the 1980's [12], [6]. It makes use of unitary superalgebras $U(n/m)$. These superalgebras can be constructed with bilinear products of creation and annihilation operators usually arranged in a matrix form

$$\begin{pmatrix} b^\dagger_\alpha b_\beta & b^\dagger_\alpha a_i \\ a^\dagger_i b_\alpha & a^\dagger_i a_j \end{pmatrix} \qquad (13)$$

The bosonic and fermionic generators are:

$$G_{\alpha\beta} = b^\dagger_\alpha b_\beta$$
$$G_{ij} = a^\dagger_i a_j$$
$$F^\dagger_{\alpha i} = b^\dagger_\alpha a_i$$
$$F_{i\alpha} = a^\dagger_i b_\alpha \qquad (14)$$

The basis $\mathcal{B}$ is:

$$b^\dagger_{\alpha_1}...b^\dagger_{\alpha_N} a^\dagger_{i_1}..a^\dagger_{i_M} \mid 0 \rangle \qquad (15)$$

The construction of the basis and of the elements of the superalgebra in terms of boson and fermion creation operators automatically insures that the spin-statistics connection is satisfied.

In the Interacting Boson-Fermion Model, the dimension n of the bosonic space is $n = 6$ and the dimension m of the fermionic space is $m = \Omega = \sum_i (2j_i + 1)$. The Hamiltonian H and other operators can then be written in terms of the elements of the superalgebra $U(6/\Omega)$. (The case discussed in 1980 [7] had $j = \frac{3}{2}$, $\Omega = 4$. This case is particularly interesting since it has a structure similar to supersymmetric gravity with $J = 0, J = 2$ (graviton), and $J = \frac{3}{2}$ (gravitino)).

A consequence of supersymmetry is that if the bosonic states and their energies are known, one can predict the fermionic states and their energies. Both are given by the same formula. For example, for $U(6/4)$ supersymmetry, the energy formula is:

$$E(\mathcal{N}, (\sigma_\infty, \sigma_\epsilon, \sigma_\ni), (\tau_\infty, \tau_\epsilon), \nu., \mathcal{J}, \mathcal{M}) = E_0$$
$$+ A \left[\sigma_1(\sigma_1 + 4) + \sigma_2(\sigma_2 + 2) + \sigma_3^2 \right]$$
$$+ B \left[\tau_2(\tau_2 + 3) + \tau_1(\tau_1 + 1) \right]$$
$$+ C J(J + 1).$$

Here $\mathcal{N}$ is the total number of bosons plus fermions and $\sigma_1, \sigma_2, \sigma_3, ...$ are the quantum numbers that label the states. [In particle physics, if one knows the mass of quarks, one can predict the mass of their supersymmetric partners, called squarks, or if one knows the properties of gluons, one can predict those of gluinos (not yet found).]

In the early 1980's, several examples of nuclear spectra with supersymmetric properties were found [13], relating spectra of even-even nuclei with those of odd-even nuclei (odd proton or odd neutron). An example is shown in Tables 1 and 2, where the observed states of the supersymmetric partners ^{190}Os and ^{191}Ir related by $U(6/4)$ supersymmetry and their electromagnetic couplings are compared with the symmetry predictions. If one defines, as a measure of supersymmetry breaking the quantity

$\phi = \frac{\sum_i |E_i^{th} - E_i^{exp}|}{\sum_i E_i^{exp}}$, where the sum goes over the observed states, one finds $\phi = 14\%$.

In view of these results, by 1984-85 it was concluded that supersymmetry in nuclei had been found (with a breaking of the order of 20%) [14].

IV   SUPERSYMMETRY IN NUCLEI CONFIRMED

Soon after the introduction of the Interacting Boson Model, it was realized that an improvement in the description of nuclear properties could be obtained by distinguishing proton pairs from neutron pairs. This model is called Interacting Boson Model-2. The building blocks here are boson operators $b_{\alpha\pi}, b_{\alpha\nu} (\alpha = 1, ..., 6; \pi =$ proton, ν = neutron). The boson operators span a (six+six)-dimensional space with algebraic structure $U_\pi(6) \oplus U_\nu(6)$. Consequently, when going to nuclei with unpaired particles, one has a model with two types of bosons (proton and neutron) and two types of fermions (proton and neutron), called Interacting Boson-Fermion Model-2.

If supersymmetry occurs for this very complex systems one expects now to have supersymmetric partners composed of a quartet of nuclei, even-even, even-odd, odd-even and odd-odd, called a magic square [15]. An example is

TABLE 1. Comparison between experimental and calculated energies in the supersymmetric partners Os and Ir. Only 21 of the known 49 states are shown.

^{190}Os		
σ_1, τ_1, J^P	$E_{\exp}(keV)$	$E_{th}(keV)$
$9, 0, 0^+$	0	0
$9, 1, 2^+$	187	180
$9, 2, 2^+$	558	360
$9, 2, 4^+$	548	500
$9, 3, 0^+$	912	540
$9, 3, 3^+$	756	660
$9, 3, 4^+$	955	740
$9, 3, 6^+$	1050	960
^{191}Ir		
$\frac{17}{2}, \frac{1}{2}, \frac{3}{2}^+$	0	0
$\frac{17}{2}, \frac{3}{2}, \frac{1}{2}^+$	82	120
$\frac{17}{2}, \frac{3}{2}, \frac{5}{2}^+$	129	200
$\frac{17}{2}, \frac{3}{2}, \frac{7}{2}^+$	343	270
$\frac{17}{2}, \frac{5}{2}, \frac{3}{2}^+$	179	360
$\frac{17}{2}, \frac{5}{2}, \frac{5}{2}^+$	351	410
$\frac{17}{2}, \frac{5}{2}, \frac{7}{2}^+$	686	480
$\frac{17}{2}, \frac{5}{2}, \frac{9}{2}^+$	502	570
$\frac{17}{2}, \frac{5}{2}, \frac{11}{2}^+$	832	680
$\frac{15}{2}, \frac{1}{2}, \frac{3}{2}^+$	539	560
$\frac{15}{2}, \frac{3}{2}, \frac{1}{2}^+$	624	680
$\frac{15}{2}, \frac{3}{2}, \frac{5}{2}^+$	748	760

TABLE 2. Comparison between experimental and calculated B(E2) values in the supersymmetric partners Os and Ir.

^{190}Os			
$(\sigma_1, \tau_1, J)_i$	$(\sigma_1, \tau_1, J)_f$	$BE(2)_{exp}(e^2b^2)$	$BE(2)_{th}(e^2b^2)$
9, 1, 2	9, 0, 0	0.478(12)	0.478
9, 2, 2	9, 0, 0	0.046(2)	0
9, 2, 2	9, 1, 2	0.259(15)	0.654
9, 2, 4	9, 1, 2	0.622(44)	0.654
9, 3, 4	9, 1, 2	0.010(2)	0
9, 3, 4	9, 2, 2	0.488(100)	0.375
9, 3, 4	9, 2, 4	0.362(72)	0.340
9, 3, 6	9, 2, 4	1.038(330)	0.715
^{191}Ir			
$\frac{17}{2}, \frac{3}{2}, \frac{1}{2}$	$\frac{17}{2}, \frac{1}{2}, \frac{3}{2}$	0.130(3)	0.425
$\frac{17}{2}, \frac{3}{2}, \frac{5}{2}$	$\frac{17}{2}, \frac{1}{2}, \frac{3}{2}$	0.640(30)	0.425
$\frac{17}{2}, \frac{3}{2}, \frac{7}{2}$	$\frac{17}{2}, \frac{1}{2}, \frac{3}{2}$	0.293(6)	0.425
$\frac{17}{2}, \frac{5}{2}, \frac{3}{2}$	$\frac{17}{2}, \frac{1}{2}, \frac{3}{2}$	0.073(13)	0
$\frac{17}{2}, \frac{5}{2}, \frac{5}{2}$	$\frac{17}{2}, \frac{1}{2}, \frac{3}{2}$	0.0111(4)	0
$\frac{17}{2}, \frac{5}{2}, \frac{7}{2}$	$\frac{17}{2}, \frac{1}{2}, \frac{3}{2}$	0.065(6)	0

$$^{194}Pt \quad ^{195}Pt$$
$$^{195}Au \quad ^{196}Au$$

Spectra of even-even and even-odd nuclei have been known for some time. However, spectra of heavy odd-odd nuclei are very difficult to measure, since the density of states in these nuclei is very high and the energy resolution of most detectors is not sufficiently good. In a major effort, led by J. Jolie, that has involved several laboratories for several years, it has very recently been possibile to measure spectra of odd-odd nuclei with unprecedented accuracy and, most importantly assign spin and parities to the individual levels. In particular, the magnetic spectrometer at the Ludwig-Maximilians Universität in München, Germany, developed by G. Graw, can separate levels only a few keV apart. It has thus been possible to measure the spectrum of ^{196}Au, the missing supersymmetric partner of ^{194}Pt, ^{195}Pt and ^{195}Au [16]. A portion of this spectrum together with the supersymmetry predictions is shown in Table 3. Here only the lowest 20 of the measured 46 states are shown. The degree of agreement for the 26 states not shown here is comparable to that of the table. Only few of the states exptected by supersymmetry (in the table the 0^- state at 221 keV and the 1^- state at146 keV) have not been conclusevely separated.

The observed spectrum of ^{196}Au meets all the criteria for supersymmetry described above: (i) to each bosonic quantum state there are fermionic partners obtained from it by a supersymmetry transformation. Here the transformation is rather complex, $U_\pi(6/4) \oplus U_\nu(6/12)$, since it involves both protons and neutrons; (ii) the energies of the states are given by a single formula and are related by supersymmetry; (iii) the measured intensities (not reported here) follow the supersymmetry predictions. In view of this new evidence, one must conclude that

TABLE 3. Comparison between experimental energies and those predicted by supersymmetry in Au.

^{196}Au			
$(\sigma_1,\sigma_2,\sigma_3),(\tau_1,\tau_2)$	J^P	$E_{\exp}(keV)$	$E_{th}(keV)$
$(\frac{13}{2},\frac{1}{2},\frac{1}{2}),(\frac{1}{2},\frac{1}{2})$	1^-	6	0
	2^-	0	18
$(\frac{3}{2},\frac{1}{2})$	0^-	...	221
	1^-	253	230
	2^-	307	303
	3^-	355	329
	3^-	375	377
	4^-	258	413
$(\frac{11}{2},\frac{1}{2},\frac{1}{2}),(\frac{1}{2},\frac{1}{2})$	1^-	167	146
	2^-	162	164
$(\frac{3}{2},\frac{1}{2})$	0^-	388	367
	1^-	348	376
	2^-	456	449
	3^-	466	475
	3^-	491	523
	4^-	520	559
$(\frac{11}{2},\frac{3}{2},\frac{1}{2}),(\frac{3}{2},\frac{1}{2})$	0^-	42	138
	1^-	...	146
	2^-	198	218
	3^-	234	245
	3^-	287	293
	4^-	213	329

supersymmetry in nuclei has been confirmed.

V IMPLICATIONS TO OTHER FIELDS

(a) Particle Physics

Supersymmetry has been sought in Particle Physics for decades. The confirmation of supersymmetry in nuclei indicates that this very complex type of symmetry can occur in Nature. It gives hope that, although badly broken, supersymmetry may occur in particle physics. However, supersymmetry in Nuclear Physics is a symmetry that relates composite objects (pairs) with fundamental objects (nucleons). Can it be the same in particle physics (Nambu, Gürsey [17],...)?

(b) Condensed matter physics

Some supersymmetric theories have been constructed in condensed matter physics (Parisi-Sourlas [18]). Nambu has suggested that supersymmetry may occur in Type II superconductors [19]. The occurrence of supersymmetry in Nuclear Physics may lead to other supersymmetric theories between composite objects and their constituents.

VI IMPLICATIONS TO SPIN-STATISTICS CONNECTION

The fact that only those representations constistent with the spin-statistics connection have been seen experimentally appears to indicate that supersymmetry occurs with the appropriate connection. However, the bosons here are composite objects of two fermions. They are obtained by making use of a bosonization of the the fermion Hamiltonian. The fermion pair operators satisfy only approximate Bose commutation relations. If one calls S the fermion creation operator for $J=0$ pairs, one finds that

$[S, S^\dagger] = 1 + O\left(\frac{1}{\Omega}\right)$ where Ω is the degeneracy of the fermion shells ($\Omega = 44$ for the $82-126$ shell). As it was observed in the sections above, supersymmetry in nuclei is broken. The approximate spin-statistics connection is one of the reasons why this breaking occurs. In view of the composite nature of the bosons, supersymmetry in nuclei cannot be used as a strict test of the spin-statistics connection. Rather, it could be used to understand other composite Bose-Fermi systems (Bose-Einstein condensates with fermionic impurities?)

VII CONCLUSIONS

Supersymmetry, one of the most fundamental types of symmetry that one may encounter in Nature, has been found and confirmed in Nuclei.

VIII AKNOWLEDGEMENTS

This work is supported in part by U.S. D.O.E. Contracts DE-FG02-91ER40608.

REFERENCES

1. Miyazawa H., *Progr. Theor. Phys.* **36**, 1266 (1966).
2. Ramond P., *Phys. Rev. D* **3**, 2415 (1971).
3. Neveu A. and Schwarz J., *Nucl. Phys. B* **31**, 86 (1971).
4. Volkov D.V. and Akulov V.P., *Phys. Lett. B* **46**, 109 (1973).
5. Wess J. and Zumino B., *Nucl. Phys. B* **70**, 39 (1974).
6. Bars I., *Physica D* **15**, 42 (1985).
7. Iachello F., *Phys. Rev. Lett.* **44**, 772 (1980).
8. Iachello F., in *Nuclear Structure and Spectroscopy*, Blok H.P. and Dieperink A.E.L., eds., (Scholar's Press, Amsterdam, 1974), p.163; Arima A. and Iachello F., *Phys. Rev. Lett.* **35**, 1069 (1975).
9. Iachello F. and Arima A., *The Interacting Boson Model*, Cambridge University Press, Cambridge, 1987.
10. Iachello F. and Scholten O., *Phys. Rev. Lett.* **43**, 679 (1979).
11. Iachello F. and Van Isacker P., *The Interacting Boson-Fermion Model*, Cambridge University Press, Cambridge, 1991.
12. Balantekin A.B., Bars I. and Iachello F., *Nucl. Phys. A* **370**, 284 (1981).
13. Iachello F., *Physica D***15**, 85 (1985); Casten R.F., *Physica D* **15**, 99 (1985).
14. Casten R.F. and Feng D.H., *Physics Today* **37**, 26 (1984).
15. Van Isacker P., Jolie J., Heyde K. and Frank A., *Phys. Rev. Lett.* **54**, 653 (1985).
16. Metz A., Jolie J., Graw G., Hertenberger R., Groger J., Gunther Ch., Warr N. and Eisermann Y., *Phys. Rev. Lett.* **83**, 1542 (1999).
17. Catto S. and Gürsey F., *Nuovo Cimento A***86**, 201 (1985).
18. Parisi G. and Sourlas N., *Phys. Rev. Lett.* **43**, 744 (1979).
19. Nambu Y., *Physica D***15**, 147 (1985).

Boson-Fermion Realization of Lie Algebras and Dynamical Supersymmetry in Fermion Systems

Hendrik B. Geyer[*], Petr Navrátil[†] and Jacek Dobaczewski[¶]

[*]*Institute of Theoretical Physics, University of Stellenbosch,
Private Bag X1, 7602 Matieland, Stellenbosch, South Africa*

[†]*Department of Physics, University of Arizona, Tucson, Arizona 85721*

[¶]*Institute of Theoretical Physics, Warsaw University,
Hoża 69 PL-00-681 Warsaw, Poland*

Abstract. Recent experimental results by Metz, Jolie and collaborators have led to renewed interest in the notion of dynamical supersymmetry in fermion many-body systems. By considering the general problem of constructing a boson-fermion realization for Lie algebras with generators expressed in terms of fermion and bifermion operators, also when favoured pairs dictate the structure of collective bosons, we demonstrate that dynamical "supersymmetry without bosons" can indeed emerge in a fermion system *without any Pauli violations*. Our construction of the general realization utilises the identification of a physical subspace and allows one to identify the images of single particle transfer operators uniquely, unlike the situation in phenomenological applications of dynamical supersymmetry.

INTRODUCTION

Supersymmetry was originally introduced into relativistic quantum field theory to exhibit an invariance with respect to the exchange of bosons and fermions (see Weinberg's recent monograph [1] for an overview), but the notion has since been successfully exploited in a variety of quantum mechanical and quantum many-body systems (see Junker's text [2] for various examples). In both cases though, whether on the microscopic or phenomenological level, the explicit presence of kinematically independent boson and fermion degrees of freedom has been considered *sine qua non* for supersymmetry.

It may therefore be somewhat surprising to discover that a fermion system on its own can also exhibit *dynamical supersymmetry* [3], in which case states with even and odd fermion numbers are then unified in a single representation of a supergroup.

In the nuclear physics context this possibility was realized by Iachello [4] in the phenomenological interacting boson-fermion model (IBFM) and subsequently shown to be applicable to various pairs of nuclei (see Ref. [5] for an introduction and review of applications). Renewed interest in this possibility has recently been created by the experimental results of Metz et al [6] where a *quartet* of nuclei have been found to fit into a single extended supersymmetric multiplet of $U_\nu(6/12) \otimes U_\pi(6/4)$ which takes both neutron (ν) and proton (π) degrees of freedom explicitly into account. (See also Iachello's contribution to these proceedings [7].)

States of even and odd nuclei are in principle eigenstates of the same Hamiltonian obtained for different particle numbers. Although there is no fundamental difference between even and odd nuclei from this point of view, their properties are, however, quite different. A unification of spectra of even and odd nuclei into a single framework is therefore indeed a challenging possibility, with the prospect of unveiling a basic underlying symmetry.

As already pointed out, the notion of supersymmetry has in this regard proved to be quite fruitful [4,5]. Properties of some neighbouring even and odd nuclei can actually be classified and understood in terms of an assumed supersymmetry within the framework of the interacting boson-fermion model (IBFM) [5]. This model introduces, together with fermions representing the single (odd) nucleon, phenomenological boson degrees of freedom (s- and d-bosons) to represent collective monopole and quadrupole fermion pairs. Dynamical supersymmetry then arises when different boson-fermion interaction strengths are related in a special way.

Although this phenomenological supersymmetry does not necessarily imply dynamical supersymmetry on a microscopic level, the IBFM [5] does achieve the type of unification of even and odd states referred to above on the phenomenological level. To be more specific: starting from a common boson-fermion Hamiltonian one finds that in some instances states of an even nucleus, described by many-boson wave functions, are linked by supersymmetry to states in a neighboring odd nucleus in which the odd fermion is treated explicitly. It is important to realize, however, that Pauli correlations between the odd particle and even core are not fully taken into account in the IBFM. In this sense the link between observed supersymmetry and an underlying microscopy is tenuous with no detailed microscopic underpinning as yet.

It is furthermore important to realize that the phenomenological supersymmetry analysis depends non-trivially on the choice which is made for single-particle transfer operators and that this choice is not necessarily (or not at all) dictated by the Hamiltonian which exhibits dynamical supersymmetry in a given case.

As discussed below, the use of an appropriate boson-fermion mapping can address both of the above phenomenological limitations. Firstly, it may be used to demonstrate that dynamical supersymmetry in a fermion system can be completely compatible with all Pauli restrictions still intact. Secondly, such a mapping achieves, in principle and often in practice, a unique prescription for the single particle transfer operators to be used in the supersymmetric analysis.

DYNAMICAL SUSY IN A FERMION SYSTEM: PHENOMENOLOGY VS MICROSCOPY

As is well known (see e.g. Ref. [2]), the simplest form of supersymmetry in quantum mechanics can be formulated in terms of nilpotent operators ("supercharges") $Q = B\alpha^\dagger$ and $Q^\dagger$ with the B's and α's the usual boson and fermion operators which are also defined to be kinematically independent, e.g. $[B, \alpha] = [B, \alpha^\dagger] = 0$. The supersymmetric Hamiltonian $\{Q, Q^\dagger\} = B^\dagger B + \alpha^\dagger \alpha$ obviously has eigenstates $|n_B, n_\alpha\rangle$, and displays the hallmark supersymmetric spectrum of a unique ground state $|0, 0\rangle$ and a set of doubly degenerate excited states. This analysis can be extended to models with more than one supercharge, $N = 2$ SUSY quantum mechanics e.g. [2], with a supersymmetric Hamiltonian which then explicitly reads $H = Q_1^2 + Q_2^2$.

In applications of dynamical supersymmetry to e.g. nuclear spectra [4–7] the appropriate Hamiltonian may also be expressed in terms of odd generators (supercharges) as well as even generators (of the type $\alpha^\dagger \alpha'$ and $B^\dagger B'$), but the nuclear interaction requires more general products than those appearing in the typical supersymmetric Hamiltonians above. While supersymmetry is therefore broken in this sense, it nevertheless remains possible to classify states of some even and odd neighbouring nuclei in terms of representations of a supermultiplet. This possibility is known as dynamical supersymmetry and is a non-trivial property of the appropriate interaction. For the whole analysis to be feasible in the first place, it is, however, clear that a Hamiltonian which reflects interactions among bosons and fermions is a prerequisite and appears in the IBFM on a phenomenological level.

What concerns us mostly in the remainder of this contribution, is the question whether dynamical supersymmetry, as summarized above, can be compatible with the Pauli principle or, alternatively, if dynamical supersymmetry can be an exact property of a fermion system. From the point of view that there are important Pauli corrections to the lowest order association between collective fermion pairs and IBM bosons [8], one might anticipate a negative answer to this question. Nevertheless, we show that the implementation of appropriate boson-fermion mappings indeed reveals instances where this compatibility holds. These mappings, introduced and refined in Refs. [9,10], and discussed and exploited below, introduce an equivalence between a system of interacting fermions on the one hand, and on the other a system of interacting bosons and fermions which are *by construction kinematically independent*. Although no guarantee in itself, this clearly fulfills the minimum requirement for a dynamical supersymmetry to exist, namely to have the appropriate degrees of freedom. Moreover, as elaborated below, one can indeed find instances where dynamical supersymmetry emerges as an exact classification scheme of fermion states. This is simply an alternative, but equivalent, classification to whatever classification scheme may have been adopted on the fermion level. What the boson-fermion mapping accomplishes is to make the inherent supersymmetric nature transparent in these instances, although they should also be identifiable on a purely algebraic level in terms of relationships between standard and supersymmetric representations.

Apart from providing a concrete link between fermion dynamics and dynamical supersymmetry, the use of boson-fermion mappings also allows one to *construct* various transition operators appropriate to the boson-fermion description, including the important single-particle transfer operator. This is in contrast to the phenomenological situation where one is obliged to truncate an infinite series of combinations of boson and fermion operators with phenomenological parameters and terms only restricted by their tensor and particle number changing properties [5]. It should be emphasized that the choice of these transition operators in phenomenological models such as the IBM or IBFM is *not* dictated by the Hamiltonian parameters in general, specifically also in the case of dynamical symmetry or supersymmetry. (See also Ref. [11] for a discussion of this point.)

Space limitations restrict our illustration of the above ideas to the familiar pairing or seniority model [12]. An extended discussion may be found in Ref. [3]

DYNAMICAL SUSY IN THE SENIORITY MODEL

The textbook SU(2) seniority model [12] has been analysed exhaustively and nothing can be added to the explicit results. Here we will show, however, how the known results may be obtained from a boson-fermion mapping and interpreted from the point of view of dynamical supersymmetry.

The SU(2) model is defined by considering in a single-j shell the monopole pair creation operator $S^+ = \sqrt{\Omega/2}(a_j^+ a_j^+)^{(0)}$ with $\Omega = j + \frac{1}{2}$. It fulfills the commutation relation $[S, S^+] = \Omega - n$, where n is the fermion number operator in the single-j shell. The SU(2) algebra can be generalized to describe odd systems by constructing the superalgebra generated by the operators S^+, S, $\Omega - n$, a_{jm}, and a_{jm}^+. The relevant commutation relation is $\left[a_{jm}^+, S\right] = -\tilde{a}_{jm}$, with $\tilde{a}_{jm} = (-1)^{j-m} a_{j,-m}$, while the single-fermion operators obey the standard anticommutation relations. Clearly this is a rather trivial superalgebra as the elements of the odd sector (single fermion operators) anti-commute only to the identity. Alternatively, by considering the *commutator* of single-fermion operators, the set of bi- and single-fermion operators may of course also be viewed as generators of a standard (orthogonal) algebra.

In the single-j shell we consider the pairing Hamiltonian $H = -GS^+S$, which has the energy spectrum [12] $E(n,v) = -\frac{1}{4}G(n-v)(2\Omega - n - v + 2)$, with n the total number of fermions and the seniority quantum number v denoting the number of fermions not coupled to angular momentum zero. This Hamiltonian describes both the even and odd systems, and the spectra in both cases are given by the same expression with v even or odd, respectively.

We can apply to this model the general Dyson boson-fermion mapping derived in Ref. [10] to find an equivalent description in the boson-fermion space. As explained there the construction, which utilises supercoherent states, is a two-step process which finally requires application of a similarity transformation, the general form of which is derived in Ref. [10]. For the SU(2) algebra, the similarity transformation involves the ideal fermion number operator $\mathcal{N} = \sum_m \alpha^\dagger_{jm} \alpha_{jm}$ with the explicit form

$$X = \frac{(\Omega - \frac{1}{2}(\mathcal{N} + \hat{\mathcal{N}}))!}{(\Omega - \hat{\mathcal{N}})!} \exp\left[\mathcal{S}^\dagger B\right]_{\widehat{}}, \qquad (1)$$

where the positional operator $\hat{\mathcal{N}}$ is explained in Ref. [10]. The ideal fermion operators $\alpha^\dagger_{jm}$ and α_{jm} commute with the ideal boson operators $B^\dagger$ and B, and the ideal fermion pair operators $\mathcal{S}^\dagger$ and $\mathcal{S}$ are obtained from S^+ and S by replacing all a's by α's. The general Dyson boson-fermion mapping obtained for the SU(2) case is

$$S^+ \longleftrightarrow \Omega B^\dagger - B^\dagger B^\dagger B - B^\dagger \mathcal{N} = B^\dagger(\Omega - N_B - \mathcal{N}) = B^\dagger(\Omega - \aleph), \qquad (2a)$$

$$S \longleftrightarrow B, \qquad (2b)$$

$$n \longleftrightarrow 2B^\dagger B + \mathcal{N} = 2N_B + \mathcal{N} = \aleph + N_B, \qquad (2c)$$

$$a^+_{jm} \longleftrightarrow \alpha^\dagger_{jm} \frac{\Omega - \aleph}{\Omega - \mathcal{N}} + B^\dagger \tilde{\alpha}_{jm} - \mathcal{S}^\dagger \tilde{\alpha}_{jm} \frac{\Omega - \aleph}{(\Omega - \mathcal{N})(\Omega - \mathcal{N} + 1)}, \qquad (2d)$$

$$a_{jm} \longleftrightarrow \alpha_{jm} + \tilde{\alpha}^\dagger_{jm} B \frac{1}{\Omega - \mathcal{N}} + \mathcal{S}^\dagger \alpha_{jm} B \frac{1}{(\Omega - \mathcal{N})(\Omega - \mathcal{N} + 1)}, \qquad (2e)$$

where $\aleph = N_B + \mathcal{N}$. We see that the single fermion images (2d) and (2e) are finite and contain terms changing the ideal fermion number by one only. Furthermore, they preserve *exactly* the anti-commutation relations on the full ideal space, i.e., as operator identities. This is guaranteed by our construction and can be verified by explicit calculation. The preservation of the commutation and anticommutation relations ensures the exact preservation of the Pauli exclusion principle once the original fermion problem is mapped into the boson-fermion space.

The mapping (2) transforms the 2-body Hamiltonian $H = -GS^+S$ into a 1-plus-2-body boson-fermion Hamiltonian of the form $H_{BF} = -GN_B(\Omega - N_B + 1 - \mathcal{N})$. H and H_{BF} have exactly the same spectrum $E(n,v) = E(N_B, \mathcal{N})$, which can be seen explicitly by equating particle numbers in the two formulations, $n = 2N_B + \mathcal{N}$, and associating $v = \mathcal{N}$. H_{BF} can also be expressed in a form which stresses its dependence on the total number of bosons and fermions $\aleph$, i.e. $H_{BF} = -G(\aleph - \mathcal{N})(\Omega + 1 - \aleph)$. Note that the boson-fermion interaction term in H_{BF}, i.e. $GN_B\mathcal{N}$, can be expressed in terms of the odd generators, $O^\dagger_m = \alpha^\dagger_{jm} B$ and $O_m = B^\dagger \alpha_{jm}$, of the U(1/2$\Omega$) superalgebra. Since the boson and ideal fermion number operators can be linked to even generators, it is possible to write H_{BF} in yet another form in terms of both even generators and *supergenerators* of U(1/2Ω):

$$H_{BF} = -G[N_B(\Omega - N_B + 1) + \mathcal{N} - \sum_m O^\dagger_m O_m]. \qquad (3)$$

On the level of group chains used to classify states, the original fermion analysis and the one which follows after mapping are of course also related. In the original fermion space of the SU(2) seniority model, the Hamiltonian eigenstates are classified according to the representations of the subgroup chain

$$SO(4\Omega+1) \supset SU(2) \otimes Sp(2\Omega) \supset U(1) \otimes SO(3) \ . \qquad (4)$$

In the boson-fermion space the appropriate classification chain starts with the supergroup $U(1/2\Omega)$ with subsequent subchain

$$\supset U_B(1) \otimes U_F(2\Omega) \supset U_B(1) \otimes U_F(1) \otimes Sp_F(2\Omega) \supset U_B(1) \otimes U_F(1) \otimes SO_F(3) \ . \qquad (5)$$

In the second chain the supergroup $U(1/2\Omega)$ is generated by $B^\dagger B$, $\alpha_{jm}^\dagger \alpha_{jm'}$, $\alpha_{jm}^\dagger B$, and $B^\dagger \alpha_{jm}$, with the first two operators belonging to the even sector and the remaining two to the odd sector of the superalgebra, respectively. The appearance of $U(1/2\Omega)$ in the chain is clearly suggested by the form (3) and also dictates that the same Hamiltonian parameters in the first expression for H_{BF} are used for both even and odd states, if the Hamiltonian should be viewed as a phenomenological boson-fermion Hamiltonian. From the mapping point of view there is of course no real choice in this matter, as the original fermion Hamiltonian pertains to any number of particles. Nevertheless, it is interesting to note already here how the phenomenological extension to superalgebras à la IBFM may be suggested from a microscopic point of view.

In the dynamical supersymmetry concept, those states of the system with the same number of *ideal* particles $\aleph$ belong to a single supersymmetric representation $[\aleph]$ of the supergroup, here $U(1/2\Omega)$. (Note that $\aleph=N_B+\mathcal{N}$ differs from the fermion number n and its ideal space image (2c), $n \longleftrightarrow 2N_B + \mathcal{N}$).

Although one can embed all states of the $SU(2)$ model in a sequence of representations of the supergroup $U(1/2\Omega)$, this does not really provide other interesting physical consequences. This is so because in the chain (5) the supergroup $U(1/2\Omega)$ is immediately split into the boson and fermion sectors which remain separate down to the bottom of the chain. As discussed in Ref. [3], potentially interesting situations which are restrictive from the supersymmetry point of view, occur if at a certain level of subgroups one may find the same subgroups in the boson and fermion sectors, and combine them together into a given boson-fermion subgroup. Such a situation arises in an $SO(8) \otimes SO(5)$ fermion model [3]. In terms of algebras appearing in the chain (5), we here only find a trivial example of the boson subalgebra $U_B(1)$, generated by the boson number operator N_B, and the fermion subalgebra $U_F(1)$, generated by the fermion number operator $\mathcal{N}$, which can be combined into the boson-fermion subalgebra $U_{BF}(1)$, generated by $\aleph=N_B+\mathcal{N}$.

At the same time, even the simple $SU(2)$ seniority model discussed here illustrates interesting consequences for the structure of the single-fermion transfer operators and the spectroscopic factors. After the mapping, the single-fermion operators acquire terms which are responsible for the Pauli correlations between the even core and the odd particle. In the phenomenological supersymmetric models these terms are postulated or motivated semi-microscopically, whereas in the supersymmetric picture derived from the boson-fermion mapping they are fixed by the mapping procedure itself. The image of the fermion annihilation operator (2e) is e.g. a combination of the ideal fermion annihilation operator and two corrective terms.

It is instructive to calculate the spectroscopic factors for the simplest states in the SU(2) model. In the original fermion space one has

$$\langle n+1, v=1, j||a_j^\dagger||n, v=0\rangle = -\langle n, v=0||\tilde{a}_j||n+1, v=1, j\rangle = -\sqrt{2\Omega - n}, \quad (6)$$

which requires one to calculate normalization factors of the fermion states. Evaluating matrix elements (6) in the mapped boson-fermion space is much simpler. In particular, one can work with the ideal boson-fermion eigenstates of the Hamiltonian H_{BF} corresponding to the states with $\mathcal{N}=0$ and 1, e.g., $|B^{N_B}\rangle$, and $|B^{N_B}, \alpha_j^\dagger\rangle$. Using Eqs. (2d) and (2e) we directly read off the corresponding result

$$(B^{N_B}, \alpha_j^\dagger||(a_j^+)_{\text{BF}}||B^{N_B})(B^{N_B}||(\tilde{a}_j)_{\text{BF}}||B^{N_B}, \alpha_j^\dagger) = -(2\Omega - n) \quad (7)$$

where $(a_{jm}^+)_{\text{BF}}$ and $(\tilde{a}_{jm})_{\text{BF}}$ are the boson-fermion images of a_{jm}^+ and $\tilde{a}_{jm}$. (Ref. [3] contains an extended discussion and closer contact with the results of Ref. [6].)

CONCLUSIONS

It has been illustrated that the construction of a general boson-fermion mapping facilitates the identification of dynamical supersymmetry in a fermion system which may be perfectly compatible with all Pauli restrictions. This mapping also leads to the unique construction of appropriate single particle transfer operators. Nontrivial occurences of dynamical supersymmetry in a fermion system are in the final instance intimately related to dynamical properties of the Hamiltonian.

REFERENCES

1. Weinberg, S., *The Quantum Theory of Fields, Vol III Supersymmetry*, Cambridge University Press, Cambridge, 2000.
2. Junker, G., *Supersymmetric Methods in Quantum and Statistical Physics*, Springer Verlag, Berlin, 1996.
3. Navrátil, P., Geyer, H. B., and Dobaczewski, J., *Nucl. Phys. A* **607**, 23-42 (1996).
4. Iachello, F., *Phys. Rev. Lett.* **44**, 772-775 (1980).
5. Iachello, F., and Van Isacker, P., *The Interacting Boson-Fermion Model*, Cambridge University Press, Cambridge, 1991.
6. Metz, A., Jolie, J., et al, *Phys. Rev. Lett.* **83**, 1542-1545 (1999).
7. Iachello, F., contribution to these proceedings.
8. Klein, A., and Marshalek, E. R., *Rev. Mod. Phys.* **63**, 375-558 (1991).
9. Dobaczewski, J., Scholtz, F. G., and Geyer, H. B., *Phys. Rev.* **C48**, 2313-2325 (1993).
10. Navrátil, P., Geyer, H. B., and Dobaczewski, J., *Phys. Rev. C* **52**, 1394-1406 (1995).
11. Geyer, H. B., Navrátil, P., and Dobaczewski, J., "Boson Mappings and Phenomenological Boson Models", in *Perspectives for the Interacting Boson Model*, edited by R. F. Casten et al, World Scientific, Singapore, 1995, pp. 189-200.
12. Ring, P., and Schuck, P., *The Nuclear Many-Body Problem*, Springer-Verlag, Berlin, 1980.

Deformed Heisenberg Algebra with Reflection, Anyons and Supersymmetry of Parabosons

Mikhail S. Plyushchay*,†

*Departamento de Física, Universidad de Santiago de Chile,
Casilla 307, Santiago 2, Chile
†Institute for High Energy Physics, Protvino, Russia

Abstract. Deformed Heisenberg algebra with reflection appeared in the context of Wigner's generalized quantization schemes underlying the concept of parafields and parastatistics of Green, Volkov, Greenberg and Messiah. We review the application of this algebra for the universal description of ordinary spin-j and anyon fields in 2+1 dimensions, and discuss the intimate relation between parastatistics and supersymmetry.

INTRODUCTION

Generalized statistics was introduced in physics in the form of parastatistics as an exotic possibility extending the Bose and Fermi statistics [1–5]. It was closely related with the discovery of color in the context of the theory of strong interactions. Nowadays generalized statistics [6–9] finds applications in the physics of the quantum Hall effect and (probably) it is relevant to high temperature superconductivity [10]. Supersymmetry, instead, unifies Bose and Fermi statistics [11–15] and its development lead to the construction of field and string theories with exceptional properties [16], that transformed the same idea of supersymmetry in one of the cornerstones of modern theoretical physics. Supersymmetry was observed in nuclear physics in the form of dynamical symmetry [17,18], whereas its manifestation as a fundamental symmetry of elementary particle physics still waits for experimental confirmation.

Though supersymmetry and generalized statistics may be unified in the form of parasupersymmetry [19], nevertheless, by the construction, the two concepts seem to be independent. Recently, the existence of intimate relation between generalized statistics and supersymmetry was established by observation of hidden supersymmetric structure in purely parabosonic [20] and purely pafermionic [21] systems. The key tool with which the observation of close relationship between generalized statistics and supersymmetry was realized is the so called deformed

Heisenberg algebra with reflection, or the R-deformed Heisenberg algebra (RDHA) [22,23]. The algebraic construction of RDHA appeared in the work of Wigner [24], where he investigated the problem of correlation of equations of motion with quantum mechanical commutation relations and proposed the generalized quantization schemes which subsequently lead to the theoretical discovery of parastatistics [1–5] (in this context, see also refs. [25–28]). RDHA represents, probably, one of the first examples of deformation of bosonic harmonic oscillator which, as it was shown recently, possesses some universality being also related to parafermions [22,23], to (2+1)-dimensional anyons [29–31], and to the bosonized form of supersymmetric quantum mechanics [32,30,33]. Besides, the RDHA structure underlies the construction of fractional supersymmetry [34]. In this talk we review the application of RDHA for the universal description of ordinary spin-j and anyon fields in 2+1 dimensions by means of first order linear differential equations [29–31], and discuss the exotic supersymmetry of purely parabosonic systems [20].

SUPERSYMMETRY OF PARABOSONS

The deformed Heisenberg algebra with reflection is generated by the creation-annihilation operators a^+, a^- and by the reflection operator R satisfying the relations

$$[a^-, a^+] = 1 + \nu R, \quad \{R, a^\pm\} = 0, \quad R^2 = 1, \qquad (1)$$

where ν is a real deformation parameter. Due to these basic relations, the creation-annihilation operators satisfy the trilinear parabosonic commutation relations $[\{a^-, a^+\}, a^\pm] = \pm 2a^\pm$. Introducing the Fock vacuum state, $a^-|0\rangle = 0$, and fixing the action of operator R on it as $R|0\rangle = |0\rangle$, we arrive at the relation $a^- a^+|0\rangle = (1+\nu)|0\rangle$, which together with trilinear commutation relation means that at $\nu = p-1$, $p = 1, 2, \ldots$, the operators $a^\pm$ have the sense of single-mode creation-annihilation operators of paraboson of order p. Vice versa, one can show that parabosonic trilinear commutation relations themselves give rise to RDHA [22]. In general case, the number operator is realized in the form $N = \frac{1}{2}\{a^-, a^+\} - \frac{1}{2}(1+\nu)$, and the reflection operator can be represented as $R = (-1)^N = \cos \pi N$.

The reflection operator introduces a natural Z_2 grading structure in the Fock space and its presence in the definition of RDHA can be considered as an indication on possible relationship between parabosons and supersymmetry. To reveal a supersymmetry of parabosons, one notes that RDHA can also be given by the relations [20]

$$a^+ a^- = F(N), \quad a^- a^+ = F(N+1), \quad [N, a^\pm,] = \pm a^\pm, \qquad (2)$$

where $F(N) = N + \nu \sin^2 \frac{\pi N}{2}$ is the characteristic function satisfying for $\nu > -1$ the relations $F(0) = 0$, $F(n) > 0$, $n = 1, 2, \ldots$. These relations mean, in particular, that for $\nu > -1$ the corresponding representations of RDHA are unitary and infinite-dimensional. On the other hand, in the case $\nu = -(2p+1)$, $p = 1, 2, \ldots$, the characteristic function possesses the property $F(2p+1) = 0$ underlying the existence of $(2p + 1)$-dimensional (non-unitary) irreducible representations of RDHA, which are associated with the deformed parafermions of order $2p = 2, 4, \ldots$ [22,23,21]. Let us restrict ourselves here by the case of unitary infinite-dimensional representations ($\nu > -1$). When $\nu = 2k + 1$, $k = 0, 1, \ldots$, the structure function satisfies the relation $F(2n + 1) = F(2n + \nu + 1)$, $n = 0, 1, \ldots$. Therefore, in the case of parabosonic systems of even order $p = 2(k + 1)$, the spectrum of the quadratic Hamiltonian $H = a^+ a^-$ (or of $H = a^- a^+$), reveals doubling of all higher-lying levels. This indicates on existence of supersymmetry in such purely parabosonic systems. As it follows from the explicit form of the characteristic function, in the case of paraboson of order $p = 2(k+1)$, $k = 0, 1, 2 \ldots$, and Hamiltonian $H = a^+ a^-$, the supersymmetry is characterized by the presence of $k + 1$ singlet states with energies $E = 0, 2, \ldots, 2k$. Therefore, only in the case $k = 0$, the spectrum has one singlet state of zero energy, whereas all other cases are characterized by the presence of k higher-lying singlet states of nonzero energy in addition to the zero energy ground state. The corresponding supercharges are the infinite series operators in the corresponding paraboson operators:

$$Q_+ = (a^+)^{2k+1} \sin^2 \pi J_0, \quad Q_- = (a^-)^{2k+1} \cos^2 \pi J_0, \quad J_0 = \frac{1}{4}\{a^+, a^-\}. \quad (3)$$

They together with the Hamiltonian satisfy the polynomial superalgebra:

$$Q_\pm^2 = 0, \quad [H, Q_\pm] = 0,$$
$$\{Q_+, Q_-\} = (H - 2k)(H - 2k + 2)\ldots(H + 2k - 2)(H + 2k), \quad (4)$$

which in the case $k = 0$ is reduced to the conventional $N = 1$ linear superalgebra. The role of the grading operator in such purely parabosonic supersymmetric systems belongs to the reflection operator $R = \cos \pi N$.

The case of paraboson system of order $p = 2$ ($\nu = 1$) given by the Hamiltonian $H = a^- a^+$ is characterized by the $N = 1$ spontaneously broken linear supersymmetry: all the states are paired in supersymmetric doublets with the lowest energy level $E = 2$. The corresponding supercharges have the form (3) with $k = 0$ and with operators a^+ and a^- changed in their places [32]. The systems of parabosons of order $p = 4, 6, \ldots$, given by the Hamiltonian $H = a^- a^+ - 2$ possess nonlinear (polynomial) supersymmetry of the form similar to (4) [20].

It was shown in ref. [20] that the supersymmetry of purely parabosonic systems can be understood as the supersymmetry of Calogero-like systems with exchange interaction and that in principle it can be realized in one-dimensional systems of identical fermions. Besides, it was demonstrated that nonlinear parabosonic supersymmetry can be obtained via appropriate modification of the classical analog of usual supersymmetric quantum mechanics.

RDHA AND ANYONS

The parabosonic supersymmetry structures corresponding to $H = a^+a^-$ and $H = a^-a^+$ can be unified and extended to the $osp(2|2)$ superalgebraic structure [32]. The operators $T_3 = \frac{1}{2}\{a^+, a^-\}$, $T_\pm = \frac{1}{2}(a^\pm)^2$ and $I = \frac{1}{2}(\nu + R)$ have a sense of even generators of $osp(2|2)$ forming $sl(2) \times u(1)$ subalgebra, whereas the operators $Q^\pm = a^\pm \Pi_\pm$ and $S^\pm = a^\pm \Pi_\mp$ are its odd generators, where $\Pi_\pm = \frac{1}{2}(1\pm R)$ are the projectors on even and odd subspaces of the Fock space. On the other hand, the operators J_μ, $\mu = 0, 1, 2$, $J_0 = \frac{1}{2}T_3$, $J_1 \pm iJ_2 = T_\pm$, and $\mathcal{L}_\alpha$, $\alpha = 1, 2$, $\mathcal{L}_1 = (a^+ + a^-)/\sqrt{2}$, $\mathcal{L}_2 = i(a^+ - a^-)/\sqrt{2}$, can be considered as even and odd generators of $osp(1|2)$ superalgebra: $[J_\mu, J_\nu] = -i\epsilon_{\mu\nu\lambda}J^\lambda$, $[J_\mu, \mathcal{L}_\alpha] = \frac{1}{2}(\gamma_\mu)_\alpha{}^\beta \mathcal{L}_\beta$, $\{\mathcal{L}_\alpha, \mathcal{L}_\beta\} = 4i(J\gamma)_{\alpha\beta}$; here (2+1)-dimensional γ-matrices appear in the Majorana representation, see ref. [29]. Since the Casimir operator $\mathcal{C} = J_\mu J^\mu - \frac{i}{8}\mathcal{L}^\alpha \mathcal{L}_\alpha$ takes the fixed value $\mathcal{C} = \frac{1}{16}(1 - \nu^2)$, this means that any irreducible representation of RDHA carries the corresponding irreducible representation of $osp(1|2)$, which, in turn, is a direct sum of two irreducible representations of $so(2,1)$ with the Casimir operator $C = J^\mu J_\mu$ taking the value $C = -\alpha_+(\alpha_+ - 1)$, $\alpha_+ = \frac{1}{4}(1+\nu)$, on even ($R = 1$) subspace of the Fock space, and $C = -\alpha_-(\alpha_- - 1)$, $\alpha_- = \alpha_+ + \frac{1}{2}$, on odd ($R = -1$) subspace. The $osp(1|2)$ structure associated with RDHA can be exploited to describe anyons by means of covariant linear differential equations. For the purpose, let us consider the field $\Psi^n(x)$ depending on space-time point x_μ in 2+1 dimensions and carrying infinite- ($\nu > -1$, $n = 0, 1, \ldots$) or finite- dimensional ($\nu = -(2p+1)$, $p = 1, 2, \ldots$, $n = 0, \ldots, 2p$) representation of RDHA. Then the following spinor set of linear differential equations describes universally ordinary spin-j fields and anyons [29–31]:

$$D_\alpha \Psi(x) = 0, \quad D_\alpha = R\mathcal{P}_\alpha + m\mathcal{L}_\alpha, \tag{5}$$

where R is the reflection operator of RDHA, m is a mass parameter, and $\mathcal{P}_\alpha = (-i\gamma_\mu \partial^\mu)_\alpha{}^\beta \mathcal{L}_\beta$. The condition of integrability of two equations (5) is equivalent to the equations $(-\partial^2 + m^2)\Psi_+(x) = 0$, $(-i\partial^\mu J_\mu - sm)\Psi_+(x) = 0$, $s = \frac{1}{4}(1+\nu)$, for the even part (in the RDHA sense) of the field, $R\Psi_+(x) = \Psi_+(x)$, whereas the solution to equations (5) in odd subspace, $R\Psi_-(x) = -\Psi_-(x)$, is trivial, $\Psi_-(x) = 0$. The parameter ν fixes the value of spin s, and one concludes that in the case of finite-dimensional representations ($\nu = -(2p+1)$), the corresponding field $\Psi_+(x)$ carries integer or half-integer spin $s = j$, whereas the case of infinite-dimensional representations of RDHA ($\nu > -1$) corresponds to the field of arbitrary spin $s > 0$ (anyon). The case of anyon with $s < 0$ can be obtained by a simple change $m \to -m$ in (5). In the case of infinite-dimensional unitary representations of RDHA, the linear differential equation $(-i\partial^\mu J_\mu - sm)\Psi_+(x) = 0$ is the (2+1)-dimensional analog of the $(3+1)D$ infinite-component Majorana equation, whose fundamental role for the description of anyons was established in ref. [35] under investigation of the $(2+1)D$ model of relativistic particle with torsion (see also refs. [36,37]).

Varying the deformation parameter in the region $\nu > -1$, one can obtain the fields of integer spin ($\nu = 4n - 1$, $s = n$, $n = 1, 2, \ldots$) as well as of half-integer spin ($\nu = 4n + 1$, $s = n + \frac{1}{2}$, $n = 0, 1, \ldots$). However, such fields of integer and half-integer spin have a nature to be essentially different from the nature of usual spin-j fields appearing in the case of finite-dimensional representations of RDHA ($\nu = -(2p+1)$) since they have hidden nonlocality. In the rest frame, the solution to the Klein-Gordon and Majorana equations has only one nontrivial component in correspondence with the pseudoscalar (helicity) nature of spin in 2+1 dimensions [38,39]. But the Lorentz boost enlivens all the infinite number of components of the field $\Psi_+(x)$ in the case $\nu > -1$ [31]. There is the analog of coordinate representation for RDHA, in which

$$a^\pm = \frac{1}{\sqrt{2}}(q \mp i\mathcal{D}_\nu), \quad \mathcal{D}_\nu = -i\left(\frac{d}{dq} - \frac{\nu}{2q}R\right), \quad (6)$$

and $R\psi(q) = \psi(-q)$. In such representation the fields $\Psi_+(x)$ have a structure of the functions to be even in continuous variable q: $\Psi_+(x,q) = \Psi_+(x,-q)$. This means that the corresponding solutions to the equations (5) in the case $\nu > -1$ have the hidden half-infinite nonlocality ($q \geq 0$) which is analogous to the string-like nonlocality of anyon fields in other approaches [40,41].

To conclude, RDHA finds various theoretical applications including the described two. It would be interesting to look for the experimental manifestation of the exotic supersymmetry of parabosons. In general, the existence of the polynomial supersymmetry is characterized by the presence of several singlet states which could be considered as an indication on parabosonic-like excitations (quasiparticles) in the system. The search for possible experimental manifestation of the nonlocal (in internal variable q) fields of integer and half-integer spin associated with parabosons seems to be another interesting problem.

ACKNOWLEDGMENTS

I am grateful to I. Bandos, M. Rausch de Traubenberg and D. Sorokin for useful discussions, and to G. Marmo and E. C. G. Sudarshan for bringing refs. [26-28] to my attention. The work was supported in part by FONDECYT (Chile) under grant 1980619 and by DICYT (USACH).

REFERENCES

1. Green, H. S., *Phys. Rev.* **90**, 270 (1953).
2. Volkov, D. V., *Sov. JETP* **9**, 1107 (1959); **11**, 375 (1960).
3. Greenberg, O. W., *Phys. Rev. Lett.* **13**, 598 (1964).
4. Greenberg, O. W., and Messiah, A. M. L., *Phys. Rev.* **B136**, 248 (1964).
5. Ohnuki, Y., and Kamefuchi, S., *Quantum Field Theory and Parastatistics*, Tokyo: University Press, 1982.
6. Leinaas, J. M., and Myrheim, J., *Nuovo Cimento* **37B**, 1 (1977).

7. Goldin, G. A., Menikoff, R., and Sharp, D. H., *J. Math. Phys.* **21**, 650 (1980); **22**, 1664 (1981).
8. Wilczek, F., *Phys. Rev. Lett.* **48**, 1144 (1982); **49**, 957 (1982).
9. Wilczek, F., and Zee, A., *Phys. Rev. Lett.* **51**, 2250 (1983).
10. Wilczek, F., *Fractional Statistics and Anyon Superconductivity*, Singapore: World Scientific, 1990.
11. Gol'fand, Y. A., and Likhtman, E. P., *JETP Lett.* **13**, 323 (1971).
12. Ramond, P., *Phys. Rev.* **D3**, 2415 (1971).
13. Neveu, A., and Schwarz, J., *Nucl. Phys.* **B31**, 86 (1971).
14. Volkov, D., and Akulov, V., *JETP Lett.* **16**, 438 (1972).
15. Wess, J., and Zumino, B., *Nucl. Phys.* **B70**, 39 (1974).
16. Green, M. B., Schwarz, J., and Witten, E., *Superstring Theory*, Cambridge, 1987.
17. Iachello, F., *Phys. Rev. Lett.* **44**, 772 (1980).
18. Metz, A., et al., *Phys. Rev. Lett.* **83**, 1542 (1999).
19. Rubakov, V. A., and Spiridonov, V. P., *Mod. Phys. Lett.* **A3**, 1337 (1988).
20. Plyushchay, M., *Hidden Nonlinear Supersymmetries in Pure Parabosonic Systems*, *Int. J. Mod. Phys.* **A**, in press, hep-th/9903130.
21. Klishevich, S., and Plyushchay, M., *Mod. Phys. Lett.* **A14**, 2739 (1999).
22. Plyushchay, M. S., *Nucl. Phys.* **B491**, 619 (1997).
23. Plyushchay, M. S., *Mod. Phys. Lett.* **A11**, 2953 (1996).
24. Wigner, E. P., *Phys. Rev.* **77**, 711 (1950).
25. Yang, L. M., *Phys. Rev.* **84**, 788 (1951).
26. Ryan, C., and Sudarshan, E. C. G., *Nucl. Phys.* **B47**, 207 (1963).
27. Mukunda, N., Sudarshan, E. C. G., Sharma, J. K., and Mehta, C. L., *J. Math. Phys.* **21**, 2386 (1980); **22**, 78 (1981).
28. Man'ko, V. I., Marmo, G., Sudarshan, E. C. G., and Zaccaria, F., *Int. J. Mod. Phys.* **B11**, 1281 (1997).
29. Plyushchay, M. S., *Phys. Lett.* **B320**, 91 (1994).
30. Plyushchay, M. S., *Ann. Phys.* **245**, 339 (1996).
31. Plyushchay, M. S., *Mod. Phys. Lett.* **A12**, 1153 (1997).
32. Plyushchay, M. S., *Mod. Phys. Lett.* **A11**, 397 (1996).
33. Gamboa, J., Plyushchay, M., and Zanelli, J., *Nucl. Phys.* **B543**, 447 (1999).
34. Rausch de Traubenberg, M., and Slupinski, M. J., *Mod. Phys. Lett.* **A12**, 3051 (1997); *J. Math. Phys.* **41**, 4556 (2000).
35. Plyushchay, M. S., *Phys. Lett.* **B262**, 71 (1991); *Nucl. Phys.* **B362**, 54 (1991).
36. Jackiw, R., and Nair, V. P., *Phys. Rev.* **D43**, 1933 (1991).
37. Plyushchay, M. S., *Phys. Lett.* **B273**, 250 (1991); *Int. J. Mod. Phys.* **A7**, 7045 (1992).
38. Cortés, J. L., and Plyushchay, M. S., *Int. J. Mod. Phys.* **A11**, 3331 (1996).
39. Plyushchay, M. S., *Monopole Chern-Simons Term: Charge-Monopole System as a Particle with Spin*, hep-th/0004032.
40. Buchholz, D., and Fredenhagen, K., *Comm. Math. Phys.* **84**, 1 (1982).
41. Fröhlich, J., and Marchetti, P. A., *Nucl. Phys.* **B356**, 533 (1991).

IV. Quantum Gravity and Related Issues

Spin and Statistics in Quantum Gravity

H.F. Dowker* and R.D. Sorkin[†]

*Department of Physics, Queen Mary and Westfield College,
London E1 4NS, UK.
[†]Department of Physics, Syracuse University,
Syracuse, NY 13244-1130, USA.

Abstract. We present a review of the spin and statistics of topological geons, particles in 3+1 quantum gravity. They can have half-odd-integral spin and fermionic statistics and since the underlying gravitational field is tensorial and bosonic, this is an example of "emergent" non-trivial spin and statistics as displayed by familiar non-gravitating objects such as skyrmions. We give the topological background and show that in a "canonical" quantization of gravity there is no spin-statistics correlation for topological geons. Allowing the topology of space to change, for example in a sum-over-histories approach, raises the possibility that a spin-statistics correlation can be recovered for geons. We review a conjectured set of rules powerful enough to give such a spin-statistics correlation for all topological geons. These would appear to rule out the possibility of parastatistics and may rule out spinorial and fermionic geons altogether.

I INTRODUCTION

In quantum mechanics and in quantum field theory in flat spacetime we have many examples of "emergent" fermionic statistics and spinorial (*i.e.*, half-odd-integral) spin for objects built from entities which are fundamentally tensorial (*i.e.*, integral spin) and bosonic. Can such a phenomenon occur in quantum gravity in which the dynamical variable, the spacetime metric, is tensorial and bosonic? The answer is yes, and in this review we will look at the best studied case, that of so-called "topological geons", particles made of non-trivial spatial topology, in 3 + 1 dimensions.

The original work on spin-half states in quantum gravity was done by Friedman and Sorkin [1,2] and on fermionic states by Sorkin [3]. In section II we recall the basic features of topological geons, particles which exist by virtue of the non-trivial topology of space, and show how they acquire their spin and statistics by a choice of unitary irreducible representation (UIR) of the so-called mapping class group (MCG). We will see that there is no spin-statistics correlation for quantum geons [3–6] and indeed there are always spin-statistics violating sectors for any species of geon [7,8]. Moreover, the presence of a particular type of symmetry between quantum geons, namely the "slides" which correspond to diffeomorphisms in which one geon slides through another, produces an extraordinary variety of quantum sectors [7,8].

The lack of a spin-statistics correlation for geons is perhaps not surprising. In the proofs of all existing spin-statistics theorems for extended objects like geons, the possibility of particle-anti-particle creation (and annihilation) is crucial. Indeed particle-anti-particle pair production and annihilation has been suggested (see [9,10]) as a unifying principle that might bring together the "topological" spin-statistics theorems, of which the Finkelstein-Rubenstein version [11] is the original, and the relativistic quantum field theory theorems (the conditions imposed — Lorentz invariance *etc.* — being supposedly just those that guarantee that antiparticles exist with the required possibilities for pair creation and annihilation).

Now, the process of geon-anti-geon pair production is a topology changing one and cannot be described within a formalism like canonical quantum gravity which assumes, a priori, that the spatial three-manifold is fixed. It has therefore been conjectured that in a formulation of quantum gravity which can accommodate topology change, the usual spin-statistics connection would be recovered for geons [4].

The sum-over-histories (SOH) for quantum gravity is such a formulation and there is indeed evidence that there's a spin-statistics theorem for geons "trying to get out" in a SOH approach [12]. To prove a general spin-statistics theorem for all geons, extra assumptions are needed and in section III we review a set of rules which would achieve this [4]. The consequences of these rules are stronger than originally envisaged and there's a possibility that they might rule out spinorial and fermionic geons altogether.

Section IV is a brief mention of some motivation for pursuing a geon spin-statistics theorem.

We will restrict ourselves to orientable three-manifolds throughout and will further assume that no handles, $S^2 \times S^1$, occur in the "prime decomposition" of the three-manifold (see section II).

II HOW THE GEON GOT ITS SPIN AND STATISTICS

Topological geons are particles made from non-trivial spatial topology. We are interested in the situation of an isolated system of these particles and thus we will be dealing with a three-dimensional manifold M which admits asymptotically flat metrics. Physically, M is three-space at a "moment of time", or, the "future boundary of truncated spacetime" [13].

A The spatial three-manifold

There is a "Three-Manifold Decomposition Theorem" that identifies candidates for elementary geons, but in order to state this theorem we must first introduce the concepts of "connected sum" (denoted #) and "prime manifold." To take the connected sum of two oriented three-manifolds M_1 and M_2, remove an open ball from each and identify the resulting two-sphere boundaries with an orientation-reversing diffeomorphism (henceforth, "diffeo"). Taking the connected sum of any three-manifold with S^3 gives a manifold diffeomorphic to the original one; taking it with $\mathbb{R}^3$ is topologically equivalent to deleting a point. A prime three-manifold, P, is a closed three-manifold that is not S^3 and such that whenever $P = P_1 \# P_2$, either P_1 or P_2 is S^3. Examples of primes are the three-torus, T^3, and the so-called spherical spaces, S^3/G, where G is some discrete subgroup of $SO(4)$ acting freely on S^3.

The M we are considering is $M = \mathbb{R}^3 \# K$ where K is a closed three-manifold. The Decomposition Theorem states that any such M can be decomposed into the connected sum of finitely many prime manifolds and this decomposition is unique:

$$M = \mathbb{R}^3 \# P_1 \# P_2 \ldots \# P_n. \tag{1}$$

We will assume that to each prime summand there corresponds an elementary quantum geon; with "correspond" being used in a suitable sense since there is a rather subtle relation between a particular piece of spatial topology and a physical particle—which subtlety has to do both with familiar "identical particle exchange effects" and unfamiliar effects due to the existence of diffeos known as "slides" [3,4].

For more details see [3,4,14]

B Wave functions

In canonical quantum gravity, for which the topology does not change, the configuration space, Q, is the space of all three-geometries on M,

$$Q = \frac{\text{Riem}^\infty(M)}{\text{Diff}^\infty(M)}, \qquad (2)$$

where $\text{Riem}^\infty(M)$ (R^∞ for short) is the space of asymptotically flat Riemannian metrics on M and $\text{Diff}^\infty(M)$ (D^∞ for short) is the group of diffeomorphisms of M that become trivial on approach to infinity. It can be shown that D^∞ acts freely on R^∞ and so Q is a manifold, R^∞ being a principal fibre bundle over Q with fibre D^∞. Thus, using the fact that R^∞ is convex and hence contractible to a point so that all its homotopy groups are trivial, we deduce that $\pi_k(D^\infty) \simeq \pi_{k+1}(Q)$.

Wave functions need not be single-valued on Q if Q contains non-contractible loops. Rather, the transformation of a wave function as such loops are traversed gives a representation of $\pi_1(Q)$. This is a special case of the general situation where wave functions are sections of a twisted vector bundle on Q. Physical observables are invariant under $\text{Diff}^\infty(M)$ which means that the quantum state space reduces to a number of inequivalent and independent sectors each transforming under a different unitary irreducible representation of the group $\pi_1(Q)$. From the above we know that $\pi_1(Q) \simeq \pi_0(D^\infty)$, where $G \equiv \pi_0(D^\infty)$ is known as the mapping class group (MCG) or the group of large diffeos of M and is the analogue for gravity of the group of large gauge transformations in a gauge theory.

C The mapping class group and UIR's

Let us restrict attention to the manifold which is a connected sum of $\mathbb{R}^3$ and a number, n, of identical primes, $M = \mathbb{R}^3 \# P \# \ldots P$. G is generated by three types of large diffeomorphism, the *exchanges*, the *internal diffeomorphisms* and the *slides*. Each of these generators can be viewed as the result of a certain *process* (a "development" [3]), with the nature of the process being suggested by the name of the category. Thus, an exchange is the result of a process in which two identical primes continuously change their positions until they have swapped places. Similarly a slide is the result of a process in which one prime travels around a closed loop threading through one or more other primes, while an internal diffeo is a diffeo whose support is restricted to a single prime.

One very important internal diffeo is the 2π-rotation of a single prime. It is the result of a process in which the prime rotates around by 2π. For particular primes, P, this diffeo is deformable to the identity and so is not a non-trivial element of G.

Geons based on these primes must have tensorial (integral) spin since a 2π rotation is equivalent to the identity and so cannot have a non-trivial effect on the quantum state. Of the known primes, only the handle (which we are explicitly excluding here) and the lens spaces are tensorial: the rest have non-trivial 2π rotation and are called spinorial primes.

The slides, internals and exchanges each form subgroups of G and the exchanges generate a subgroup isomorphic to the permutation group, $\mathcal{S}_n$. In [7,8] the literature on the MCG was sublimated into the following result

$$G = (slides) \ltimes (internals) \ltimes \mathcal{S}_n, \qquad (3)$$

where the symbol $\ltimes$ denotes semidirect product (with the normal subgroup on the left). What this says is that every element of G is uniquely a product of three diffeomorphism-classes, one from each subgroup, and that each subgroup is invariant under conjugation by elements of the subgroups standing to its right in equation 3. See [15–17] for more details on the MCG.

The fact that G is a semidirect product allows us to analyze its UIR's in terms of representations of its factor groups and their subgroups. Let

$$G = N \ltimes K \qquad (4)$$

be a semidirect product with N being the normal subgroup. A finite dimensional UIR of G is then determined by the following data

- Γ = a UIR of N
- T = a PUIR of $K_0 \subseteq K$,

where K_0 is the subgroup of K that remains "unbroken by Γ" (K_0 is "the little group") and "PUIR" stands for "projective UIR", *i.e.*, representation up to a phase. (The Schur multiplier for T is determined by Γ.)

In seeking the UIR's of this group there are two situations, depending on whether the slide subgroup is represented trivially or not. The results in the following two subsections are contained in [7,8].

D The sectors with trivial slides

In the simpler case of UIR's of G which annihilate the slides, in effect a complete classification is possible. In this case, the mathematical problem is reduced to finding the UIR's of the quotient group, $G/(slides)$, which by equation 3 is just the semidirect product

$$(internals) \ltimes \mathcal{S}_n. \tag{5}$$

Let us find the finite dimensional UIR's of this group (sometimes called the "particle group" [17]). The normal group of internal diffeos is the direct product of n copies of the MCG group, H, for a single prime P:

$$N = (internals) = H \times H \times \cdots \times H. \tag{6}$$

The most general UIR of this normal subgroup N is itself a product, namely the tensor product,

$$\Gamma = \Gamma_1 \otimes \Gamma_1 \cdots \Gamma_2 \otimes \Gamma_2 \cdots \Gamma_r, \tag{7}$$

of UIR's of H, where the first n_1 factors are Γ_1, the next n_2 factors are Γ_2, etc.. Physically a given Γ_i specifies a certain "internal structure" for the corresponding geon, and is therefore a "species parameter" or "quantum number". Different Γ_i's mean that the corresponding geons are not identical particles but are of different physical species. So here can we have a quantum breaking of indistinguishability.

With respect to the choice (7), the unbroken subgroup $K_0 \subseteq \mathcal{S}_n$ reduces to a product of permutation groups,

$$K_0 = \mathcal{S}_{n_1} \times \mathcal{S}_{n_2} \times \cdots \times \mathcal{S}_{n_r}. \tag{8}$$

The statistics is then given by a UIR T of K_0, that is to say by an independent UIR $T_1, T_2, \cdots$ for each of the subgroups $\mathcal{S}_{n_1}, \mathcal{S}_{n_2}, \cdots$. Each of these T's in turn, can be specified by a choice of a Young tableau, and determines whether the corresponding geons will manifest Bose statistics, Fermi statistics or some particular parastatistics. Since there is no restriction on the choice of T, there is no restriction on which combinations of these possible statistics can occur.

To summarize, all possible sectors with trivial slides are accounted for by specifying

- a *species* for each geon (i.e. a UIR of H)

- a *statistics* for each resulting set of identical geons

E Some sectors with nontrivial slides

When the slide subgroup is represented nontrivially, a full classification of the possible UIR's of G does not exist. It is clear, however, that there is an extraordinary variety of possible UIR's and in order to give an idea of this richness we give an example of a UIR in a special case which avoids most of the complications which obstruct the full classification.

The prime in this special case is $\mathbb{R}P^3$, which can be visualized as a region of M produced by excising a solid ball and then identifying antipodal pairs of points on the resulting S^2 boundary. The internal group is trivial for this prime. For each pair of $\mathbb{R}P^3$'s, one can slide one through the other, with the square of this slide being trivial (since $\pi_1(\mathbb{R}P^3) = \mathbf{Z}_2$), making a total of $n(n-1)$ independent order 2 generators. The complete group $(slides)$ is then generated by products of these elementary slides.

Since for $\mathbb{R}P^3$, $(internals)$ is trivial, the MCG reduces to

$$G = (slides) \ltimes \mathcal{S}_n. \tag{9}$$

Hence, according to the general scheme outlined earlier, we get a UIR of G by choosing first a UIR, Γ, of $N = (slides)$, and then a suitable PUIR, T, of the resulting unbroken subgroup $K_0 \subseteq \mathcal{S}_n$. As before, we may interpret K_0 as describing the surviving indistinguishability of the geons, and T as describing the statistics within each set of identical geons.

Consider abelian UIRs of N in which, in effect, each ordered pair of primes is assigned a ± 1. We can represent the various such UIR's pictorially by drawing n dots to represent the primes and an arrow to represent each ordered pair that receives a minus sign (meaning the a slide of the first prime through the second produces a phase-factor of -1). Each distinct diagram of this type then gives rise to a different class of UIR's of $(slides)$, and therefore furnishes a different building block for constructing UIR's of G.

If we make a particular choice of UIR of $(slides)$ by choosing a diagram in which the dots and arrows form a circle and the arrows point anticlockwise, say, this pattern leaves $\mathbf{Z}_n \subseteq \mathcal{S}_n$ as the unbroken subgroup K_0. We acquire distinct UIR's of G corresponding to the possible UIR's of $\mathbf{Z}_n$. Here, geon identity is expressed not by a permutation group, $\mathcal{S}_n$, at all, but by the cyclic group $\mathbf{Z}_n$. With this new type of group comes a new type of statistics, in which a cyclic permutation of the geons produces the complex phase q or $\bar{q}$, $q = 1^{1/n}$ being an nth-root of unity.

Another interesting possibility, arising when the UIR, Γ, of $(slides)$ is non-abelian, is that of "projective statistics," meaning a type of statistics expressed by a properly projective representation of the permutation group or one of its subgroups. To construct such an example, we would need at least four geons because $\mathcal{S}_n$ possesses properly projective representations only for $n \geq 4$.

F No spin-statistics correlation

All the UIRs of the MCG are on a equal footing as far as the canonical theory goes. Given a fixed spatial three-manifold, there appears in the theory a set of, in principle unpredictable, new parameters, some discrete and some continuous, corresponding to Nature's choice of a particular UIR amongst all the possibilities. Some of these UIRs violate the usual spin-statistics correlation in the grossest possible way. Subsection II D shows that if the basic prime is tensorial (the 2π rotation is trivial) then there is a UIR in which the permutation group is represented by the totally antisymmetric UIR. If the basic prime is spinorial then the quantum geon can be tensorial or spinorial and the permutation group be represented either by the totally symmetric or antisymmetric UIR. In other words there are UIRs in which the geons are spinorial bosons or tensorial fermions.

This lack of a correlation can be attributed to the fact that in a theory in which the topology is fixed (such as canonical quantization) there is no allowance for geon-anti-geon production. This is because a process in which a geon and anti-geon are created from $\mathbb{R}^3$ is a topology changing one (we know this from the decomposition theorem: one piece of non-trivial topology can't "cancel" another). The known spin-statistics theorems for objects such as skyrmions and other kinks which have these emergent properties of spin and statistics all require, for their proofs, that the process of pair creation and annihilation be describable as a path in the configuration space ([18] is a general such proof). This leads one to expect that in a formulation of quantum gravity in which topology change is naturally accommodated the spin statistics correlation might be recovered. The SOH is such a formulation and we therefore turn now to that approach.

III SPACETIME APPROACH

In [12] it was shown that two non-chiral geons (*i.e.*, one which are their own antiparticle) which are pair created in a topology changing process, satisfy a limited spin-statistics theorem which eliminates some of the spin-statistics violating sectors. Instead of reviewing that work, we will turn to the more general question of how a general spin-statistics theorem could be proved for all geons.

In the SOH framework the fundamental dynamical input is a rule attaching a quantum amplitude to each pair of truncated histories which "come together" at some "time" [19,13,20,21]. Let us call such a pair a "Schwinger history" and its underlying manifold a "Schwinger manifold". In the case of quantum gravity, a truncated history is a Lorentzian manifold with final boundary (and possibly initial boundary depending on the physical context), and the "coming together" means the identification or "sewing together" of the final boundaries. In a framework

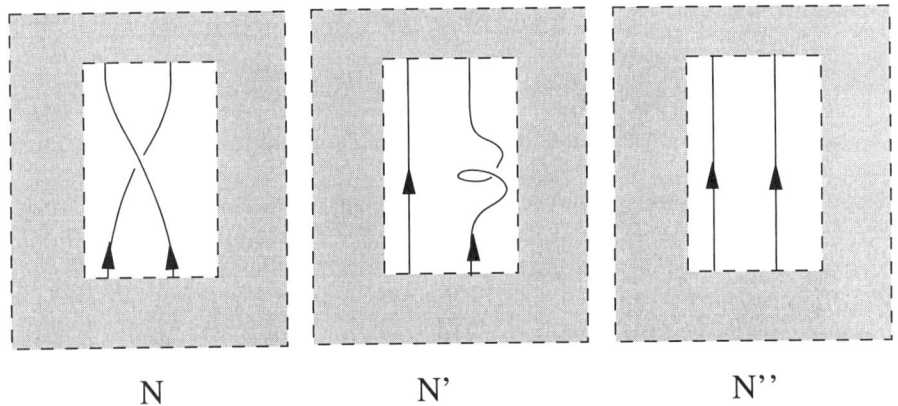

N   N'   N''

FIGURE 1. Three manifolds which differ only in the unshaded compact region

in which topology change is allowed, all four-manifolds and all metrics on each manifold are summed over. Now, without disturbing the classical limit of the theory or the local physics, we can multiply the amplitude of each Schwinger history by a "weight" w depending only on the topology of the underlying manifold (and on the two initial metrics, if initial boundaries are present). We would like, somewhat analogously to [22], to argue that these weights carry some unitary representation of G, and that sets of weights belonging to disjoint representations "do not mix". We cannot do so in the present case for several reasons. We do not know what G is in a topology changing scenario precisely because the topology of space changes and so the MCG changes with it. Even if we could overcome this, we would be faced with the question of how to implement the higher-dimensional non-abelian UIR's.

We will ignore these detailed questions here and instead sketch what a spin-statistics theorem might look like in a SOH framework.

Consider two Schwinger manifolds, N and N', that contribute to the SOH and which differ only in a compact region in which, in N two identical primes swap places and in N' one prime stays put and the other rotates around by 2π. This is illustrated in figure 1 in which the "environment" depicted in grey is common to N and N'. The lines represent the "worldlines" of the primes and are in reality framed curves in a four-dimensional manifold. (Given a framed curve in a four-dimensional space, there is a topologically unique way of attaching a prime to each point of that curve.) The corkscrew effect in N' represents the frame rotating by 2π.

The geons are bosons (fermions) if the weight of N is equal to (minus) the weight of a third manifold, N'' which differs from N only in that the two primes do not move at all in that same compact region. The geons are tensorial (spinorial) if the

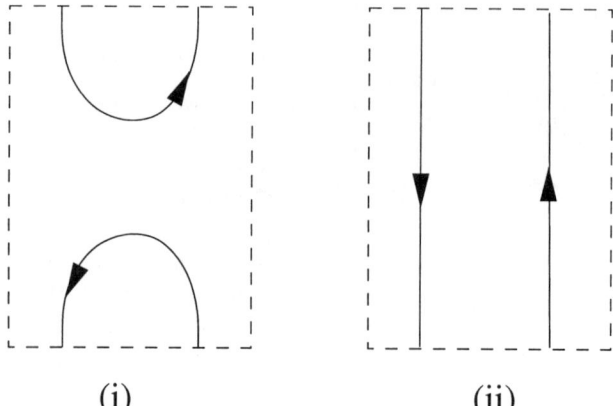

(i)   (ii)

FIGURE 2. The annihilation of a prime-anti-prime pair followed by a pair creation is congruent to the trivial propagation in time of a prime-anti-prime pair

weight of N' is equal to (minus) the weight of N''. A spin statistics theorem would consist of the result that given any two Schwinger manifolds such as N and N', they must have the same weight in the SOH.

The rules that would achieve this result are the following [4]. We dub manifolds with the same weight "congruent."

- Manifolds which are diffeomorphic, via a diffeomorphism which is the identity on the initial boundaries and at spatial infinity if these are present, are congruent.

- Let two manifolds, N_1 and N_2 differ only in a certain "compact region," so $N_1 = R_1 \cup E$ and $N_2 = R_2 \cup E$, where E is the "environment". If N_1 and N_2 are congruent then $N_1' = R_1 \cup E'$ and $N_2' = R_2 \cup E'$ are also congruent.

- If N is a manifold which admits a spacelike hypersurface which divides N in half so that the second half is the time reverse of the first, then N is congruent to the product manifold $V_0 \times [0,1]$ where V_0 is the initial boundary of N. (If there's no initial boundary then N is congruent to S^4).

Now, for any geon topology, P, there is a canonical manifold, X, which mediates geon-anti-geon annihilation. The third rule has the consequence that the manifold which consists of X followed by its time reverse (pair annihilation followed by pair creation) is congruent to the manifold in which a prime and an anti-prime just sit there. This is illustrated in figure 2 where the lines again represent the worldlines of the primes and the arrows remind us that they are framed curves. The "anti-prime" is the mirror image, or "chiral conjugate," of the prime.

This can be used to prove a spin-statistics theorem as illustrated in figure 3 which shows a sequence of congruent compact regions, to be thought of as embedded in

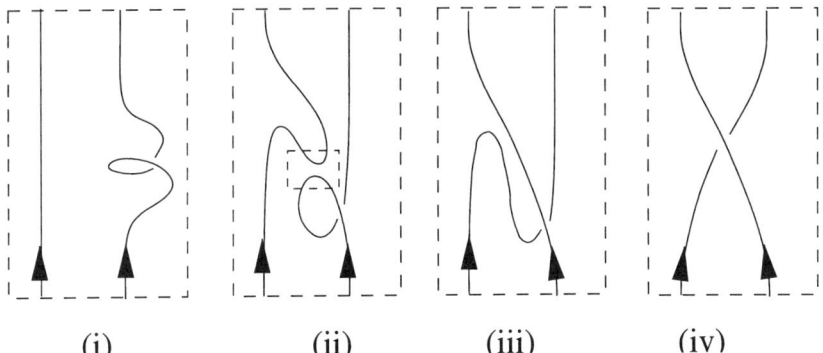

FIGURE 3. A sequence of congruent manifolds which establish a spin-statistics correlation for geons

some common environment. Again, the lines represent the (framed) worldlines of two identical primes. Region (i) contains a static prime and one which rotates by 2π. "Untwisting" the rotation gives region (ii) which is diffeomorphic to (i). Notice that region (ii) contains a small region which is a pair-annihilation-pair-creation event and replacing that with the prime-anti-prime product, as we may do by rule 3, gives region (iii). This is in turn diffeomorphic to (iv). So (i) is congruent to (iv) which is the statement of a spin-statistics correlation in this framework.

It has now been realised that these rules are much more powerful than originally envisaged. In particular they allow only abelian UIR's as potential weights and thus eliminate the possibility of parastatistics. (This seems to be a consequence of "topological" spin-statistics theorems in general [9,10].) A sketch of the proof of this result is given in figure 4 which shows a sequence of (compact regions of) congruent manifolds – the surrounding "boxes" are to be imagined.

In (i) is shown a manifold in which at early times there are a number of primes sitting in fixed positions, propagating in time: that's denoted by the single vertical line which now stands for a collection of a number of framed worldlines. In the intermediate region a large diffeo, g, an element of the MCG of the three-manifold comprising those prime summands, is developed: that's denoted by the shaded box labeled g. It's useful to imagine an example in which, say, there are two identical primes and g is the exchange diffeo so that diagram (i) is actually the manifold N in figure 1. (i) is diffeomorphic to (ii) which is congruent to (iii) by rule 3. The line on the right in (iii) is now a spectator for the next few steps. Into the "closed loop" we may introduce the development of any diffeo h followed by its inverse. h is an element of the MCG not of the original multi-prime manifold but one which is its mirror image in which each original prime is replaced by its chiral conjugate. This can be seen by following carefully what is created and destroyed at the pair creation and annihilation events. Then the diffeos h and h^{-1} are eased along the

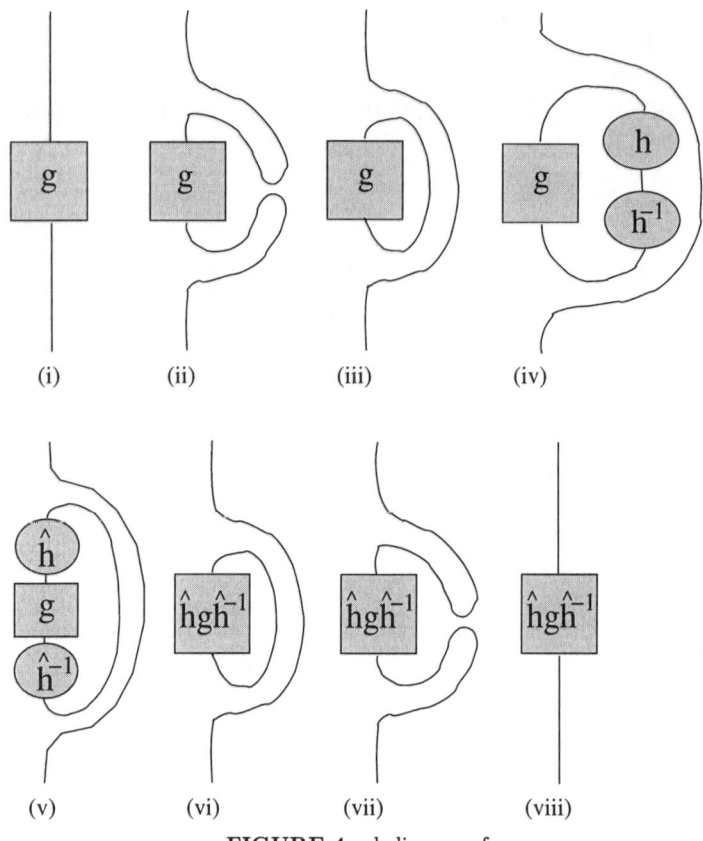

FIGURE 4. abelian proof

loop in opposite directions until they are adjacent to g as shown in (v). In doing so h becomes a new diffeo, $\hat{h}$, an element in the same MCG as g, related to h in the obvious way. Similarly for $\hat{h}^{-1}$. Composing $\hat{h}$, g and $\hat{h}^{-1}$ gives (vi) and the reverse of the first two steps results in (viii). This may be done for any h and so (i) is congruent to (viii) for any $\hat{h}$ in the MCG.

This means that if we are able, in some effective sense, to implement the Laidlaw-Morette-DeWitt type of scheme, the weight of manifold (i) would have to be the same as that of manifold (viii). Thus, the weights can only depend on the conjugacy class of the elements of the mapping class group and so can only carry abelian UIR's.

It might be seen as an advantage that only abelian UIR's need be considered – the representation theory is radically simplified in this case and we feel that we understand how to attach mere phases as weights to the manifolds in the SOH.

But there is a potential, serious drawback. All spinorial geons known to us are non-abelian. That is, we do not know of a prime whose 2π rotation is non-trivial in an abelian UIR of the MCG. If no such primes do in fact exist, the restriction to abelian UIR's would eliminate spinorial and fermionic geons altogether and this seems much to high a price to pay for a spin-statistics theorem.

IV MOTIVATION

Why should we care whether or not topological geons obey a spin-statistics theorem? We can hope that by formulating rules that would give us a spin-statistics theorem we may glean clues about the nature of the underlying, more fundamental theory of quantum gravity which should eventually give rise to such rules. It is clear that the spin-statistics question is intimately connected to the question of the weights attachable to the different manifolds in the SOH, whose arbitariness in the absence of something like these ideas seems a problem. Geons contain the potential to be the mechanism of the ultimate unification between spacetime and matter: perhaps spacetime is all there is. Bosonic fields could arise from Kaluza-Klein reductions and fermions could be quantum geons. Indeed Friedman and Higuchi showed that in pure Kaluza-Klein gravity there will be stable geons with the kinematical quantum numbers of the standard model, although obtaining chiral fermions is problematic [23]. (Showing that geons have the standard model masses is a problem of a different order, though one potential difficulty has been slightly alieviated since the discovery of a neutrino mass. There is still an enormous number, $m_{planck}/m_{neutrino}$, to explain but at least it's finite.) As unlikely as this might seem, the payoff, were it to be true, is so high that it is worth bearing the possibility in mind. Then a geon spin-statistics theorem would give us the fundamental explanation of the perfect spin-statistics correlation we see in nature.

V ACKNOWLEDGMENTS

R.D.S would like to thank the Aspen Center for Physics for its hospitality and to acknowledge support from NSF grant PHY-9600620. H.F.D. is supported in part by an EPSRC Advanced Fellowship.

REFERENCES

1. Friedman, J.L., and Sorkin, R.D., *Phys. Rev. Lett.* **44**, 1100 (1980).
2. Friedman, J.L., and Sorkin, R.D., *Gen. Rel. Grav.* **14**, 615 (1982).

3. Sorkin, R.D., "Introduction to topological geons," in *Proceedings of the NATO Advanced Study Institute on Topological Properties and Global Structure of Space-Time, Erice, Italy, May 12-22, 1985*, edited by P.G. Bergmann and V. De Sabbata, Plenum Press, New York, 1986.
4. Sorkin, R.D., "Classical topology and quantum phases: Quantum geons," in *Proceedings, Geometrical and algebraic aspects of nonlinear field theory, Amalfi, Italy, May 1988*, edited by G. Marmo, S. De Filippo, M. Marinaro and G. Vilasi, North Holland, Amsterdam, New York, 1989.
5. Aneziris, C., et al, *Mod. Phys. Lett.* **A4**, 331 (1989).
6. Aneziris, C., et al, *Int. J. Mod. Phys.* **A4**, 5459 (1989).
7. Sorkin, R.D., and Surya, S., *Int. J. Mod. Phys.* **A13**, 3749 (1998).
8. Sorkin, R.D., and Surya, S., "Geon statistics and UIR's of the mapping class group," in *Proceedings: First Latin American Symposium on High Energy Physics and VII Mexican School of Particles and Fields, Mérida, México, Nov. 1996*, edited by Juan Carlos D'Olivo, Martin Klein-Kreisly, Héctor Méndez, AIP Conference Proceedings 400, New York, 1997.
9. Balachandran, A.P., et al, *Mod. Phys. Lett.* **A5**, 1575 1586 (1990).
10. Balachandran, A.P., et al, *Int. J. Mod. Phys.* **A8**, 2993–3044 (1993).
11. Finkelstein, D., and Rubinstein, J., *J. Math. Phys.* **9**, 1762 (1968).
12. Dowker, H.F., and Sorkin, R.D., *Class. Quant. Grav.* **15**, 1153 (1998).
13. Sorkin, R.D., *Int. J. Theor. Phys.* **33**, 523–534 (1994).
14. Friedman, J.L., "Space-time topology and quantum gravity," in *Conceptual problems of quantum gravity: Proceedings of the 1988 Osgood Hill Conference, North Andover, Massachusetts, 15-19 May 1988*, edited by A. Ashtekar and J. Stachel, Birkhäuser, Boston, 1991.
15. Giulini, D., *Int. J. Theor. Phys.* **33**, 913–930 (1994).
16. Giulini, D., *Helv. Phys. Acta* **68**, 86–111 (1995).
17. Giulini, D., *Banach Center Publ.* **39**, 303–315 (1997).
18. Sorkin, R.D., *Commun. Math. Phys.* **115**, 421 (1988).
19. Bialynicki-Birula, I., "Transition amplitudes versus transition probabilities and a reduplication of space-time" in *Quantum Concepts in Space and Time*, edited by R. Penrose and C.J. Isham, Clarendon Press, Oxford, 1986.
20. Louko, J., Daughton, A., and Sorkin, R.D., In *Proceedings of the 5th Canadian Conference on General Relativity and Relativistic Astrophysics*, edited by R.B. Mann and R.G. McLenaghan, World Scientific, Singapore, 1994.
21. Sorkin, R.D., "Quantum measure theory and its interpretation," in *Quantum Classical Correspondencer: Proceedings of the 4th. Drexel Symposium on Quantum Non-integrability, Philadelphia, USA, September 8-11, 1994*, edited by D.H. Feng and B-L Hu, International Press, Cambridge Mass., 1997.
22. Laidlaw, M.G.G., and Morette-DeWitt, C., *Phys. Rev.* **D3**, 1375–1378 (1971).
23. Friedman, J.L., and Higuchi, A., *Nucl. Phys.* **B339**, 491 (1990).

Quantum Field Theory on Noncommutative Space-Time and its Implication on Spin-Statistics Theorem

M. Chaichian[a], A. Demichev[a,b] and P. Prešnajder[a,c]

[a] High Energy Physics Division, Department of Physics,
University of Helsinki
and
Helsinki Institute of Physics,
P.O. Box 9, FIN-00014 Helsinki, Finland
[b] Nuclear Physics Institute, Moscow State University, 119899, Moscow, Russia
[c] Department of Theoretical Physics, Comenius University, Mlynská dolina, SK-84215 Bratislava, Slovakia

Abstract. We study properties of a scalar quantum field theory on noncommutative space-times. We show that field theories on a noncommutative plane have ultraviolet divergences, while the theory on a noncommutative cylinder is ultraviolet finite. Thus, ultraviolet behaviour of a field theory on noncommutative spaces is sensitive to the topology of the space-time, namely to its compactness. We present general arguments for the case of higher space-time dimensions and as well discuss the implication of the noncommutativity on the spin-statistics theorem.

I INTRODUCTION

The standard concept of a geometric space is based on the notion of a manifold $\mathcal{M}$ whose points $x \in \mathcal{M}$ are locally labelled by a finite number of real coordinates $x^\mu \in \mathbb{R}^4$. However, it is generally believed that the picture of space-time as a manifold $\mathcal{M}$ should break down at very short distances of the order of the Planck length $\lambda_P \approx 1.6 \times 10^{-33}\,cm$. This implies that the mathematical concepts for high energy (small distance) physics have to be changed, or more precisely, our classical geometrical concepts may not be well suited for the description of physical phenomena at small distances. No convincing alternative description of physics at very short distances is known, though different routes to progress have been proposed. One such direction is to try to formulate physics on some noncommutative space-time [1–3].

If the notions of the noncommutative geometry are used directly for the description of the space-time, the notion of points as elementary geometrical entity is lost and one may expect that an ultraviolet (UV) cutoff appears. The simplest model of this kind is the fuzzy sphere [4], *i.e.* the noncommutative analog of a two-dimensional sphere. As is well known from the standard quantum mechanics, a quantization of any *compact* space, in particular a sphere, leads to finite-dimensional representations of the corresponding operators, so that in this case any calculation is reduced to manipulations with finite-dimensional matrices and thus there is simply no place for UV-divergences [5]. Things are not so easy in the case of non-compact manifolds. The quantization leads to infinite-dimensional representations and we have no guarantee that noncommutativity of the space-time coordinates removes the UV-divergences. In this work we show that field theories on a noncommutative plane with the most natural Heisenberg-like commutation relations among coordinates or even on a noncommutative quantum plane with $E_q(2)$-symmetry have ultraviolet divergences, while the theory on a noncommutative cylinder is ultraviolet finite. Thus, ultraviolet behaviour of a field theory on noncommutative spaces is sensitive to the topology of the space-time, namely to its compactness.

II TWO-DIMENSIONAL QUANTUM FIELD THEORY ON NONCOMMUTATIVE SPACE-TIME WITH HEISENBERG-LIKE COMMUTATION RELATIONS

The noncommutative version $P_\lambda^{(2)}$ of a two-dimensional plane is obtained via replacing the commuting coordinates by the hermitian operators $\hat{x}_i$, $(i,j = 1,2)$ satisfying the commutation relations

$$[\hat{x}_i, \hat{x}_j] = i\lambda^2 \varepsilon_{ij}, \qquad i,j = 1,2, \qquad (1)$$

where λ is a positive constant of the dimension of *length*. The noncommutative analogs of field derivatives $\partial_i \hat{\varphi}(\hat{x}_i)$, $i = 1, 2$ are defined as

$$\partial_i \hat{\varphi}(\hat{x}_i) = \varepsilon_{ij} \frac{i}{\lambda^2}[\hat{x}_j, \hat{\varphi}(\hat{x}_i)], \qquad i = 1,2. \qquad (2)$$

They satisfy the Leibniz rule and reduce to the usual derivatives in the commutative limit.

The free part of the field action has the form

$$S_0^{(\lambda)}[\hat{\varphi}, \hat{\varphi}^\dagger] = \text{Tr}\left[(\partial_i \hat{\varphi})^\dagger (\partial_i \hat{\varphi}) + m^2 \hat{\varphi}^\dagger \hat{\varphi}\right], \qquad (3)$$

where the operator trace substitutes an integral used in the case of commutative geometry. The calculation of Green functions and other quantities in NC-QFT can be carried out in simple and natural way by the use of operator symbols. An operator symbol, as it is defined in the ordinary Quantum Mechanics, is a function on the phase space which is constructed by a definite rule from a given operator. Different rules for the construction produce different symbols (*e.g.*, Weyl, normal symbols *etc.*). Sets of such functions endowed with a so-called star-product form algebras which are isomorphic to the initial operator algebra. Knowledge of explicit forms of star-products ($\star$-products) allows to make calculations in an easy and short way.

The Weyl symbol ψ_W has some special properties which makes it convenient for the calculations. In particular, the explicit form of the $\star$-product which makes the algebra of Weyl symbols isomorphic to the operator algebra is defined by the expression

$$(\varphi_W \star \psi_W)(x) = \varphi_W(x) \exp\left\{i\frac{\lambda^2}{2} \overleftarrow{\partial}_i \, \varepsilon^{ij} \, \overrightarrow{\partial}_j\right\} \psi_W(x) \qquad (4)$$

so that

$$\int d^2x \, (\varphi_W \star \psi_W)(x) = \int d^2x \varphi_W(x)\psi_W(x) \ , \qquad (5)$$

because ε^{ij} is antisymmetric. Therefore, the free action of the NC-QFT in terms of the Weyl symbols has the same form as usual QFT on commutative space. Higher order (interaction) terms contain non-locality, but analysis shows (see also [6]) that they do not remove UV-divergences.

If the interaction is switched on, there naturally appears the problem of a perturbative determination of the full Green function $G_\lambda = \langle \varphi_N(x)\varphi_N(y) \rangle$. Within the perturbation theory the problem is reduced to the calculations of free field averages of the type $\langle \varphi_N(x)\varphi_N(y)S^n_{int} \rangle_0$. However, now the problem of a noncommutative generalization of the interaction term arises. If we choose as a commutative prototype the $(\varphi^*\varphi)^2$-interaction, the most direct noncommutative generalization is

$$S^\lambda_{int}[\hat{\varphi},\hat{\varphi}^\dagger] = g\text{Tr}\,[\hat{\varphi}^\dagger\hat{\varphi}\hat{\varphi}^\dagger\hat{\varphi}] = g\int d^2x\, \varphi^*_N(x) \star \varphi_N(x) \star \varphi^*_N(x) \star \varphi_N(x) \ . \qquad (6)$$

This action produces vertices containing factors $e^{\lambda^2 k^2/2}$ on each leg with the momentum k_i, $i=1,2,3,4$, plus additional phase factors $\exp\{\pm i\lambda^2(k_1 \times k_2 + k_3 \times k_4)/2\}$ (here $k \times p \stackrel{\text{def}}{=} \varepsilon_{ij}k_ip_j$). The Gaussian factor $e^{-\lambda^2 k^2/2}$ from the propagators are cancelled in Feynman diagrams and the UV-divergences appear. However, this is not the only possibility. Insisting only on a commutative limit condition $\lim_{\lambda \to 0} S^\lambda_{int}[\hat{\varphi},\hat{\varphi}^\dagger] = S_{int}[\varphi,\varphi^*]$, the integrand in the noncommutative integral $I[\hat{\varphi}^\dagger\hat{\varphi}\hat{\varphi}^\dagger\hat{\varphi}]$ is defined up to the *operator ordering*. There is no problem to modify the operator ordering of the generators $\hat{x}_1$ and $\hat{x}_2$ in the integrand $\hat{\varphi}^\dagger\hat{\varphi}\hat{\varphi}^\dagger\hat{\varphi}$ in such a way that the vertices will not contain the exponential factors $\exp\{\lambda^2 k_i^2/2\}$ on legs. For example, one can use the normal symbols for the construction of the free action but the Weyl symbols for the interaction part. The resulting action will lead to UV-regular Feynman diagrams. However, besides this pragmatic point of view, we have not been able to find any deeper principle preferring such different ordering.

III QUANTUM FIELD THEORY ON A NONCOMMUTATIVE CYLINDER

A standard commutative cylinder with radius ρ can be identified with the set of points

$$C_\rho = \{(x,t),\ t \in \mathbb{R},\ x = \rho e^{i\phi},\ \rho = \text{const}\}\ . \tag{7}$$

We shall interpret C_ρ as a space-time manifold. It is convenient to introduce $x_+ = x$, $x_- = x^*$.

The starting point for development of a field theory on a non-commutative cylinder is the replacement of commutative variables $x_0, x_\pm$ by the noncommutative ones $\hat{x}_0, \hat{x}_\pm$, satisfying the $e(2)$ Lie algebra commutation relations

$$[\hat{x}_0, \hat{x}_\pm] = \pm \lambda \hat{x}_\pm\ , \qquad [\hat{x}_+, \hat{x}_-] = 0\ . \tag{8}$$

The field $\hat{\varphi} = \varphi(\hat{x}_0, \hat{x}_\pm)$ is an operator in $\mathcal{L}^2(S^1, d\phi/2\pi)$ and we suppose that it possesses the expansion

$$\hat{\varphi} = a_0(\hat{t}) + \sum_{k=1}^{\infty} [a_k^{(+)}(\hat{t})\hat{x}_+^k + a_k^{(-)}(\hat{t})\hat{x}_-^k]\ , \tag{9}$$

The free field action has the form similar to that in the commutative case:

$$S_0[\hat{\varphi}, \hat{\varphi}^*] = \lambda \text{Tr}\left[-\varphi^*(\Box_\lambda + m^2)\varphi\right]\ , \tag{10}$$

where the noncommutative analog $\Box_\lambda$ of the operator $\Box$ is defined as

$$\Box_\lambda \hat{\varphi} = -\frac{1}{\lambda^2}[x_0, [x_0, \varphi]] + \frac{1}{\lambda^2 \rho^2}[x_+, [x_-, \varphi]]\ . \tag{11}$$

In the noncommutative case the interaction term we take in the form $V = \frac{g}{4} : (\varphi^* \varphi)^2 :$, where the ordering $: \cdots :$ means that all expansion coefficients of fields $a_k(\hat{t}), a_k^\dagger(\hat{t})$ are collected as a right (or equivalently left) factor. The tadpole contribution is *finite* [7]:

$$G_0(x,t;x,t) = \int_0^{\pi/\lambda} d\omega\, \frac{\cot\left(\sqrt{\Omega_\lambda^2(\omega) - m^2 + i\varepsilon}\right)}{\sqrt{\Omega_\lambda^2(\omega) - m^2 + i\varepsilon}} < \infty\ . \tag{12}$$

In the two-dimensional scalar field theory on commutative space the tadpole is the only divergent contribution to Feynman diagrams. We have shown that transition to the noncommutative cylinder leads to *UV-finite field theory*.

IV QUANTUM FIELD THEORY ON A NONCOMMUTATIVE QUANTUM PLANE WITH $E_Q(2)$-SYMMETRY

In this section we consider Quantum Mechanics on a plane $P_q^{(2)}$ induced by the quantum group $E_q(2)$ [8] (see also [9] and refs therein):

$$\bar{v}v = v\bar{v} = 1, \qquad t\bar{t} = q^2 \bar{t}t,$$
$$vt = q^2 tv, \qquad \bar{v}t = q^{-2} t\bar{v} \qquad q \in \mathbb{R}. \tag{13}$$

Other commutation relations follow from the involution: $v^\dagger = \bar{v}$, $t^\dagger = \bar{t}$.

The elements $\bar{z}$, z and $\bar{\partial}_q$, ∂_q defines the q-deformed algebra of functions on $P_q^{(2)}$ together with the q-deformed left-invariant vector fields (derivatives). Its defining relations read

$$z\bar{z} = q^2 \bar{z}z, \qquad \partial_q \bar{\partial}_q = q^2 \bar{\partial}_q \partial_q$$
$$\partial_q z = 1 + q^{-2} z \partial_q, \qquad \bar{\partial}_q \bar{z} = 1 + q^2 \bar{z} \bar{\partial}_q, \tag{14}$$
$$\bar{\partial}_q z = q^2 z \bar{\partial}_q, \qquad \partial_q \bar{z} = q^{-2} \bar{z} \partial_q.$$

In the case $q > 1$, the tadpole contribution proves to be proportional to the sum [10]

$$\sum_{N=-\infty}^{\infty} \frac{q^2 - 1}{q^4 \sqrt{[(q^N - 1)^2 + (q^2 - 1)^2 \sigma^2/q^4][(q^N + 1)^2 + (q^2 - 1)^2 \sigma^2/q^4]}}. \tag{15}$$

Since the terms in this series tends at $N \to -\infty$ to a constant which is independent of N, the series has the *linear* divergence at the lower limit (the IR-regularization parameter σ again, as in the nondeformed case, is inessential in the $N \to \pm\infty$ limit). Notice that in the nondeformed limit $q \to 1$ the series (15) coincides with the nondeformed result (logarithmic divergence). Quite similar calculation in the case $q < 1$ shows that the tadpole diagram has the linear divergence at the upper limit.

Thus the perturbation theory for the φ^4-model on the noncommutative plane $P_q^{(2)}$ with the action constructed with the help of the $E_q(2)$–quantum group invariant measure contains the UV-divergences and, hence, can not be considered as a regularization of the usual scalar field theory on the commutative plane.

V CONCLUSION

We have shown that transition to a noncommutative space-time does not necessarily lead to an ultraviolet regularization of the quantum field theory constructed in this space, at least in the most natural way of introducing noncommutativity as we have performed in this paper. However, in general, theories which have the same UV-behaviour on classical spaces may acquire essentially different properties after the quantization. The reason is that quantization procedure is highly sensitive to the topology of the manifold under consideration. Thus, while in the case of classical space-time the theories on a sphere, cylinder or plane have UV-divergences, in the case of noncommutative space-time the two-dimensional theories on the fuzzy sphere and on the quantum cylinder do not have divergences at all. This can be traced to the compactness properties of the space-time in question:

It is of great interest to investigate the problem further in order to find out whether there exists any other, than the one presented in this paper, way of introducing noncommutativity so that it would remove the UV-divergences even in the case of fully noncompact space-times.

Another interesting area of investigations in noncommutative field theories concerns the problems of causality and the relation between spin and statistics. Since observables on noncommutative space-times are constructed by using the star-product, they prove to be nonlocal objects and the standard argumentation about the spin-statistics relation (see, e.g., [11]) should be reconsidered. The preliminary results show [12] that modifications of the usual commutation relations for creation and annihilation operators, which are no more excluded, alter the causality properties of the field theories on noncommutative space-times.

Acknowledgements The financial support of the Academy of Finland under the Projects No. 163394 is greatly acknowledged. A.D.'s work was partially supported also by RFBR-00-02-17679 grant and P.P.'s work by VEGA project 1/4305/97

REFERENCES

1. Connes A., *Noncommutative Geometry*, Academic Press, New York, 1994.
2. Doplicher S., Fredenhagen K. and Roberts J. E., *Phys. Lett.* **B331**, 39 (1994);
   Doplicher S., Fredenhagen K. and Roberts J. E., *Comm. Math. Phys.* **172**, 187 (1995).
3. Seiberg N. and Witten E., **JHEP** 9909: 032 (1999).
4. Hoppe J., *Elem. Part. Res. J.* **80**, 145 (1989);
   Madore J., *Journ. Math. Phys.* **32**, 332 (1991).
5. Grosse H., Klimčik C. and Prešnajder P., *Commun. Math. Phys.* **185**, 155 (1997).
6. Filk T., *Phys. Lett.* **B376**, 53 (1996).
7. Chaichian M., Demichev A. and Prešnajder P., *Nucl. Phys.* **B567**, 360 (2000).
8. Vaksman L.L. and Korogodsky L.I., *Dokl. Akad. Nauk SSSR* **304**, 1036 (1989).
9. Chaichian M. and Demichev A., *Introduction to Quantum Groups* (World Scientific, Singapore, 1996).
10. Chaichian M., Demichev A. and Prešnajder P., *J. Math. Phys.* **41**, 1647 (2000).
11. Streater R.F. and Wightman A.S., *PCT, Spin and Statistics, and All That*, Benjamin, New York, 1964.
12. Chaichian M., Demichev A., Prešnajder P. and Tureanu A. (in preparation).

V. Experimental Tests of the Spin-Statistics Connection and Symmetrization Postulate

How We Know That Photons Are Bosons: Experimental Tests of Spin-Statistics for Photons

D. DeMille,[1] D. Budker,[2] N. Derr,[3*] and E. Deveney[3†]

[1] *Physics Department, Yale University, New Haven, CT 06520 USA; david.demille@yale.edu*
[2] *Physics Department, Univ. of California, Berkeley, CA 94720 USA; budker@socrates.berkeley.edu*
[3] *Physics Department, Amherst College, Amherst, MA 01002 USA*
[*] *Present Address: MIT Lincoln Laboratory, Lincoln, MA 02420 USA; ndderr@ll.mit.edu*
[†] *Present Address: Bridgewater State College, Bridgewater, MA 02325 USA; edeveney@bridgew.edu*

Abstract. We discuss theoretical models and experimental data that shed light on possible small violations of the spin-statistics relation for photons. Particular emphasis is given to our recent experimental search for non-symmetric photon states, using atomic two-photon transitions.

I. INTRODUCTION

Photons are interesting particles with which to search for small violations of the usual spin-statistics relation, for a number of reasons. The photon is of course the only *fundamental* boson that can be readily produced in states with many identical particles present. In addition, some explicit theoretical models for a spin-statistics violation have been formulated for photons, but have been absent for other particles.[*]

With this in mind, we review here the experimental data that place limits on possible small violations of the spin-statistics relation for photons. We consider three experimental signatures of such a violation: deviations from the Planck blackbody spectrum; non-standard behavior at large single-mode occupation number; and the existence of non-permutation-symmetric states. We conclude with a description of our recent experiment to search for non-symmetric states using two-photon transitions in atoms. Before moving to the experiments, however, we begin with a brief discussion of the available models for interpreting each of these types of data.

Theoretical Models for Spin-Statistics Violation for Photons

We know of two theoretical models that have been used to describe possible spin-statistics violations for photons; both are based on deformed commutation relations. The first of these is the "quon algebra,"[1] where

---

[*] Only recently has a description been formulated, using the quon algebra, of experiments searching for permutation symmetry-forbidden states in molecules and in atomic He. See the contribution by R. Hilborn in these proceedings.

$$a_k a_l^\dagger - q a_l^\dagger a_k = \delta_{kl}, \qquad (1)$$

with q real and $|q| \leq 1$. It has been shown that photons with deformation parameter $q < 1$ can exist in non-permutation-symmetric states,[2] and that certain microwave cavity states are observably altered.[3] Limits on the output power of lasers (i.e., a soft upper bound on the single-mode occupation number N) were suggested to exist,[4] but this argument was shown to be invalid.[5] No further work on possible non-standard behavior of quons at high N has been reported, and the situation with regards to such behavior is thus unclear. The status of blackbody radiation in the quon theory is similarly unclear. Though the relevant partition function has been given,[6] we have not found any explicit derivation of the blackbody spectrum for quons. Any such distribution may be difficult to interpret, since the statistical mechanics of quons suffers from the Gibbs paradox (for $|q| < 1$), and thus may not be tenable at all.[7]

The behavior of photons which obey the "Q-oscillator" algebra,[8] where

$$a_k a_k^\dagger - Q a_k^\dagger a_k = Q^{-N}, \qquad (2)$$

has also been considered. (We use capital Q to distinguish from the quon algebra.) Several versions of a deformed Planck distribution have been derived,[9] and a model has been developed in which this Q-deformation leads to a blue shift of the mode frequency at high N.[10] In this algebra, there is no specified relation between different modes; thus, this deformation need not lead to the existence of non-symmetric states.

II. PREVIOUS EXPERIMENTAL LIMITS

Deviations from the Planck Blackbody Spectrum

It is reasonable that a deviation from the usual spin-statistics relation for photons could lead to a deformation of the Planck blackbody spectrum, since the derivation of this spectrum explicitly invokes Bose-Einstein (BE) statistics. To our knowledge, no analysis of blackbody spectral data has been performed to search for small, systematic deviations from the Planck formula. However, it appears that the best absolute measurements of blackbody spectra have uncertainties of no better than about 1%.[11]

We note in passing that recent data on the cosmic microwave background (CMB) are stated to have deviations of <50 ppm from the Planck distribution.[12] However, this is based on comparison of the CMB with a reference blackbody (with an assumed Planck distribution), rather than on absolute measurements. Since the reference is held at a temperature very close to that of the CMB, such a comparison is insensitive to any general deformation of the blackbody spectrum. We conclude that deviations from the Planck distribution of $\leq 1\%$ cannot be ruled out by any existing data.

Non-Standard Behavior At Large Occupation Number N

One obvious distinction between bosons and fermions is that bosons can have unlimited occupation number in a single mode, while fermions can have at most one particle in any given state. Thus, it seems sensible that a breakdown of the spin-statistics relation could give some upper limit on the single-mode occupation number N, or at least some unusual behavior in the limit $N \to \infty$. With this in mind, it is interesting to point out what is known about electromagnetic fields with large N.

It is commonly understood that the existence of powerful lasers shows that large occupation numbers are possible. From a pulsed laser with wavelength $\lambda \sim 1$ μm, $N \sim 10^{19}$ has been demonstrated.[13] Even more striking results come from high-power microwave resonance cavities: with $\lambda \sim 10$ cm, $N \gtrsim 10^{23}$ has been achieved.[14] These bounds improve even further at lower frequencies. Indeed, the existence of *any* finite energy stored in a *static* electromagnetic field would indicate $N \to \infty$, since the energy per photon $h\nu \to 0$ in this case. It is thus difficult to imagine any way to reconcile the very existence of static fields with an upper bound on N. In reality, of course, no field can ever be truly static (because of power outages, battery failure, etc.). Nevertheless, the laboratory limits from "nearly-static" fields are very strong: we estimate that $N \gtrsim 10^{40}$ has been achieved without any special effort.[†]

In the Q-oscillator model, a deviation from BE statistics is manifested by a blue-shift of any mode frequency ω, at high values of N. The fractional shift is $\delta\omega/\omega \approx \lambda^2 N^2/2$, where $\lambda = \ln Q$ and $|\lambda| \approx 1-Q \ll 1$.[10] A strict bound has been placed on λ, by comparing the frequency difference between beams of light from two lasers as a function of the power in the beams.[15] No frequency difference was observed at the level $\delta\omega/\omega \approx 10^{-14}$, despite changing the occupation number $N \sim 10^{10}$ by nearly its full value. This null result implies that in the Q-oscillator model, $1-Q \lesssim 10^{-17}$.

Existence of non-permutation-symmetric states

According to the usual spin-statistics relation, any multi-photon wavefunction must be symmetric under interchange of particle labels. Thus, the existence of any non-symmetric state would indicate a violation of BE statistics for photons. How could such states be distinguished from the usual symmetric states? Answers to the analogous question for particles such as electrons and nuclei are familiar. For example, every physicist is taught that the exchange-symmetric $(1s)^2$ $^3S_1$ state of the two electrons in atomic helium does not exist. It is far less familiar that, because photons obey BE statistics, there are "missing" states among the possible wavefunctions of two photons. The result has nevertheless been known for some time, and is referred to in particle physics as the Landau-Yang (LY) theorem.

The LY theorem states that a particle with angular momentum $J = 1$ cannot decay into two photons.[16] A simple proof shows that this result is simply a manifestation of the fact that the two-photon state is symmetric under interchange.[17] The kinematics of the hypothetical two-photon decay of a J=1 (vector) particle, V, are shown in Fig. 1. The decay amplitude $\mathcal{A}$ for this process can be written in terms of the only quantities

---

[†] This estimate is based on a large NMR magnet with magnetic field $B \sim 5$ T in volume $V \sim (10 \text{ cm})^3$, for a time $T \sim 1$ week.

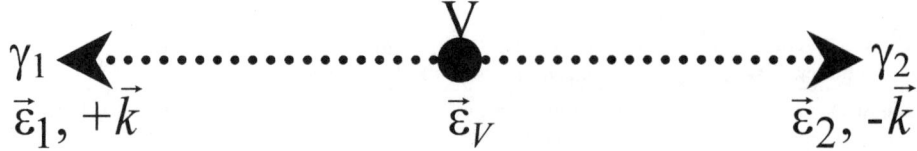

FIGURE 1. Kinematics of the decay V → γγ, in the rest frame of V.

available: the polarizations of the two photons, $\vec{\varepsilon}_{1,2}$; the polarization of V, $\vec{\varepsilon}_V$; and the outgoing photon momentum, $\vec{k} = \vec{k}_1 = -\vec{k}_2$. In quantum field theory the creation and annihilation operators for V and γ carry with them the corresponding polarization vectors. Thus, in order to destroy one V and create two γ's, each of the polarization vectors must appear exactly once in the expression for $\mathcal{A}$. (The momentum vector $\vec{k}$ can appear an arbitrary number of times.) With the requirement of gauge invariance on the photons (i.e., $\vec{\varepsilon}_{1,2} \cdot \vec{k} = 0$), only three forms of $\mathcal{A}$ are possible:

$$\mathcal{A}_a = (\vec{\varepsilon}_1 \times \vec{\varepsilon}_2) \cdot \vec{\varepsilon}_V \, F_a(k^2); \tag{3a}$$

$$\mathcal{A}_b = (\vec{\varepsilon}_1 \cdot \vec{\varepsilon}_2) \cdot (\vec{\varepsilon}_V \cdot \vec{k}) F_b(k^2); \tag{3b}$$

$$\mathcal{A}_c = [(\vec{\varepsilon}_1 \times \vec{\varepsilon}_2) \cdot \vec{k}] (\vec{\varepsilon}_V \cdot \vec{k}) F_c(k^2). \tag{3c}$$

Interchange of the photons leads to the transformations $\vec{\varepsilon}_1 \leftrightarrow \vec{\varepsilon}_2$ and $\vec{k} \leftrightarrow -\vec{k}$. It can be easily verified that each of the forms written in Eqns. (3) is *odd* under interchange. Therefore, these amplitudes vanish exactly because photons obey BE statistics.

It is useful to look at the LY theorem from a different point of view. Suppose we try to calculate the decay rate of a particle into two photons, using Feynman diagrams. The total decay amplitude will be the sum of amplitudes from two diagrams, with photon labels exchanged: $\mathcal{A}_{total} = \mathcal{A}_{12} + \mathcal{A}_{21}$. The LY theorem shows that for the process V → γγ, it must be that $\mathcal{A}_{12} = -\mathcal{A}_{21}$. This leads to a way to search for the existence of exchange-*antisymmetric* two-photon states. For such states, the amplitude $\mathcal{F}$ for decay into "fermionic photons" should be $\mathcal{F} = \mathcal{A}_{12} - \mathcal{A}_{21}$, so that in general $\mathcal{F} \neq 0$. If we assume that there is a BE statistics-violating probability v for any pair of photons to be in an antisymmetric state, then the decay rate Γ_v for the process V → γγ will be $\Gamma_v \propto v|\mathcal{F}|^2$. If $\mathcal{F}$ can be calculated, one can set limits on v from experimental upper limits on the partial width $\Gamma(V \to \gamma\gamma)$.

We have compiled here all experimental limits known to us on the decay rates of vector particles into two photons. For each particle, we have also estimated the spin-statistics violating decay rate $\Gamma_v(v=1)$ and the corresponding branching ratio (B.R.) for V → γγ, in order to set limits on the BE statistics-violation parameter v. The manner of obtaining estimates for Γ_v differs between cases, and bears some explanation.

For $\chi_{c1} = c\bar{c}$ 1p $^3P_{1(+)}$, we estimate that the partial width for decay to two photons is $\Gamma_v(\chi_{c1} \to \gamma\gamma) \sim v \cdot 1$ keV. This value is obtained by assuming that $\Gamma_v(v=1)$ will lie

TABLE 1. Limits on two-photon antisymmetric states from particle physics.

Particle, State, $^{2S+1}L_{J(P)}$	Observed B. R. → γγ (ppm)	B.R. → antisymmetric γγ, with ν=1 (ppm)	Limit on ν	References
ortho-Ps = e^+e^- 1s $^3S_{1(-)}$	<200	0?	none?	[19]
J/Ψ = $c\bar{c}$ 1s $^3S_{1(-)}$	<500	0?	none?	[20]
Ψ(2S) = $c\bar{c}$ 2s $^3S_{1(-)}$	<200	0?	none?	[20]
χ_{c1}(1P) = $c\bar{c}$ 1p $^3P_{1(+)}$	<1500	≲1000	ν ≲ 1	[20]
Z^0	<50	≲3	none	[20]

between the measured rates for the allowed decays $\chi_{c0} \to \gamma\gamma$ and $\chi_{c2} \to \gamma\gamma$. The decay $Z^0 \to \gamma\gamma$ could occur only through loop diagrams, since—like all neutral particles—the Z^0 has no direct coupling to photons. Each such diagram is thus suppressed by two electromagnetic vertex factors (i.e., by $\sim q^2\alpha^2$ for a fermion of charge q), even before the cancellation due to Bose statistics. We have estimated $\mathcal{F}$ based on the known couplings of the Z^0 to all fermions (and assuming a typical, additional factor of π for the "geometric" suppression of the loop-diagram amplitude). We stress that in both of these cases, a careful calculation of $\mathcal{F}$ could be of interest. Finally, we set $\mathcal{F} = 0$ for decays of the form $f\bar{f}$ $^3S_{1(-)} \to \gamma\gamma$. Unlike for the previous two cases, these decays are forbidden not only because photons are in symmetric states, but also because charge conjugation (C) symmetry is violated by this decay. Since there is no *a priori* reason to expect these two symmetries to be broken together, these decays could be absent even for $\nu \sim 1$.‡

The data and our conclusions are summarized in Table 1. Remarkably, it appears that little can be learned about ν from the existing data. We note in this context that Ignatiev *et al.* previously discussed the decay $Z^0 \to \gamma\gamma$, as a means for studying violation of BE statistics for photons.[18] They emphasize that one may naively expect that any such violation would be enhanced at high energies. However, they did not attempt to derive a value for the parameter ν as we have done here. Unfortunately, our estimate indicates that the experimental limits on $\Gamma_\nu(Z^0 \to \gamma\gamma)$ must be improved dramatically, in order to place any quantitative limit on ν in the high-energy regime.

III. USING ATOMIC TWO-PHOTON TRANSITIONS TO SEARCH FOR NON-SYMMETRIC PHOTON STATES

We have recently used a generalization of the ideas outlined above, to search for photon states that are not symmetric under interchange. By searching for certain forbidden two-photon transitions in atoms, we have placed an upper limit $\nu < 1.2 \times 10^{-7}$ on the BE statistics violation parameter for visible photons.[21] The

---

‡ Note, however, that the intimate relationship between the spin-statistics connection and the discrete space-time symmetries such as C, make it entirely plausible that a violation of BE statistics could carry with it a violation of C as well. In this case, the limit on ν from , e.g., the ortho-Ps $\to \gamma\gamma$ B.R., could be quite strong.

remainder of this paper will discuss the principle and execution of this experiment, as well as some subtleties in its interpretation.

Review of Two-Photon Transitions in Atomic Physics

We begin with a brief review of the properties of two-photon transitions in atomic physics. Imagine we shine two plane-wave beams of light on an atom, as shown in Fig. 2. The photons in beam 1(2) have polarization $\vec{\varepsilon}_{1(2)}$, momentum $\vec{k}_{1(2)}$, and frequency $\omega_{1(2)}$. Consider a two-photon transition between initial ($|i\rangle$) and final ($|f\rangle$) atomic states, which have well-defined energies $E_{i,f}$, angular momenta $J_{i,f}$, and parities $P_{i,f}$. In general, such a transition could be possible whenever energy conservation is satisfied, i.e. when $\hbar\omega_1 + \hbar\omega_2 = E_f - E_i$. Conceptually, the two-photon transition can be thought of as proceeding stepwise via two single-photon absorptions, with the first photon absorption resulting in a virtual excited atomic state (see Fig. 3). There are two possible paths (i.e., amplitudes) along which the two-photon transition can proceed, corresponding to the order in which the two photons are (virtually) absorbed. If photons obey BE statistics, the amplitudes for these two paths must be *added*. The amplitude for the two-photon process is enhanced when the virtual intermediate state is near in energy to a real atomic state ($|n\rangle$). This means that if the two absorbed photons have different frequencies (i.e., $\omega_1 \neq \omega_2$), then the amplitudes for the two excitation paths will in general be different.

We will find the series expansion of the plane wave of each light beam useful:

$$e^{i\vec{k}\cdot\vec{r}} = 1 + i(\vec{k}\cdot\vec{r}) - (\vec{k}\cdot\vec{r})^2/2 + ... \qquad (4)$$

This expansion is sensible because $(\vec{k}\cdot\vec{r})$ is small for typical atomic transitions, where $\lambda = 2\pi/k$ is in the visible and $r \sim a_0$ (the Bohr radius): $(\vec{k}\cdot\vec{r}) \sim 2\pi a_0/\lambda \lesssim \alpha/2 \ll 1$. Atomic transitions are usually classified according to their multipolarity, i.e., according to the leading term in Eqn. (4) from which they arise. This expansion can be used to clarify the connection between selection rules for atomic transitions and the quantum numbers of the photon participating in the transition. This connection is not usually emphasized in atomic physics, so we re-derive it here.[22]

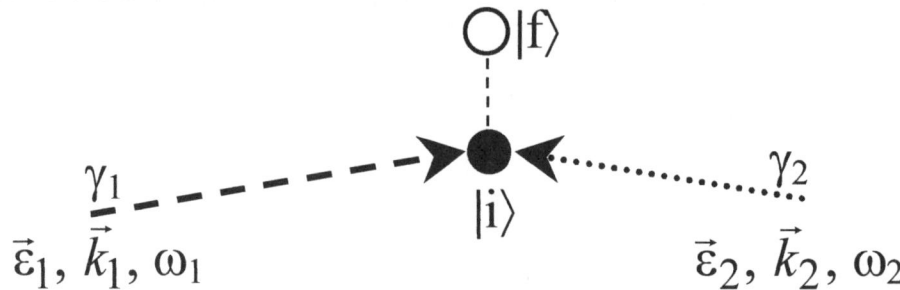

FIGURE 2. Kinematics of an atomic two-photon transition.

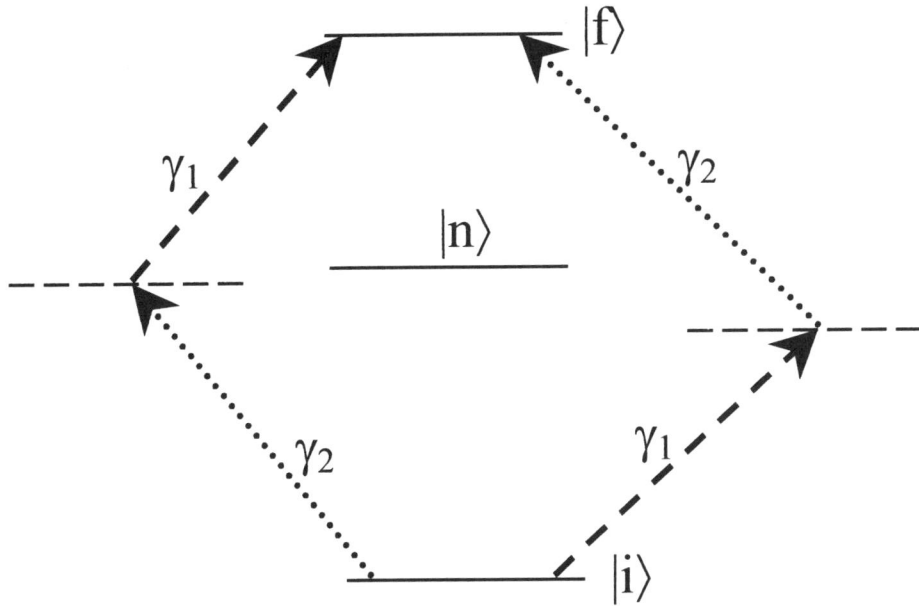

FIGURE 3. Energy level diagram for an atomic two-photon transition. The solid horizontal lines indicate real atomic states; the dashed horizontal lines are virtual intermediate states.

The term with power $|\vec{k}|^n$ in Eqn. (4) arises only from photons with orbital angular momentum $l = n$, $n-2$, $n-4$, etc. [This can be seen easily from the fact that the l-wave radial wavefunctions for free particles are the spherical Bessel functions of order l: $\psi_l \propto j_l(kr)$]. Since photons have spin $s = 1$, the total angular momentum of the photon can only be $j = l\pm 1$ or l. By convention, magnetic multipoles have $l = j$; electric multipoles have $l = j\pm 1$. Photons have odd intrinsic parity, so the total parity is $P_\gamma = -(-1)^l$. As summarized in Table 2, these properties correspond exactly to the standard atomic transition selection rules.

TABLE 2. Photon Quantum Numbers and Atomic Selection Rules.

Multipole Term	E1	M1; E2	M2; E3
Power of k in Eqn. (4)	0, 2, ...	1, 3, ...	2, 4, ...
Relative Amplitude	1	$ka_0 \sim \alpha/2$	$(ka_0)^2/2 \sim \alpha^2/8$
photon l	0, 2	1; 1, 3	2; 2, 4
photon s	1	1	1
photon j	1	1; 2	2; 3
atom ΔJ	± 1, 0 ($0 \nrightarrow 0$)	± 1, 0 ($0 \nrightarrow 0$); ± 2, ± 1, 0 ($0 \nrightarrow 0, 1$)	± 2, ± 1, 0 ($0 \nrightarrow 0, 1$); ± 3, ± 2, ± 1, 0 (etc.)
photon $P = -(-1)^l$	-1	$+1$	-1
atom ΔP	-1	$+1$	-1

Analogue of the Landau-Yang Theorem in Atomic Physics

It seems reasonable that atomic transitions of the form J=0 + 2γ → J=1 might be forbidden, as were particle decays of the form J=1 → 2γ. However, the analogy is not completely general. The kinematics of the transition have been complicated by the atomic recoil, which in general allows absorption of two photons with $\vec{k}_1 \neq -\vec{k}_2$; this in turn invalidates the earlier proof. In order to recover the LY result, let us for now confine ourselves to the case where the two beams of light are both *collinear* (so that $\vec{k}_1 \| \vec{k}_2$) and *degenerate* (so that $\omega_1 = \omega_2$). This situation is almost identical to that in the earlier proof. The only difference is that there is now a possibility for the atom to absorb two photons from the *same* beam, so that $\vec{k} = \vec{k}_1 = \vec{k}_2$ (rather than $\vec{k}_1 = -\vec{k}_2$ as was automatically satisfied in the particle decay). Indeed, this seems to invalidate our earlier proof: for such co-propagating photons, the amplitude of Eqn. (3b) need not vanish. However, even this amplitude can be eliminated if we add one additional condition. Under a parity (P) transformation, Eqn. (3b) is odd, while (3a) and (3c) are even. Thus, the amplitude of Eqn. (3b) vanishes for transitions between atomic states of the same total parity. Therefore, our earlier proof can still be applied, to yield the following generalization of the LY theorem: *atomic transitions of the form J=0 + 2γ → J=1 (with ΔP = 0) are forbidden for degenerate, collinear photons.*

This result suggests that we can search for evidence that photons are in a non-permutation-symmetric state, by looking for these forbidden atomic transitions. Indeed, this is the principle of our experiment, which is described below.

Transition Amplitudes, Selection Rules, and Bose-Einstein Statistics

The hierarchy of atomic transition multipoles plays an important role in understanding the J=0 + 2γ → J=1 transitions we are considering. Under ordinary conditions, E1 amplitudes are dominant, so we will focus exclusively on these from here on. We note in passing that higher multipoles in principle can lead to very small rates for these transitions, which might be significant in a sensitive search for transitions induced by violation of BE statistics. However, the effect of the higher multipoles can be made extremely small by good experimental design.[§] Likewise, atomic parity violation gives nonzero, but negligibly small, transition rates.[**]

From the point of view of selection rules for atomic transitions, the LY result looks rather surprising. Transitions of the form J=0^P + 2γ → J=1^P (ΔP=0) obey all selection rules for a two-step E1-E1 process: the first step is J=0^P + γ → J=$1^{(-P)}$, and the second is J=$1^{(-P)}$ + γ → J=1^P. Let us look in detail at the amplitude for such a process. The general expression for the atomic E1-E1 resonant transition rate W_{if} is:[23]

[§] Such terms can enter because of imperfect collinearity of the light beams. However, the largest components involve two M1/E2 type one-photon steps, and thus have amplitude in atomic units of $\leq \theta\alpha^2$, where θ is a small misalignment angle.
[**] Parity violation induces a tiny E1-E2 or E1-M1 component to the transitions of interest.

$$W_{if}(\Omega_1,\Omega_2) \propto |\varepsilon_{1a}\varepsilon_{2b}\langle f|Q_{ab}|i\rangle|^2 \frac{dI_1}{d\Omega_1}\frac{dI_2}{d\Omega_2}\delta(\omega_{if}-\Omega_1-\Omega_2); \quad (5)$$

$$Q_{ab}(\Omega_1,\Omega_2) = d_a\left(\sum_n \frac{|n\rangle\langle n|}{\omega_{ni}-\Omega_1}\right)d_b + d_b\left(\sum_n \frac{|n\rangle\langle n|}{\omega_{ni}-\Omega_2}\right)d_a. \quad (6)$$

Here $dI_j/d\Omega_j$ is the spectral distribution of light intensity for light beam j; ω_{jk} is the frequency of the atomic transition $|j\rangle \to |k\rangle$; $\vec{d} = e\vec{r}$ is the atomic dipole operator; and the subscripts a, b refer to Cartesian components. Eqns. (5) and (6) exhibit many of the features we have previously discussed qualitatively. The two terms in Eqn. (6) correspond to the two alternative absorption paths in Fig. 3, i.e., to the direct and exchange Feynman diagrams in a two-photon decay. Because photons obey BE statistics, the amplitude $\varepsilon_{1a}\varepsilon_{2b}\langle f|Q_{ab}|i\rangle$ must be symmetric under interchange of the photon labels (1↔2); this is the origin of the plus sign between the terms in Eqn. (6).

For the specific case of a J=0 → J=1 transition, only the irreducible rank-1 component of Q_{ab} can contribute to the matrix element in Eqn. (5). Thus, Q_{ab} reduces to its antisymmetric part, $Q_{ab}^{(1)}$:[24]

$$Q_{ab}(\Omega_1,\Omega_2) = Q_{ab}^{(1)} = \frac{Q_{ab}-Q_{ba}}{2} = \frac{(\Omega_1-\Omega_2)}{2}\sum_n \frac{d_a|n\rangle\langle n|d_b - d_b|n\rangle\langle n|d_a}{(\omega_{ni}-\Omega_1)(\omega_{ni}-\Omega_2)}. \quad (7)$$

Eqn. (7) also illustrates several of our earlier points. The transition amplitude $\varepsilon_{1a}\varepsilon_{2b}\langle f|Q_{ab}|i\rangle$ is antisymmetric under interchange of the polarization vectors, and in fact takes the form of the rotational invariant in Eqn. (3a): $(\vec{\varepsilon}_1 \times \vec{\varepsilon}_2)\cdot\vec{\varepsilon}_\gamma$. (That is, this amplitude requires the two light beams to have different polarizations.) The overall exchange symmetry of the amplitude is maintained because of the additional antisymmetric term $(\Omega_1-\Omega_2)$ that appears in Q_{ab}. This term makes it explicit that the $J=0^P + 2\gamma \to J=1^P$ transitions are completely forbidden if (and only if) the two photons are degenerate (with $\Omega_1 = \Omega_2$).

The vanishing of the transition amplitude, specifically in the degenerate case, leads to another way of understanding the role played by BE statistics for photons in these transitions. Consider the wavefunction of the two-photon state. For an E1-E1 transition, we take each photon to have $l = 0$. (This is valid up to a correction of order $(\vec{k}\cdot\vec{r})^2$; see Table 2.) Then the entire angular momentum of the photons arises from spin, and the transition J=0 + 2γ → J=1 requires $S_{tot} = 1$ for the two-photon system. Since each photon has $s = 1$, the $S_{tot} = 1$ combination is antisymmetric under interchange (as can be seen e.g. from the Clebsch-Gordon coefficients for $1\otimes 1 \to 1$). Thus, in order to satisfy BE statistics, the spatial part of the two-photon wavefunction must also be antisymmetric. The spatial part of each photon's wavefunction can be described by an s-wave: $\psi_j(\vec{r}) \propto j_0(k_j r)$. Of course, then, the antisymmetrized two-photon wavefunction will vanish if and only if $|k_1| = |k_2|$ (i.e., $\Omega_1 = \Omega_2$).

It may be instructive to consider the meaning of the factor $(\Omega_1 - \Omega_2)$ that appears in Q_{ab}. As is suggested in Fig. 3, the presence of the atomic intermediate state $|n\rangle$ allows the atom to *distinguish* between two photons with different energies. We mean this in the same sense that one can completely distinguish between an electron in Anacapri and one in New Haven, because the overlap between their wavefunctions is negligible.[25] In our case, the relevant wavefunction overlap is best considered in momentum space. Here, the atom evidently presents an effective potential function for the photons [see Eqn. (6)]. Only when the two photons are completely indistinguishable (i.e., when they are degenerate) is there perfect cancellation of the spatial-momentum part of the wavefunction, due to its antisymmetrization.

Searching for Spin-Statistics Violation with J=0 → J=1 Transitions

We have shown that for exactly degenerate photons, the usual transition amplitude Q_{ab} vanishes. However, the analogous amplitude for non-symmetric photons should *not* vanish. We construct this amplitude $\mathcal{F}_{ab}$, as before, by replacing the plus sign in Eqn. (6) with a minus sign; we then obtain

$$\mathcal{F}_{ab}(\Omega_1, \Omega_2) = \sum_n \left(\omega_{ni} - \frac{\Omega_1 + \Omega_2}{2} \right) \frac{d_a |n\rangle\langle n| d_b - d_b |n\rangle\langle n| d_a}{(\omega_{ni} - \Omega_1)(\omega_{ni} - \Omega_2)}, \qquad (8)$$

which is in general non-zero for $\Omega_1 = \Omega_2$. [A similar result is obtained in the quon algebra, with the amplitude $\mathcal{F}_{ab} \propto (1-q)$].[2] Returning to the general case where $\Omega_1 = \Omega_2$ is not required, we take the transition rate to be the sum of two probabilities:

$$W_{if}(\Omega_1, \Omega_2) \propto \left\{ |\varepsilon_{1a}\varepsilon_{2b}\langle f|Q_{ab}(\Omega_1,\Omega_2)|i\rangle|^2 + v|\varepsilon_{1a}\varepsilon_{2b}\langle f|\mathcal{F}_{ab}(\Omega_1,\Omega_2)|i\rangle|^2 \right\} \\ \times \frac{dI_1}{d\Omega_1} \frac{dI_2}{d\Omega_2} \delta(\omega_{if} - \Omega_1 - \Omega_2). \qquad (9)$$

Note that here we have assumed that the normal and BE-violating amplitudes do not interfere; this is expected as a consequence of the superselection rule for transitions between states with different representations of the permutation group.[26]

Eqns. (7), (8), and (9) summarize the central principle of our measurement. That is: for monochromatic light, the degenerate J=0 + 2γ → J=1 transition rate is due *entirely* to violation of BE statistics: i.e., $W_{if}(\Omega_1 = \Omega_2) \propto v$. We can obtain a value (or upper limit) for v by searching for these degenerate transitions, so long as we know the amplitude $\mathcal{F}_{ab}$ and can calibrate the sensitivity of our apparatus.

The nature of the photon states that can be detected using this principle is implicit in Eqn. (9). Note that $W_{if} \propto I_1 I_2$, (where I_j is the intensity in light beam j), for both the allowed and the BE statistics-forbidden processes. The intensity $I \propto N$, where N is the number of photons in the beam. Thus, using the language of photons to describe the beams of light, $W_{if} \propto N_1 N_2$. This can be understood by a counting argument. Each transition requires one photon from each beam, in order to absorb two photons with orthogonal polarizations. For the allowed process, there are N_1 choices for the first photon and N_2 for the second, and thus a total of $N_1 N_2$ pairs available. For the BE-statistics violating process, at least one of the photons must be antisymmetric with the rest. However, there are N_j choices for the antisymmetrized photon in beam j; with any given choice, this photon can pair with any of the N_k photons in beam k. Therefore, again we conclude that $N_1 N_2$ pairs are available. Note that this argument implies that we are sensitive to photon states with Young tableaux of the form

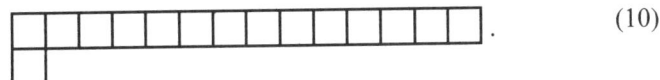 . (10)

In this sense, our experiment is sensitive not only to violations of the spin-statistics relation, but also to violations of the more general permutation symmetry postulate.[27]

Specific Features of Our Experiment

All relevant details of our experiment were given in Ref. [21]; here we give only a brief outline of our scheme. We searched for the transition $6s^2\ ^1S_0 + 2\gamma \rightarrow 5d6d\ ^3S_1$ in atomic Ba; this transition has an unusually large BE-violating amplitude. Light from a dye laser was split into two beams with orthogonal linear polarizations. These beams counterpropagated through a Ba vapor cell. The laser was tuned around the required frequency for the degenerate two-photon transition ($\lambda = 549$ nm). Transitions were detected by observing fluorescence at $\lambda_{fl} = 436$ nm, accompanying the decay $5d6d\ ^3S_1 \rightarrow 6s6p\ ^3P_2$. Excess signal within a narrow tuning range of the laser wavelength would indicate a violation of BE statistics. The sensitivity of the experiment was calibrated with the same detection system, using non-degenerate photons ($\lambda_1' = 532$ nm and $\lambda_2' = 566$ nm) to drive the same transition. That is, we measured the ratio of signals for the degenerate and calibration transitions:

$$S \equiv \frac{W_{if}(\Omega_1 = \Omega_2 = \omega_{if}/2)}{W_{if}'(\Omega_1', \Omega_2')}, \qquad (11)$$

where the primed quantities correspond to the non-degenerate transition.

Using the expression for W_{if} from Eqn. (9), it can be seen that the value of S determines v (assuming $v \ll 1$ and no sources of background signals):

$$v = S \frac{\delta\Omega_1}{\delta\Omega_1'} \frac{\delta\Omega_2}{\delta\Omega_2'} \frac{I_1' I_2'}{I_1 I_2} \frac{1}{R^2}, \qquad (12)$$

where $\delta\Omega_j$ and I_j are the spectral width and total intensity, respectively, of the laser

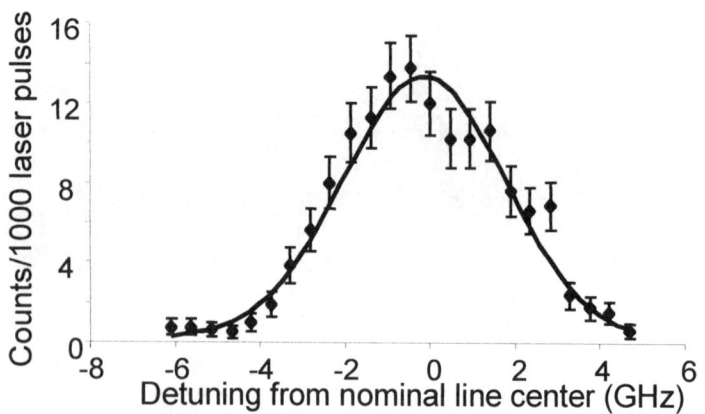

FIGURE 4. Typical scan through the calibration transition and fit to peak height and linewidth. Taken with ~0.4 μJ/pulse at $\lambda_1'=532$ nm and ~200 μJ/pulse at $\lambda_2'=566$ nm

at frequency Ω_j; and R^2 is the ratio of matrix elements of two-photon operators: $R^2 \equiv \left|\mathcal{F}_{if}(\Omega_1 = \Omega_2 = \omega_{if}/2)/Q_{if}(\Omega_1',\Omega_2')\right|^2$. $R^2 = 10 \pm 2$ was determined from known atomic transition energies and dipole matrix elements. All other quantities in Eqn. (12) were measured individually. Uncertainties from each of these measurements led to an overall uncertainty of ~70% in the calibration constant used to determine the value of v from the measured value of S. Fig. 4 shows a typical scan of the laser through the nondegenerate calibration transition.

A typical scan through the degenerate transition is shown in Fig. 5. In all data sets, there is evidence for a statistically significant peak above the background. We believe these peaks are due to the finite bandwidth of the laser. For light from a laser of *finite* spectral width, the transition probability of Eqn. (9) does not vanish for $v=0$, even though $Q_{ab}(\Omega_1 = \Omega_2) = 0$. This is because an atom can absorb two photons from opposite sides of the laser spectral distribution; then, from the atom's point of view, $\Omega_1 = \Omega_2$ is not strictly satisfied. From a crude model for our laser spectra, we predicted the size of the residual signal S due to this "bandwidth effect." Averaging over all data gave the result:

$$\frac{S(\text{observed})}{S(\text{predicted})} = 1.5 \pm 0.6. \tag{13}$$

That is, the observed resonances were consistent with those expected for purely bosonic photons, due to the finite bandwidth of the dye laser.

It should be noted that these residual peaks are extremely small. This can be verified simply by looking at the data in Figs. 4 and 5. For the calibration transition, with $I_1'I_2' \sim 10^{-4}$ (arb. units), the signal is $W_{if}' \sim 0.1$. By contrast, for the degenerate transition, we typically saw signals $W_{if} \sim 3 \times 10^{-3}$, with $I_1 I_2 \sim 3$. Using Eqn. (12) (with

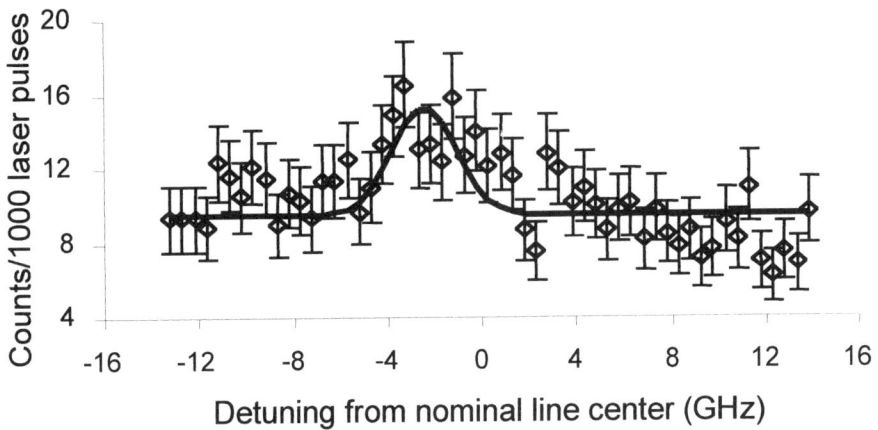

FIGURE 5. A typical scan through the degenerate transition, with the fit to peak plus background. Taken with ~1.5 mJ/pulse in each beam, at $\lambda_1 = \lambda_2 = 549$ nm.

all $\delta\Omega_j$ approximately the same), we see that the bandwidth-effect peaks have a size similar to that which would occur for a BE-statistics violation, with $v \sim 10^{-7}$.

A violation of BE statistics would appear as a resonant signal *in excess* of the peak due to the bandwidth effect [see Eqn. (9)]. Thus, in order to determine an upper limit on v, we should subtract this background. Unfortunately, although the observed peak is consistent with our crude predictions, we also find that the size of the bandwidth-effect peak is quite sensitive to details of the laser spectra. Thus, for determination of v, we take a conservative approach and assume that the *entire* observed peak could be due to violation of BE statistics. In this case, we can use Eqn. (12) to find v. This allows us to set a limit on the BE statistics violation parameter for photons:

$$v < 1.2 \times 10^{-7} \ (90\% \text{ c.l.}). \tag{14}$$

This represents the first result based on a new principle, which in the ideal case gives a background-free signal arising from violation of BE statistics for photons. We believe that the limit on v can be decreased by several orders of magnitude, with an experiment based on this same principle but using improved apparatus. Such an experiment is now underway, and is described in another contribution to these proceedings.[28]

ACKNOWLEDGMENTS

We thank C. Bowers, D. Brown, E. Commins, O. Greenberg, R. Hilborn, L. Hunter, K. Jagannathan, S. Rochester, M. Rowe, M. Suzuki, and M. Zolotorev for useful discussions; and R. Hilborn for equipment loans. This work was supported by funds from Amherst College, Yale University, and NSF (grant PHY-9733479).

REFERENCES

1. Greenberg, O. W., *Phys Rev. D* **43**, 4111 (1991).
2. Hilborn, R. C., and Greenberg, O. W., unpublished manuscript; see also the contribution by R. C. Hilborn in these proceedings.
3. Gerry, C. C. and Hilborn, R. C., *Phys. Rev. A* **55**, 4126 (1997).
4. Fivel, D. I., *Phys. Rev. A* **43**, 4913 (1991).
5. Greenberg, O. W., "Quons, An Interpolation Between Bose and Fermi Oscillators," in *Workshop on Harmonic Oscillators*, edited by D. Han, Y. S. Kim, and W. W. Zachary, NASA Conf. Pub. 3197, Greenbelt, MD: NASA, 1993, p. 5.
6. See e.g. Scipioni, R., *Mod. Phys. Lett. B* **7**, 1911 (1993); Inomata, A., *Phys. Rev. A* **52**, 932 (1995); and Ref. [7]. A partition function based on an interpretation of quons as arising from "ambiguous statistics" was given in: Medvedev, M.V., *Phys. Rev. Lett.* **78**, 4147 (1997), but was criticized in: Meljanac, S., Milekovic, M., and Ristic, R., *Mod. Phys. Lett. A* **14**, 2413 (1999).
7. This statement holds in three spatial dimensions: Werner, R. F., *Phys. Rev. D* **48**, 2929 (1993); Goodison, J. W., and Toms, D. J., *Phys. Lett. A* **195**, 38 (1994). For two dimensions, see Inomata, A., and Kirchner, S., *Phys. Lett. A* **222**, 213 (1996), **231**, 311 (1997), and in these proceedings.
8. Biedenharn, L. C., *J. Phys. A* **22**, L873 (1989); Macfarlane, A. J., *J. Phys. A* **22**, 4581 (1989).
9. See e.g. Martin-Delgado, M. A., *J. Phys. A* **24**, L1285 (1991); Neskovic, P. V., and Urosevic, B. V., *Int. J. Mod. Phys. A* **7**, 3379 (1992); Song, H. S., Ding, S. X., and An, I., *J. Phys. A* **26**, 5197 (1993).
10. Man'ko, V. I., Marmo, G., Solimeno, S., and Zaccaria, F., *Phys. Lett. A* **176**, 173 (1993).
11. See e.g. Sapritsky, V. I., *Metrologia* **32**, 411 (1995/96).
12. Fixsen, D. J., et al., *Ap. J.* **473**, 576 (1996).
13. Patterson, F. G., and Perry, M. D., *J. Opt. Soc. Am. B* **8**, 2384 (1991).
14. Zolotorev, M., private communication. This is based on cavities with volume $V \sim \lambda^3$ and internal electric fields $E \sim 100$ kV/cm.
15. Man'ko, V. I., and Tino, G. M., *Phys. Lett. A* **202**, 24 (1995).
16. Landau, L. D., *Dokl. Akad. Nauk SSSR* **60**, 207 (1948); Yang, C. N., *Phys. Rev.* **77**, 242 (1950).
17. See e.g. Sakurai, J. J., *Invariance Principle and Elementary Particles*, Princeton: Princeton University Press, 1964, pp. 15-16.
18. Ignatiev, A. Yu., Joshi, G. C., and Matsuda, M., *Mod. Phys. Lett. A* **11**, 871 (1996).
19. Gidley, D. W., Nico, J. S., and Skalsey, M., *Phys. Rev. Lett.* **66**, 1302 (1991).
20. Caso, C., et al. (Particle Data Group), *Eur. Phys. J. C* **3**, 1 (1998).
21. DeMille, D., Budker, D., Derr, N., and Deveney, E., *Phys. Rev. Lett.* **83**, 3978 (1999).
22. Blatt, J. M., and Weisskopf, V. F., *Theoretical Nuclear Physics*, New York: John Wiley & Sons, 1952, pp. 796-815.
23. Loudon, R., *The Quantum Theory of Light*, New York: Oxford University Press, 1983.
24. This result has been independently (re)derived several times. See e.g. Grynberg, G., and Cagnac, B., *Rep. Prog. Phys.* **40**, 791 (1977); Bonin, K. D., and McIlrath, T. J., *J. Opt. Soc. Am. B* **1**, 52 (1984); and Bowers, C. J., et al., *Phys. Rev. A* **53**, 3103 (1996).
25. For a standard explanation of the relationship between distinguishability (in our sense) and the observability of exchange symmetry, see e.g.: Sakurai, J.J., *Modern Quantum Mechanics*, Menlo Park, CA: Benjamin/Cummings, 1985, pp. 365-366.
26. Amado, R. D., and Primakoff, H., *Phys. Rev. C* **22**, 1338 (1980).
27. For a discussion of this distinction, see Tino, G. M., *xxx.lanl.gov*, quant-ph/9907028 (1999).
28. Brown, D. E., Budker, D., and DeMille, D., in these proceedings.

High Precision Experimental Tests Of The Symmetrization Postulate For Fermions

J. D. Gillaspy

Atomic Physics Division, National Institute of Standards and Technology
Gaithersburg, MD 20899-8421

Abstract. A review is given of experiments that have been used to infer an upper bound on the degree (β^2) to which the symmetrization postulate might be violated in nature. A proposed grouping of the experiments into seven different categories is put forth, based on different conceptual difficulties associated with interpreting them. Some comments are made about the strengths and weaknesses of representative experiments from several of the categories. It is suggested that there may be a considerable body of existing experimental data that could be analyzed to infer additional limits on SP-violation; as an example of this I consider the suppression of molecular recombination in spin-polarized atomic hydrogen by strong magnetic fields. While this article is limited to experiments involving fermions, an extensive bibliography on both theory and experiments related to the spin-statistics connection in general is referred to.

INTRODUCTION

The Symmetrization Postulate

In this paper, I use the term "symmetrization postulate" (SP) in the original sense described by Messiah [1], namely that multiparticle states are either all symmetrical or all antisymmetrical with respect to the permutation of any two identical particles. Under this restriction, wavefunctions that predict identical values for all observables can differ by no more than a complex phase. This essentially limits the mathematics of quantum mechanics to 1-dimensional Young diagrams, and rules out the possibility that there can be a "small" violation of the Pauli Exclusion Principle or SP. As reviewed in detail at this conference[2] the inclusion of higher dimensional representations of the permutation group allows more flexibility, and brings up the possibility that "every once in awhile" one will find a wavefunction which has an abnormal permutation symmetry. Such a violation of the symmetrization postulate is not in conflict with proofs of the spin-statistics theorem which neglect higher dimensional representations of the permutation group from the onset. A bibliography [3] described below contains several hundred references which delve into the theoretical, physical, and philosophical aspects of the issues mentioned in this paragraph.

Why Fermions?

At the most fundamental level, all matter particles are fermions (leptons and quarks), so the neglect of bosons in this paper is less restrictive than may be apparent at first glance. Furthermore, an approximate expression recently derived by Greenberg and Hilborn [4] implies that (for fundamental particles) ". . . almost all particles obey Bose or Fermi statistics to a precision comparable to the precision with which electrons obey Fermi statistics".

THE EXPERIMENTS

Overview

Figure 1 gives an overview of selected experimental results considered in this paper. The data were obtained from 3 sources: (1) published references known to me from my earlier involvement in this field around 1990, (2) published references found through a computerized literature search carried out in April 2000, and (3) work that is in preprint stage and known to me through direct contact with the authors. The data shown in figure 1 are identified in table 1a, where they are also assigned a "category" (see table 1b) and "retro-analysis" marker, as described below. In addition, table 1a indicates whether the experiment is a test for electrons or nucleons.

The recent literature search (item 2 above) was made using a database in the NIST Virtual Library [5]. The database encompasses 5,300 major scientific journals from 1983 onward. The search was made on (a) authors known to have worked in this field, (b) papers that contain references to these authors, and (c) key phrases such as "symmetrization postulate", "spin-statistics", and "pauli exclusion principle" from the title, abstract and keyword list. After discarding references that appeared from the title to be irrelevant, nearly 300 references were retained. These references have been combined with additional references collected by R. C. Hilborn (including some from professional journals within the philosophy community) and made available on the web [3].

The data shown in figure 1 are replotted as function of publication date in figure 2. Two features are striking: (1) the burst of activity after the 1987 publication of Greenberg and Mohapatra [6] in which it was emphasized that "no high precision tests of the Pauli Principle have been made" and (2) the lack of correlation of β^2 with time. This brings up two interesting questions: (a) why were the two relatively high precision experiments before 1987 discarded and (b) why are relatively low precision experiments still being carried out? One possible answer to both of these questions lies in the following key issue: the difference between precision (or claimed sensitivity) and accuracy.

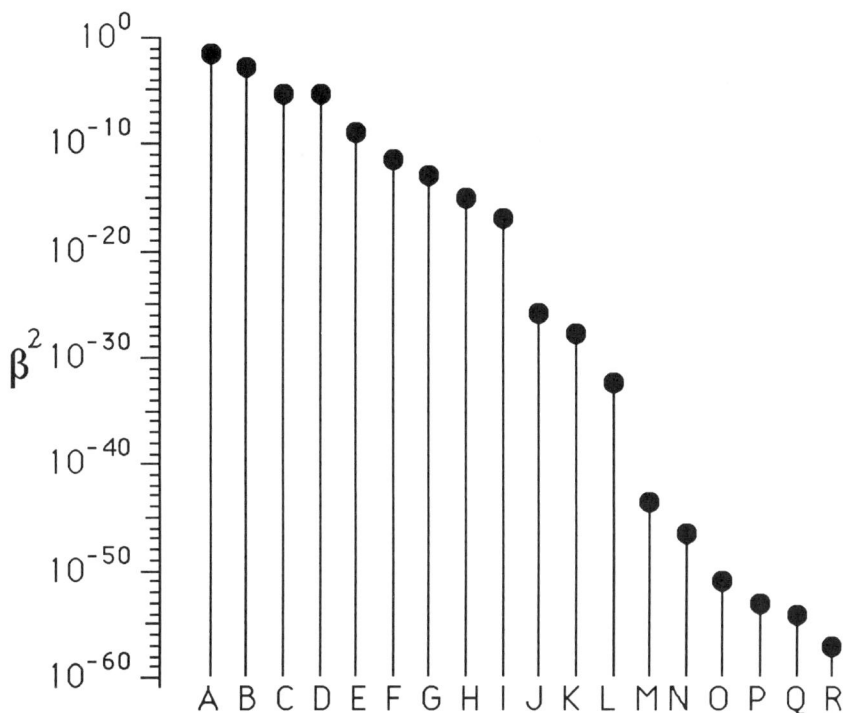

FIGURE 1. Values of β^2 from selected experiments listed in table 1a (see text).

TABLE 1a. Identification of data shown in figure 1.

Label	Selected author and reference	Category	Retro	Electron	Nuclear
A	Goldhaber [7]	II	R	X	
B	Fischbach [8]	VI	R	X	
C	Fischbach [9]	VI		X	
D	Gillaspy [10]	V		X	
E	Gillaspy [11]	V	R	X	
F	Kekez [12]	IV			X
G	Nolte [13]	V			X
H	Plaga [14]	VII	R		X
I	Mohapatra [15]	VI	R		X
J	Ramberg & Snow [16]	II		X	
K	Nolte [17]	VI			X
L	Nolte [18]	VI	R		X
M	Reines & Sobel [19]	I	R	X	
N	Logan & Ljubicic [20]	I			X
O	Kishimoto [21]	I			X
P	Ejiri [22]	III			X
Q	Bernabei [23]	III			X
R	Ejiri [24]	III	R		X

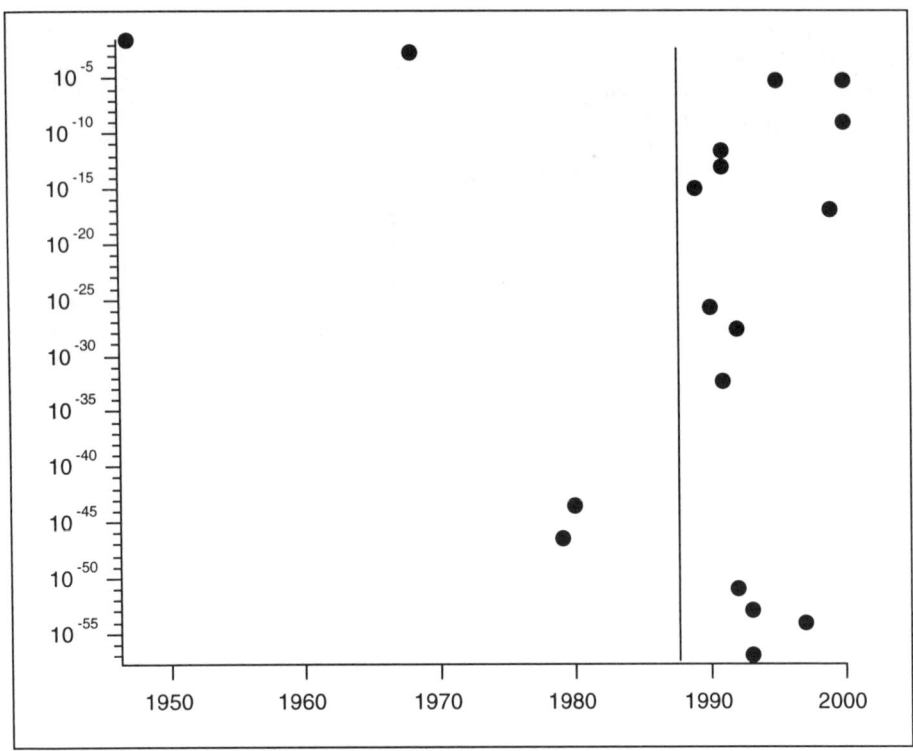

FIGURE 2. Data shown in figure 1, plotted as function of publication date. The vertical line marks the publication of Greenberg and Mohapatra's "paron" paper in 1987 [6].

When an experiment yields a null result, it is sometimes particularly difficult to assess with high confidence what the signal *would have been* if the sought after effect were present. Neglected numerical factors in the data reduction, hidden assumptions, or changing theoretical interpretations can change the results significantly (I will discuss several examples below). Particularly for experiments seeking to detect a violation of one of the fundamental laws of physics, there can be a tradeoff between choosing an experiment that has some chance of seeing a violation if it is very small, or an experiment that has a high chance of seeing a violation if it is relatively large. The former (high precision) type of experiment may miss detecting the violation due to a failure in accuracy, while the latter (possibly higher accuracy) type of experiment may in fact see the violation first. In the absence of a credible theoretical prediction for the expected magnitude of the violation, the two experiments can be complementary.

In addition to concerns about accuracy, there is another reason why a variety of types of experiments might be carried out even if they have a precision less than that of some previous experiments. Just as in the case of parity-violation, it may be that an SP-violation can be more readily detected in some physical systems (types of atoms, electronic levels, etc.) than in others. Indeed, Okun has argued that the Pauli exclusion principle should be tested throughout the periodic table [25]. Similarly, D.E. Kelleher has suggested that the near degeneracy of the states of allowed and forbidden symmetry in metastable excited helium may make them more sensitive to SP-violation than other states [10].

Categorization

At first glance, the experiments fall into three categories, based on the primary way in which the results have been expressed: (1) "β^2", the probability that an unbiased sampling of two particle states yields one with unusual statistics (2) "lifetime", and (3) "absolute abundance". The experiments that fall into category (2) or (3) have also been expressed, at one point or another, in terms of a probability, hence I have used "β^2" to plot them in comparison to each other in figure 1. I propose an alternative categorization that is based on more underlying conceptual features of the experimental design. Specifically, I divide the experiments into the seven types listed in table 1b. Each type is discussed briefly below.

TABLE 1b. Types of experiments that test the symmetrization postulate.

Type	Descriptor	Notes (see text)
I	Spontaneously Relaxing Blocks	Photons out
II	Absorbing Blocks	Fresh electrons in
III	Decaying Blocks	Fresh electrons out; fermion number change
IV	Internally Absorbing Blocks	Electron capture
V	Collisions in Vacuum	2-body systems
VI	Ground State Accumulation	Mass Spectrometry
VII	Miscellaneous	Everything else

Type-I Experiments: Spontaneously Relaxing Blocks

This is the classic experiment of the Reines and Sobel type [19] in which a small violation of the exclusion principle is assumed to lead to the slow relaxation of bulk matter, with eventually all of the electrons of all of the atoms in the universe ending up in the innermost shell (n=1, or "K-shell). Thus, if one watches a block of matter, one should occasionally see some photons spontaneously being emitted as the electrons fall into the innermost shell from time to time. A similar situation could occur for nuclear shells. The problem with this type of experiment is that it fails to take into account the effect of a superselection rule which prevents transitions between ordinary and SP-violating states [26], even if the SP-violating states can exist. Transitions are only allowed between two SP-violating states, and those occur at the ordinary (typically rapid) rates. It was for this reason that the two relatively high precision experiments plotted in figure 2 for dates near 1980 were considered invalid (zero accuracy) by Greenberg and Mohapatra.

A Second Look At The Superselection Rule

While the superselection rule erodes the basis of type-I experiments, such experiments can still be interpreted in terms of a "superficial" violation of the symmetrization postulate (and/or the associated Pauli exclusion principle) if one accepts the possibility that electrons are not elementary particles, but themselves have substructure [27]. This is not a strict violation because the quantum fields that form the substructure obey bose and fermi statistics in the usual way, but the experimental signature of such a superficial violation would appear identical to the sort of SP-violation originally envisioned for the type-I experiments. Indeed, according to Akama et al. [27] "such superficial violation of the Pauli principle must exist, no matter how small it is, if the electron has any substructure at all". It has been emphasized elsewhere [17] that low-energy SP-violation experiments can be used in this way to infer limits on electron substructure that are competitive with TeV accelerator experiments which test the radius of the electron directly.

Type-II Experiments: Absorbing Blocks

These types of experiments are similar in practice to type-I experiments, except that "fresh" electrons are brought into the system in some way (by flowing an electrical current through the system, for example). The scenario is that occasionally one of these fresh electrons may be in an anomalous symmetry state with respect to the observed system, so once this electron is brought into contact with the system it can bind in an excited state and fall down to the ground level, emitting a photon in the process. I call these experiments "absorbing" because they accumulate extra electrons in the ground levels. Conceptual issues concerning the nature of "fresh" electrons continue to be addressed.

The most precise experiment of this type is by Ramberg and Snow [16]. Issues concerning the accuracy of this experiment, however, have not been discussed in detail in the literature. As an example, I note that even if an electron entered the experiment in an anomalous symmetry state with respect to a copper atom, and was then captured into an excited state, it would most likely decay to the ground state non-radiatively (blind to the detectors) because the K-shell fluorescence yield is less than 50%, even for elements as heavy as copper [28]. This correction reduces the sensitivity of the experiment, but does not invalidate it, unlike the case of neglecting the superselection rule for type-I experiments. The fluorescence yield is one of several correction factors that should be applied to the Ramberg and Snow experiment in order to determine its quantitative accuracy.

Type-III Experiments: Decaying Blocks

"Decaying blocks" are to be distinguished from "relaxing blocks" in that the former involves a nuclear reaction in which the identity of the particles is changed. Frequently, an electron or positron is spontaneously ejected from the material, so in some sense one could say this is an "inverse current" experiment (fresh electrons come out, instead of being put in). Associated with the ejected particle is another particle (remaining in the block) that has changed its identity and therefore may have formed an anomalous symmetry that would allow it to relax down to a normally full ground level and emit a photon (or other particle). In this way, it has been claimed, the superselection rule can be avoided.

The most precise experimental test of the symmetrization postulate to date is of this type. It was carried out with the large Kamiokande detector (roughly 16 meters across, consisting of 3,000 tons of pure water, located 1,000 meters underground, and observed with 1,000 photomultiplier tubes). After collecting data for 3 years, a limit of $\beta^2 < 2.3 \times 10^{-57}$ was obtained [24].

Type-IV Experiments: Internally Absorbing Blocks

There is only one experiment of this type that I know of [12], but because of its rather unique conceptual aspects, I put it in a class of its own. I call this type of experiment "internally absorbing " because a current (of sorts) is involved, much in the sense of the type II experiments, but in this case the electrons come from an electronic shell to a nuclear shell. The process involved is called "electron capture decay" and essentially involves an atomic electron being pulled into the nucleus and captured by a proton, which then changes into a neutron. This "fresh" neutron is presumed to have some probability of being in a SP-violating symmetry so that it can relax to a ground level, emitting a photon or other particle that can be detected. The internal current between the electronic and nuclear regions of the atom is a distinguishing feature of this type of experiment.

Type-V Experiments: Collisions In Vacuum

This type of experiment is the atomic analogue of the "condensed matter" type-II experiment. Instead of flowing a macroscopic current through piece of bulk material, individual electrons (or other particles) are projected through a vacuum towards atoms and analyzed for their effect, often on a case-by-case basis. Experiments of this type will produce a relatively modest number of events, but each event will be more easily interpretable. While sensitivity is therefore sacrificed in these types of experiments, it is done so (hopefully) for the sake of accuracy.

Type-VI Experiments: Ground State Accumulation

These experiments typically use mass spectroscopy techniques to search for atoms which have an anomalous number of electrons (or nucleons) in their ground levels, thereby directly violating the Pauli exclusion principle. They are distinguished from the other experiments in that they do not depend on a *transition* (accompanied by the emission of photons or other particles) for detection.

I point out that experiment-K in figure 1 above is based on a reanalysis of the same author's earlier data (experiment-L), considering instead a different channel for SP-violation to occur in the history of the universe. The more recent analysis yields a considerably *less* stringent limit on SP-violation, and illustrates well how the accuracy of the results interpreted from a measurement depends on more than just the precision of the experimental data.

Note that many of these types of experiments depend on assuming that if a valence electron is moved to the ground level, the remaining valence electrons see a perfectly screened nucleus so that the modified atoms behaves chemically as if it were the next lowest atom on the periodic table. Atomic structure theorists [29] have indicated that screening corrections should be quite significant, however. Presumably, the anomalous atoms could have a very different chemical reactivity than presumed. Because the effect of an additional electron in the ground level of an atom has never been calculated, assumptions about the chemical reactivity are an open question.

Type-VII Experiments: Miscellaneous

This category is reserved for experimental limits that do not fit into any of the other categories.

Notes On Adjustments

The comparative data I have plotted in this paper have been adjusted from the published results in several instances. The experiments of Fischbach et al. [8] have been converted into β^2 by Greenberg [30]. A more recent measurement by Fischbach and colleagues has a factor of 300 improved sensitivity [9] so I have decreased the corresponding β^2 factor by that same amount. Note that these experiments involve, in some sense, a double violation of the exclusion principle (4 Beryllium electrons in the ground level). In the presence of a continuously evolving theoretical framework, the strict interpretation and intercomparison of these (and other) experiments, is uncertain. As emphasized by Hilborn at this conference, each experiment should be taken to stand on its own.

The paper of Kishimoto [21] does not explicitly present a β^2 factor, but notes a 7.2 MeV width (corresponding to a lifetime of 6×10^{-22} s) of the corresponding allowed transition, so the ratio of their forbidden to allowed lifetimes gives a β^2 factor of 1×10^{51}.

Logan and Ljubicic [20] quote a forbidden lifetime of 5×10^{27} s, and note that the corresponding allowed transition is $\sim 10^{-18}$ s, but they appear to have dropped a factor of 10 in taking the square root of the ratio, so they obtain $\beta < 1.3 \times 10^{-22}$ instead of $\beta < 1.3 \times 10^{-23}$. I therefore adjust their value, and use the more precise number given by Kishimoto [21] for the corresponding allowed transition, to obtain $\beta^2 < 4 \times 10^{-47}$.

RETRO-ANALYSIS

Not all of the published limits discussed in this paper have been the result of experiments designed and constructed specifically for the purpose of testing SP. The data indicated in table 1a with an "R" in the column labeled "Retro" were those that were clearly the result of a retrospective analysis of data originally taken for another purpose. This "retro-analysis" of existing data for a possible limit on SP-violation can be a very valuable source of information; contributions along these lines by researchers from a wider variety of sub-disciplines of physics would be welcome.

A New (Old) Type IV Experiment: Retro-Analysis Of The Decay Of Spin-Polarized Atomic Hydrogen

As an example of a limit on SP violation from retro-analysis of existing data, I present here a discussion of the decay of spin-polarized atomic hydrogen. Although the combined effect of electronic and nuclear spin makes atomic hydrogen a composite boson, the fundamental fermionic character of the constituent components becomes felt in collisions. Indeed, it is the antisymmetry of the final state 2-electron wavefunction that prevents H_2 molecules from forming if the constituent electrons have their spins polarized in the same direction. By experimentally forcing the electron spins in a gas of hydrogen to be aligned, one can suppress molecular recombination by many orders of magnitude. This was the basis of the "stabilization" of atomic hydrogen by Silvera and Walraven in 1980 [31], in which they cooled the gas to a fraction of a degree above absolute zero in the presence of a strong (several tesla) magnetic field. In this configuration, the Zeeman spin-flip energy is much larger than kT, so the gas becomes highly polarized. In such an environment, a gas sample that would ordinarily recombine in a fraction of a second can be make to stay around for hours. Basically, the Pauli exclusion principle is functioning to keep the electrons from being in the same place at the same time, so collisions that would tend to produce localized bound states (molecules) become ineffective.

If the exclusion principle were to be violated during a small-fraction of collisions of spin-polarized hydrogen atoms, the recombination would no longer be suppressed, and the gas would decay a bit faster than expected. Since spin-polarization can be used to suppress the recombination by a factor of 3×10^{-6}, a significant limit on β^2 can be obtained, (compare B=0 data and B=11 tesla data for 2-body surface recombination at T<300mK in [32], figure 5.9). The dominant residual decay is due to the fact that the lowest energy hyperfine state retains a small admixture of depolarized electronic spin, even at high magnetic field (proportional roughly to the inverse magnetic field), can be further suppressed by purifying the gas in the second lowest hyperfine state. The associated "nuclear polarization" has been produced as high as 99.8%, corresponding to factor of 500 suppression of residual decay rate, and a corresponding better limit on β^2. Finally , if we combine these two factors and claim that the measured decay rates can be accounted for to an accuracy of 15%, then we can infer a bound of $\beta^2 < 1 \times 10^{-9}$.

This type of experiment falls into the type-V category, although it is different from the rest in that it does not rely on the emission of photons or other particles for detection. Because atomic hydrogen is the only material that is thermodynamically stable as a gas at temperatures less than a degree above absolute zero, the atomic-to-molecular transition is followed by freezing of the molecules into a solid, and so a direct measurement of the decay of the macroscopic gas pressure (or volume, at fixed hydrostatic pressure) is used as the measurement variable.

A Reconsideration Of Anecdotal Limits From Atomic Spectroscopy

Greenberg has stated that "Atomic spectroscopy is the first place to search for violations of the exclusion principle since that is where Pauli discovered it"[33]. Greenberg and Mohapatra [30] have furthermore suggested that if β^2 were greater than about 10^{-7}, then atomic spectroscopists would have detected unexplained emission lines long ago. This is what I call an "anecdotal limit". I suggest that this limit is several orders of magnitude too stringent, and illustrate the reasons why by discussing a NIST experiment [10] that was designed specifically to search for SP-violations.

The first point to be made is that there *do* exist spectroscopic lines that have been recorded in laboratory experiments but not explained, despite considerable effort by leading spectroscopists to do so [34]. Most such lines probably go unreported and undiscussed in the literature. It is perhaps amusing to note that several papers reporting limits on SP-violation have reported positive events that correspond exactly to the predicted signal for an SP violation [10,21,24]. For most of these cases, the authors provided a convincing alternative explanation, but one of the cases (with a 10-sigma discrepancy) remains unaccounted for [21]. It is likely that researchers who did not originally set out to search for SP-violations would be even more likely to find alternative explanations for anomalous observations, and less likely to elaborate on them in their publications. It is reasonable that there would be considerable resistance to any publication claiming evidence for a violation of SP or the Pauli exclusion principle. Plaga's 1989 explanation of the solar neutrino problem in terms of a small violation of the Pauli exclusion principle [14], for example, did not receive wide discussion.

A second point to be made is that precision tests of SP require a very different type of resolution from that of conventional high precision spectroscopy. Tests of SP operate along the intensity axis, rather than the wavelength axis, so the conventional sort of resolution (wavelength) is only relevant indirectly in that overlapping lines may contribute intensity noise to each other. Many of the powerful techniques of modern spectroscopy, therefore, are of limited value in testing SP. As an example, I note that in the experiment C.J. Sansonetti and I performed [35] to measure the binding energy of helium to a few parts in 10^{10}, a relatively large SP-violation (several percent) could have occurred and we would have not observed it. Nearly all spectroscopic measurements are carried out with signal-to-noise ratios of less than 10,000 along the intensity axis, corresponding to a relatively modest $\beta^2 > 1\times10^{-4}$.

Greenberg and Mohapatra have suggested that astrophysical spectra would be a valuable source of data to infer a limit on SP-violation, but that Doppler shifts and collisional broadening will tend to obscure the results [30]. In the following I illustrate how these problems also plague laboratory measurements in a way that may not have been fully appreciated for measurements concerned with the intensity axis rather than the wavelength axis.

In the NIST SP-experiment, the laser system was frequency-locked to a table-top low pressure helium discharge[10], which provided a reference point to the ordinary (SP-allowed) spectroscopic lines. This system operates at room temperature (discharge tube cold to the touch). Nevertheless, a direct laser scan over the SP-allowed line used in that work shows a roughly 5 GHz Doppler linewidth, broad enough to totally mask the corresponding weak SP-forbidden line which is only 0.5 GHz away from the allowed line.

A saturated absorption technique was then used to measure a "Doppler-free" line profile that was approximately 100 times narrower. One of the infrequently mentioned limitations of this technique, however, is that the Doppler-free line rests on a small but significant "Doppler pedestal" induced by collisions. The profile of this pedestal will typically not be apparent in any published data because of the huge difference between the two linewidths--any plot that shows one in detail will obscure the other. It is there, nonetheless, and restricts the ability to obtain a limit on an SP-violation to $\beta^2 \sim 10^{-3}$. It should be apparent from this example, that Doppler effects and collisions can severely degrade the ability to test SP not only in astrophysical observations, but also in laboratory measurements using modern high resolution techniques.

While I admit that the best available spectroscopic techniques *could* detect an SP-violation at a level of 10^{-6}-10^{-8}, I claim that it does not follow that such a violation *would* have been detected, let alone reported in the literature or otherwise widely discussed. I thus claim that anecdotal evidence must be replaced with direct experimental data from specific experiments.

One of the most powerful spectroscopic techniques for testing SP is the "photon burst" method applied to an atomic beam. This essentially realizes a gedanken experiment in which individual atoms are analyzed one at a time to confirm whether or not any of them are in an SP-forbidden state. Using the photon burst method in this way, a limit of $\beta^2 < 10^{-10}$ should be attainable [10]. Such an effort is currently being carried on in H. Batelaan's group at the University of Nebraska.

CONCLUSION

Greenberg has summarized the situation from a theoretical perspective in one of his papers: "It is not easy to violate the exclusion principle" [36]. I conclude with an experimentalist's followup: It is not easy to detect such a violation either!

ACKNOWLEDGMENTS

I thank Dan Kelleher for introducing me to this research topic, attracting me to NIST to do experiments on it, and serving as my research advisor; the National Research Council for the fellowship that paid my salary; the management (Jim Roberts, Wolfgang Wiese, Bill Ott, and Katharine Gebbie) for providing the money needed to purchase/build most of the apparatus needed for my experiments (and for allowing me to still spend a bit of my time thinking about such crazy things), Kavoos Deilamian for daring to do his PhD thesis on this topic with me, and Craig Sansonetti for lending me critical apparatus and knowledge and inviting me to participate in some "non-null" measurements along the way.

REFERENCES

1. Messiah, A. M. L., *Quantum Mechanics*, John Wiley and Sons, New York, 1976.
2. International Conference on Spin-Statistics Connection and Commutation Relations: Experimental Tests and Theoretical Implications, Anacapri, Italy, May 31-June 3, 2000.
3. http://physics.nist.gov/MajResFac/EBIT/peprefs.html.
4. Greenberg, O. W., and Hilborn, R. C., *Foundations of Physics* **29**, 397-407 (1999).
5. http://nvl.nist.gov.
6. Greenberg, O. W., and Mohapatra, R. N., *Physical Review Letters* **59**, 2507-2510 (1987).
7. Goldhaber, M., and Scharff-Goldhaber, G., *Phys. Rev.* **73**, 1472-1473 (1948).
8. Fischbach, E., Kirsten, T., and Schaeffer, T., *Physical Review Letters* **20**, 1012-1014 (1968).
9. Javorsek, D., Bourgeois, M., Elmore, D., Fischbach, E., Hillegonds, D., Marder, J., Miller, T., Rohrs, H., Stohler, M., and Vogt, S., to be published (2000).
10. Deilamian, K., Gillaspy, J. D., and Kelleher, D. E., *Phys. Rev. Lett.* **74**, 4787-4790 (1995).
11. Gillaspy, J. D., this paper (2000).
12. Kekez, D., Ljubicic, A., Kaucic, S., and Logan, B. A., *Nuovo Cimento Della Societa Italiana Di Fisica a-Nuclei Particles and Fields* **104**, 607-609 (1991).
13. Miljanic, D., Ljubicic, A., Nolte, E., Faesterman, T., Gail, H., Gillitzer, A., Korschinek, G., Muller, D., Scheuer, R., Calvi, G., Lattuada, M., Spitaleri, C., Zadro, M., and Logan, B. A., *Nuclear Instruments & Methods in Physics Research Section B- Beam Interactions With Materials and Atoms* **56-7**, 508-510 (1991).
14. Plaga, R., *Zeitschrift Fur Physik a-Hadrons and Nuclei* **333**, 397-403 (1989).
15. Baron, E., Mohapatra, R. N., and Teplitz, V. L., *Phys. Rev. D* **59**, 036003 (1999).
16. Ramberg, E., and Snow, G. A., *Physics Letters B* **238**, 438-441 (1990).
17. Thoma, M. H., and Nolte, E., *Physics Letters B* **291**, 484-487 (1992).
18. Nolte, E., Faestermann, T., Gillitzer, A., Korschinek, G., Muller, D., Novikov, V. M., Pomansky, A. A., Ljubicic, A., and Miljanic, D., *Zeitschrift Fur Physik a-Hadrons and Nuclei* **340**, 411-413 (1991).
19. Reines, F., and Sobel, W. H., *Phys. Rev. Lett.* **32**, 954- (1974).
20. Logan, B. A., and Ljubicic, A., *Phys. Rev. C* **20**, 1957-1958 (1979).
21. Kishimoto, T., Shibata, T., Imamura, M., Shibata, S., and Uwamino, Y., *Journal of Physics G-Nuclear and Particle Physics* **18**, 443-448 (1992).
22. Ejiri, H., and Toki, H., *Physics Letters B* **306**, 218-223 (1993).
23. Bernabei, R., Belli, P., Montecchia, F., DeSanctis, M., DiNicolantonio, W., Incicchitti, A., Prosperi, D., Bacci, C., Dai, C. J., Ding, L. K., Kuang, H. H., and Ma, J. M., *Physics Letters B* **408**, 439-444 (1997).
24. Suzuki, Y. et al., *Physics Letters B* **311**, 357-361 (1993).
25. Okun, L. B., "On the experimental basis of the pauli exclusion principle," in *Festi-Val--Festschrift for Val Telegdi*, edited by K. Winter, Elsevier Science Publishers, B.V., 1988.
26. Amado, R. D., and Primakoff, H., *Phys. Rev. C* **22**, 1338-1340 (1980).
27. Akama, K., Terazawa, H., and Yasue, M., *Physical Review Letters* **68**, 1826-1829 (1992).
28. Krause, M. O., *J. Phys. Chem. Ref. Data* **8**, 307 (1979).
29. Kim, Y.-K., Private communication.
30. Greenberg, O. W., and Mohapatra, R. N., *Physical Review D* **39**, 2032-2038 (1989).
31. Silvera, I. F., and Walraven, J. T. M., *Phys. Rev. Lett.* **44**, 164-168 (1980).
32. Silvera, I. F., and Walraven, J. T. M., "Spin-polarized Atomic Hydrogen," in *Progress in Low Temperature Physics Vol. X*, edited by D.F. Brewer, Elsevier Science Publishers, B.V., 1986.
33. Greenberg, O. W., "Small Violations of Statistics," in *Orbis Scientiae 1998*, Plenum Press, 1999.
34. Reader, J., Private communication.
35. Sansonetti, C. J., and Gillaspy, J. D., *Phys. Rev. A* **45**, R1-R3 (1992).
36. Greenberg, O. W., *Nuclear Physics B (Proc. Suppl.)* **6**, 83-89 (1989).

CP Violation and Symmetry of the Neutral Kaon Wave Function

Italo Mannelli

Scuola Normale Superiore
I-56126 PISA

Abstract. The interpretation in terms of CP violation of the experimental results concerning the neutral kaon system requires, in general, assumptions about the validity of quantum mechanics, about the coherence of the wave function and of its behavior under particle exchange. A few examples are discussed where these assumptions can be tested. An update is also given of the present experimental situation concerning in particular direct CP violation in two-pion decay.

Gell-Mann and Pais were first to point out in 1955 [1] that it was meaningful to consider states with wave functions expressed in terms of coherent linear combinations of neutral kaon and antikaon and predicted very interesting phenomenological consequences. In particular, assuming charge conjugation(C) invariance of the weak interactions, the symmetric and antisymmetric combinations would propagate in vacuum independently of each other. Since the most important decay modes accessible to the C eigenstates with opposite eigenvalue had very different phase space available, a different lifetime was expected for them. This fact was soon established experimentally by observing the distributions of decay times for 2-pion and 3-pion final states. Another striking consequence of the mixing between neutral kaon and antikaon is the phenomenon of coherent regeneration of the short-lived eigenstate in a beam of a long-lived one propagating in matter, as a consequence of the difference in forward scattering amplitudes.

After the discovery of parity(P) and charge conjugation(C) non-invariance of the weak interactions, the description of the phenomenology of the neutral kaon system remained essentially unchanged, although in terms of CP eigenstates(K1 and K2) rather than C eigenstates.

In 1964 the experiment by Christenson, Cronin, Fitch and Turlay [2] demonstrated that the long lived eigenstate KL was actually decaying, with a branching ratio of the order of 3×10^{-3}, in the same 2-pion final state which is the preferred decay mode of the short lived KS. This result, coupled with the measurement of a charge asymmetry in the semileptonic decays of KL, implies CP violation in the mixing. This manifests itself in particular with the presence of a K1 component in KL (and of K2 in KS) and hence of 2-pion decays of KL, even in the absence of a direct CP violation in the decay.

In case CP violation occurred only in K-antiK mixing, then from a generic neutral kaon state, only the component with CP eigenstate +1(-1) K1(K2), would decay into a final state with the same CP +1(-1) eigenvalue. This would imply that such a final state would be identically the same whether it is originating from a KS or from a KL decay. On the contrary, in presence of direct CP violation in the decay process, this would not be the case and distinctive features would allow in general to distinguish between the two cases.

It was soon realized [3] that the early experimental results could very well be described, within their quoted accuracy, on the basis of the so called superweak model, which limited the occurrence of CP violation to DeltaS=2 transitions. However, the extension to three families of quarks of the weak quark mixing mechanism allows, as first shown by Kobayashi and Maskawa [4],for the introduction of a non-trivial phase in the mixing matrix, with consequent expectation of CP violation effects, not only in DeltaS=2 transition via the so-called box diagram, but also, barring accidental cancellations, in DeltaS=1 decay processes to which quarks from all three families can contribute. This realization, together with the relevance of CP violation for cosmology, has provided a strong motivation to extend the search for CP violation to the charm and bottom sector and to devise and perform experiments with the best statistical and systematic accuracy made possible by improvements in accelerator and detection techniques. Although no clear signal of CP violation has yet been found outside of the neutral kaon system, an all-out effort is underway exploiting, in addition to fixed-target accelerators, new dedicated electron-positron colliders, covering also the neutral b-meson for which a rich and well constrained phenomenology is predicted according to the standard model. The potential in this field of the Fermilab proton-antiproton collider and of LHC at CERN will also be exploited in the future.

Some basic properties are naturally (often implicitly) assumed in the planning and in the interpretation of the experiments, such as the validity of quantum mechanics and the coherence and symmetry character of the wave function.

In this respect it should be noted that the fact that the ratios of the decay rates (KS into $\pi^+\pi^-$)/(KS into $\pi^0\pi^0$) and (KL into $\pi^+\pi^-$)/(KL into $\pi^0\pi^0$) are not identical, a fact which needed more than one decade of dedicated experiments to be firmly established, can actually be interpreted in two radically different ways, as remarked by Greenberg and Mohapatra [5].

The argument goes as follows: the wave function of a two π^0 final state is necessarily symmetric under particle exchange; assuming the validity of the generalized Bose-Einstein statistics for pions, a $\pi^+\pi^-$ final state with spin J=0 is also necessarily in an eigenstate of CP with eigenvalue +1. With this hypothesis, different ratios of decay rates mean unequivocal evidence for direct CP violation. However, if the generalized Bose-Einstein statistics is slightly violated, a $\pi^+\pi^-$ pair with J=0 could exist with isospin I=1, i.e. in an antisymmetric CP=-1 eigenstate. There could then be a

CP conserving decay amplitude for the K2 component of KL to end up in such state. This being possible for $\pi^+\pi^-$ but not for $\pi^0\pi^0$ could be the reason, rather then direct CP violation as commonly claimed, for the observed difference between KS and KL decay ratios. Along the above lines an upper limit for the violation of the generalized Bose-Einstein statistics was in fact derived by Greenberg and Mohapatra from preliminary data by the NA31 collaboration, actually reported at the International Symposium on Lepton and Photon Interactions at High Energies in 1987 [6] as first evidence for direct CP violation in KL to two-pion decay.

As mentioned above, in the standard model, as a consequence of the non-trivial phase in the quark mixing matrix, CP violation is expected both in K-antiK mixing and in the direct decay transition to two-pions. Since the latter effect results from partial cancellation of rather large amplitudes, its size cannot be reliably predicted, but the establishment of its very existence provides strong support for the extension of the standard model to include CP violation. To reach the accuracy in the comparison of the branching ratios necessary to establish the presence of direct CP violation, two dedicated experimental programmes have been in progress at CERN and Fermilab for more than 15 years and have recently published their updated results [7].

An outline of the main features of the CERN NA48 experiment is given below:

According to the usual phenomenological description, where ε characterizes the CP violation in K-antiKmixing, direct CP violation is induced by a difference in phase between the kaon transition amplitudes into two-pion in isotopic spin 0 and 2. This difference is parametrized by ε' directly linked to the measurable double ratio R:

$$\eta_{+-} \equiv \frac{A(K_L \to \pi^+\pi^-)}{A(K_S \to \pi^+\pi^-)} \approx \varepsilon + \varepsilon'$$

$$\eta_{00} \equiv \frac{A(K_L \to \pi^0\pi^0)}{A(K_S \to \pi^0\pi^0)} \approx \varepsilon - 2\varepsilon'$$

$$\text{Re}(\varepsilon'/\varepsilon) \approx \frac{1}{6}\left\{1 - \frac{\Gamma(K_L \to \pi^0\pi^0)}{\Gamma(K_S \to \pi^0\pi^0)} \Big/ \frac{\Gamma(K_L \to \pi^+\pi^-)}{\Gamma(K_S \to \pi^+\pi^-)}\right\} = \frac{1}{6}(1-R)$$

$$R = \frac{N_L^{00}}{N_S^{00}} \Big/ \frac{N_L^{+-}}{N_S^{+-}} \quad \text{i.e. the double ratio of the number of detected events,}$$

provided at least K_S/K_L or $\pi^+\pi^-/\pi^0\pi^0$ are collected at the same time.

To detect a Re(ε'/ε) of $O(10^{-3})$ and to determine its size, a dedicated double beam and detector has been built, with which the double ratio R is measured. The Kaons range in momentum between 70 and 170 Gev/c and the KS beam is produced by a fraction of the 450Gev primary protons impinging on the KL target, which are channelled by a bent silicon crystal and then brought to the KS target, located 120m downstream and displaced 7.2cm vertically relative to the KL beam axis. A KS decay is identified by the time coincidence of the event with the signal of a so called tagging hodoscope located on the attenuated proton beam line leading to the KS target.

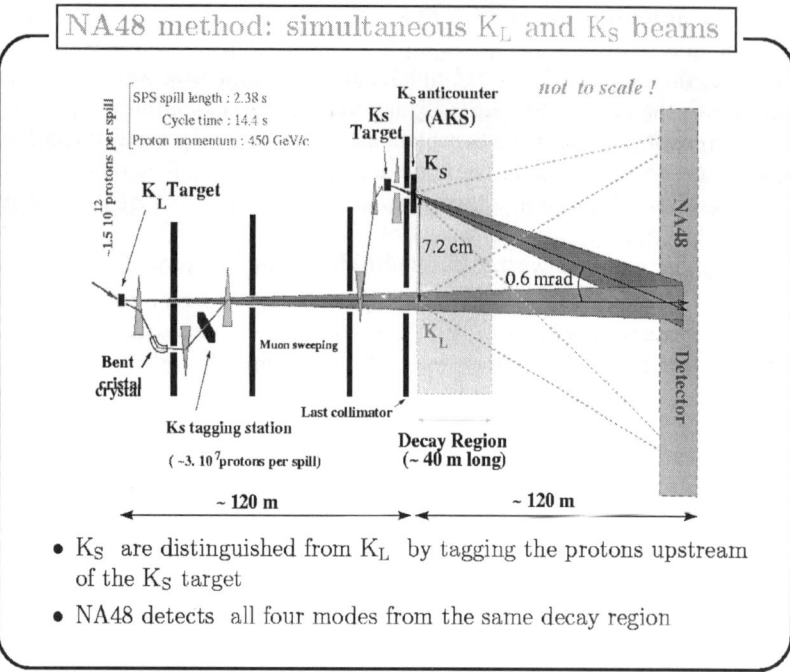

An iridium crystal, following a beam dump and collimator 6m after the KS target, converts gamma rays in $\pi^+\pi^-$ pairs and a scintillation counter in anticoincidence defines precisely the beginning of the acceptance region for the decays, which is taken, for both KS and KL, 3.5 times as long as the KS decay length.

The detector setup consists basically of two sets of minidrift wire chambers, with a magnet in between, providing a transverse momentum kick of 265MeV/c, which measure the angle of emission of the charged pion and their momenta. A fast pretrigger is provided by a scintillation counter hodoscope which follows the magnetic spectrometer. The detection of the photons produced by the π^0 decays is done in a purpose-built 20ton liquid Krypton ionization calorimeter. The active volume, which is 27radiation length thick, is subdivided in 13212 2x2cm2 towers with their axis pointing to the middle point of the fiducial decay region. The photon energy resolution obtained is:

$$\sigma(E)/E = \frac{3.2\%}{\sqrt{E}} \oplus \frac{100\text{MeV}}{E} \oplus 0.5\%$$

and the time resolution has an RMS of 277ps with tails extending beyond 2ns not exceeding the level of $9*10^{-5}$.

The calorimeter is followed by a Fe–scintillator hadronic calorimeter and by a muon veto system for background rejection.

The most recent, but still preliminary result, is based on the analysis of 1.14million

K_L in $\pi^0\pi^0$ reconstructed events. This is the smallest of the four number of events appearing in R.

The determination of R is done in bins of energy and after weighting the KL events according to the ratio of the KS/KL longitudinal decay distribution. In this way only small corrections for relative acceptance for the different decay modes, originating mostly from the slightly different transverse profiles of the KS and KL beams, need to be evaluated by Montecarlo simulation. In the next figures theNA48 result for 1998 date and the most significant results of the CERN and Fermilab collaborations are shown together with the averages from different combinations.

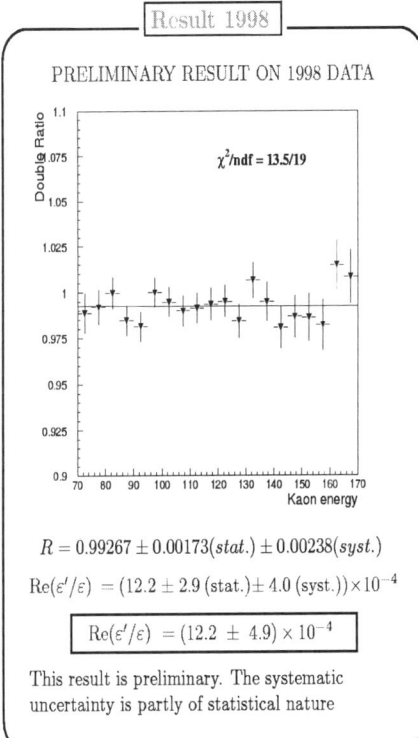
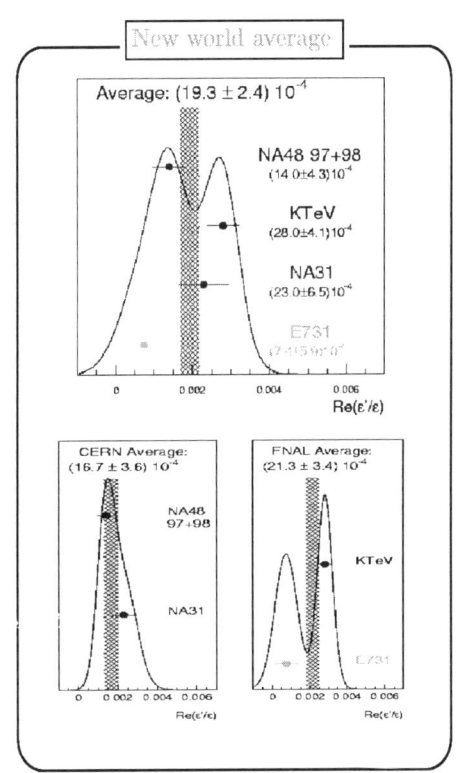

The overall world average is:

$$Re(\varepsilon'/\varepsilon) = (19.3 +- 2.4) * 10^{-4}$$

An interesting different approach has been exploited by the CPLEAR collaboration, identifying the K or antiK initial state by observing the sign of the charge of the other Kaon produced in the same proton-antiproton annihilation. The results obtained are relevant in particular for T violation and for setting upper limits to CPT non-invariance at the level of $10^{-3} \div 10^{-4}$ for many of the amplitudes describing neutral kaon decays to different final states and at the level of $2*10^{-18}$ Gev/c**2 for the difference between the diagonal elements of the neutral kaon mixing matrix [8].

They also set limits [9] to the possible existence of a dissipative decoherence of the wave function [10] which could indicate failure of quantum mechanics or e.g. an anomalous interaction of the kaon system with a gravitational background [11] and also exclude the presence of a breakdown of the equivalence principle resulting in a difference in gravitational interaction of the K relative to the antiK which had been considered by several authors, in particular e.g. by G.Chardin and J-M.Rax [12] who proposed such an effect as an alternative explanation for the CP violation in mixing.

Recently, at the INFN Frascati National Laboratory, the DAFNE-KLOE collider and detector has come in operation. In this case the neutral K antiK are produced in a state with the well defined quantum number of the Phi-meson. Very distinctive features are predicted which depend critically on the quantum mechanical description of the system. For example a coincident decay in a $\pi^+\pi^-$ pair on one side and a $\pi^0\pi^0$ pair on the other would indicate by itself the presence of direct CP violation while the observation of a coincident decay in two equal pairs would mean a loss of coherence or, particularly in the case of two pair of $\pi^0\pi^0$, an even deeper breakdown of basic principles. Of course even in the absence of hypothetical signals of entirely new physics, a number of very interesting results are expected from this experiment to help complete at a high level of precision the description of the neutral kaon system in many of its most interesting features.

Even more recently, at Stanford and KEK, new electron-positron colliders have been put in to operation, coupled with specialized detectors capable of detecting decays of b-mesons and of resolving their vertices. They are expected to be operated most of the time at the CM energy corresponding to the formation of the Y^{4S} resonance. An important point is that, given the already known properties of the Y^{4S} and of the neutral b-mesons, to detect CP asymmetries, which would vanish in case the sign of the time difference between the two decays from a given production event is not measured, it has been necessary to produce the Y^{4S} using collisions between electron and positron beams of sufficiently different energies such as to give a large boost to the Y^{4S} in the laboratory system. This fact has cost a large amount of money and technical complexity and is a direct consequence of the entanglement of the wave functions of the two b-meson, as determined by the quantum numbers of the parent Y^{4S}. There is little doubt in my opinion that the results will justify the costly choice of design for these colliders.

To conclude, I would like to point out that to overcome any problem with the possible ambiguity in the interpretation of the experimental results on neutral kaon two-pion decays, mentioned in the first part of the talk, attention has been recently focussed again on obtaining evidence for direct CP violation by comparing, at a greatly enhanced level of accuracy, the Dalitz plot distribution for K^+ $\pi^+\pi^+\pi^-$ with the corresponding one for K^- $\pi^-\pi^-\pi^+$.

Any difference among the two distributions would represent, by definition, unequivocal evidence for direct CP violation.

ACKNOWLEDGMENTS

It is a pleasure to aknowledge the stimulating friendly atmosphere of the Spin 2000 Conference and the very kind hospitality by the local organizers.

REFERENCES

1. Gell-mann, M.,and Pais, A., *Phys.Rev* **97**, 1387-1389 (1955).
2. Christenson, J.H., Cronin, J.W., Fitch, V.L., and R. Turlay R., *Phys.Rev.Lett.* **13**, 138-140 (1964).
3. Wolfenstein L., *Phys.Rev.Lett.* **13**, 562-564 (1964).
4. Kobayashi, M., and Maskawa, T., *Prog. Theor. Phys.* **49**, 652-657 (1973).
5. Greenberg, O.W., and Mohapatra, R.N., *Phys.Rev D* **39**, 2032-2038 (1989).
6. Mannelli, I., *Nucl.Phys. B* **Proc.Suppl 3**, 367-388 (1988)
7. Fanti, V., et al, *Phys.Lett. B* **465**, 335-348 (1999),
   Alavi-Harati, A., et al., *Phys.Rev.Lett*. **83**, 22-27 (1999).
8. Angelopoulos, A., et al, *Phys lett B* **471**, 332-338 (1999).
9. Adler, R., *Phys lett B* **364**, 239-245 (1995).
10. Ellis, J., et al., *Nucl Phys B* **241**, 381-405 (1984).
11. Hawking, S., *Comm.Math.Phys.* **87**, 395-415 (1982).
12. Chardin, G., and Rax, J-M., *Phys.Lett.B* **282**, 256-262 (1992).

Testing the Symmetrization Postulate and the Spin-Statistics Connection for Nuclei by Molecular Spectroscopy

Guglielmo M. Tino

Dipartimento di Scienze Fisiche dell'Università di Napoli "Federico II" and INFM, Complesso Universitario di Monte S. Angelo, via Cintia, I-80126 Napoli, Italy

Abstract. I present the idea underlying the experiments on molecules to test the validity of the symmetrization postulate of quantum mechanics and the spin-statistics connection for nuclei. I describe in some detail recent experiments on O_2 and CO_2 molecules and show that these should be considered as tests of the spin-statistics connection for the spin-0 oxygen nuclei. I discuss the possibility of new experiments to test the spin-statistics connection and the symmetrization postulate.

INTRODUCTION

The symmetrization postulate establishes that in a system containing identical particles the only possible states are either all symmetrical or all antisymmetrical with respect to permutations of the particles. In the first case, the particles are called bosons and follow Bose-Einstein statistics; in the second case they are called fermions and follow Fermi-Dirac statistics. Experiments indicate that particles with integral values of spin are bosons, while particles with half-integral spin are fermions. The reason why only symmetric and antisymmetric states seem to occurr in nature and the connection with the spin of the particles has been a puzzle since the early days of quantum mechanics [1,2]. The spin-statistics theorem, proved by W. Pauli [3] from the basic principles of quantum field theory and special relativity, states that given the choice between Bose and Fermi statistics, integral-spin particles must obey Bose statistics and half-integral-spin particles must obey Fermi statistics. Proofs of the spin-statistics theorem are discussed in [4] and in contributions to this volume. Quantum mechanics would nevertheless allow also symmetries different from those imposed by the symmetrization postulate, and theories have been developed allowing for small deviations from conventional symmetry which might have been masked in the experiments performed so far. It is worth noting that no theory so far predicts the possibility of observing a violation in a particular system or in some specific condition. Consistent theories can be formulated, however, which would lead to different symmetry properties. Experiments are needed then to discriminate between these theories by imposing constraints which are the more stringent the higher is the experimental precision.

In this paper, after a brief account of the theoretical background, the experimental tests on molecules including identical spin-0 nuclei are presented, trying to clarify in a consistent manner their meaning. In the spectra of these molecules, some lines are missing. Their absence can be understood when the requirement of symmetry of the molecular wave function under exchange of the identical nuclei is considered, which prevents the occupation of some of the states. The basic idea of the experiments presented here is to search for molecules in such "wrong-symmetry" states. It is shown that the experiments performed so far should be considered as tests of the spin-statistics connection for ^{16}O nuclei. The possibility of experiments to search for violations of the symmetrization postulate for nuclei is also discussed.

THEORETICAL BACKGROUND

If a system includes N particles which have the same intrinsic properties (mass, charge, spin,...), the states represented by vectors differing only by a permutation of identical particles cannot be distinguished by any observation. All the physical observables must be invariant under permutation of identical particles. The Hilbert space of the complete system can be decomposed into orthogonal sub-spaces which correspond to the irreducible representations of the permutation group. The irreducible eigensubspaces may have dimension greater than one. If a representation is of order one, the level is not degenerate; the consequence of a permutation of identical particles is only to multiply the eigenfunction by a phase factor. If the representation is of order $h > 1$, the associated eigenvalue is h-fold degenerate. Although it is not possible in this case to associate a single vector to a physical state, this degeneracy does not cause any difficulty: the measurable results on a state do not depend on which vector is chosen in the h-dimensional subspace to represent the state [5].

The symmetrization postulate admits only the representations of order one, that is only the completely symmetric and antisymmetric functions. The connection with the spin of the particles (taken as an experimental evidence or demonstrated from basic assumptions) completely defines the symmetry allowed for each type of particle.

Two important points must be noted to avoid confusion in the interpretation of experiments searching for violations of the symmetries dictated by the symmetrization postulate and by the spin-statistics connection. The first is that in a system including only two identical particles, the symmetric and antisymmetric representations are the only irreducible representations. Only two "entangled" states are then possible:

$$\Psi_S(1,2) = \tfrac{1}{\sqrt{2}}[\psi_a(1)\psi_b(2) + \psi_b(1)\psi_a(2)]$$
$$\Psi_A(1,2) = \tfrac{1}{\sqrt{2}}[\psi_a(1)\psi_b(2) - \psi_b(1)\psi_a(2)]$$
(1)

where ψ_a and ψ_b are single-particle wave functions. In this case, the symmetrization postulate does not play a relevant role, the question being only the connection between the intrinsic spin of the particles and the symmetry of the two-particle state. This is not the case for systems including three or more identical particles. For three identical particles, for example, the state of the system could be represented by any of six vectors. The permutation group has two one-dimensional representations (the symmetric and the antisymmetric one) and two two-dimensional representations. In this case, the symmetrization postulate limits the possible states to the symmetric and antisymmetric ones and the connection to the particle spin determines which of the two states is occupied by each type of particle.

The second important remark is that if a system of N particles is in a particular state belonging to a given irreducible representation of the permutation group, it cannot be brought by any perturbation into another state belonging to a different representation. The perturbation operator must indeed be symmetric under particles exchange so that matrix elements between the initial state and states corresponding to a different representation vanish. The symmetry character of the wave function must then remain the same in time (the case of a variable number of particles requires special consideration). This "superselection rule" is a rigorous selection rule holding also in the presence of perturbations such as collisions or electric fields [1,6].

A consequence of the superselection rule is that it is not possible to consider states given by a coherent superposition of states with different permutation symmetries. Therefore symmetry violations in systems including identical particles can only be described in terms of an incoherent mixture which is represented by a density matrix. In the case of two integral-spin particles, for example, the density matrix taking into account small symmetry violations is:

$$\rho_2 = \left(1 - \tfrac{1}{2}\beta^2\right)\rho_s + \tfrac{1}{2}\beta^2 \rho_a \tag{2}$$

where $\rho_{s(a)}$ is the symmetric (antisymmetric) two-particle density matrix. A pair of particles will be found in the normal symmetric state with probability $(1 - \beta^2/2)$ and in the anomalous antisymmetric state with probability $\beta^2/2$. In the case of particles with half-integral values of spin, ρ_s and ρ_a are exchanged.

Using the notation adopted in the literature, in the following $\beta^2/2$ indicates the "symmetry-violation" parameter. Its real meaning needs to be specified, however, for the particular physical system and the theoretical model considered.

It is not the purpose of this paper to discuss in detail the theories allowing for symmetry properties different from the ones which are peculiar to bosons and fermions. A survey can be found in [7] and in contributions to the present volume. It is of interest, however, to put in evidence the possibility of theories allowing for small deviations from the usual symmetry relations, whose search is the subject of the experiments discussed in this paper. Such deviations can be expressed as a different symmetry of the state under particle exchange or, in Fock-space representation, as a deformation of the algebra of the creation and annihilation operators. A statistics intermediate between Bose and Fermi cases was first proposed in [8] considering the possibility that at most n identical particles could occupy the same quantum state. This idea led to a generalized field theory [9], called parastatistics, in which the field operators obey trilinear commutation relations instead of the usual bilinear relations. These theories predict, however, gross violations of statistics which are immediately excluded by experimental evidence. The possibility of a continuous interpolation between bosonic and fermionic behaviours is given by "quons" [10]. The commutation and anticommutation relations are replaced by generalized bilinear commutation relations depending on a parameter q (q-mutators):

$$a_k a_l^+ - q a_l^+ a_k = \delta_{kl} \qquad -1 \leq q \leq 1 \qquad (3)$$

with the vacuum condition $a_k|0\rangle = 0$. As q varies between -1 and 1, the symmetry changes continuously from the completely antisymmetric case (fermions) to the completely symmetric case (bosons). In this frame, the value of $\beta^2/2$ can be related to the value of the q parameter: for small violations of Bose statistics, $\beta^2 = 1-q$; for small violations of Fermi statistics, $\beta^2 = 1+q$. It can be shown that this interpolation preserves positivity of norms and the non-relativistic form of locality [11]. Other aspects are still doubtful such as the possibility of accounting for local observables in a relativistic theory or for the existence of antiparticles. Statistics other than Fermi and Bose have also been investigated for one- and two-dimensional systems [12] and in connection with anyon high-temperature superconducting systems [13].

EXPERIMENTAL TESTS

In this section, experiments performed to search for violations of the symmetrization postulate and/or of the spin-statistics connection are discussed. After a brief review of experiments on electrons and photons, recent experiments on integral-spin nuclei in molecules are described in detail. In fact, no accurate test for integral-spin particles had been reported until recently. This is due to the fact that while there are several systems in which a violation of the Pauli exclusion principle would be detected as a signal on a zero background, the effect of a small violation for particles following Bose-Einstein statistics would usually manifest itself as a small change in the properties of a many-particle system. This obviously limits the achievable accuracy. In [14], a bound to a possible violation of the generalized Bose statistics for pions was inferred considering the $K_2 \rightarrow \pi^+\pi^-$ decay, which is usually considered as due to CP violation (see also the contribution of I. Mannelli in this book).

For a review of the experiments before 1989, the reader is referred to [14]. Some of the initial experiments suffered from a misunderstanding of the constraints imposed by the "superselection rule". It is shown here that also the interpretation given for some recent experiments is questionable. In fact, although the papers published on this subject usually present their "null results" as a confirmation of the validity of the symmetrization postulate, most of them should be considered as tests of the spin-statistics connection. An experiment that would allow a genuine test of the symmetrization postulate for nuclei is discussed in the following.

Experiments on electrons

A few experiments have been performed to test the validity of the Pauli principle for half-integral-spin particles. In particular, a high precision test on electrons was performed in [15] by running a current through a copper bar and searching for X-rays that would be emitted if some of the electrons introduced in the sample were captured by a copper atom and cascaded down to the 1S state, which is already filled with two electrons. No signal was found and this was interpreted as giving a limit $\beta^2/2 \leq 1.7 \times 10^{-26}$ to the probability that a new electron introduced into copper would form a mixed-symmetry state with respect to the other electrons already present in the copper sample. In this experiment, the large number of electrons in the system was important to reach such a high sensitivity but, on the other hand, makes the interpretation of the result more complicated. Conclusions may depend, for example, on whether we consider the symmetry of the system composed by the injected electron plus the electrons already present in the copper bar, or we consider a model in which the electron collides with a copper atom and is captured. A simpler two-electron system was investigated in [16]. A spectroscopic test was performed on helium atoms, searching for a transition involving the permutation symmetric 1s2s $^1S_0$ state. An upper bound $\beta^2/2 \leq 5 \times 10^{-6}$ was set to a violation of the Pauli principle. In spite of the lower sensitivity, the interpretation of this result is simpler. Since only two identical particles are involved, this should be considered as a test of the spin-statistics connection for electrons. Doubts can be raised, however, about what would be the chemical stability of "paronic" atoms in ordinary samples. In [16], this was taken into account by having the atoms ionize and recombine in a discharge before entering the detection region.

Experiments on Photons

Several papers have been published recently reporting or proposing experiments to set a limit to possible violations of Bose statistics for photons [17,18,19,20,21,22]. The fundamental nature of the photon and its peculiar properties make it very interesting to investigate this particle in this context. It is hard, however, to find an experiment that would give a direct evidence of a violation with a significant sensitivity. This is one case, in fact, in which a small deviation from normal statistics would usually produce only a small signal over a large background. An attempt to set a limit to a possible violation of Bose statistics was made in [17], based on light intensities attainable in laser systems. In [18], a possible dependence of the frequency of light on its intensity was searched for. This effect is expected if a q-nonlinearity is introduced in the description of the electromagnetic field. Since nonlinearities in the commutation relations give rise to mixed-symmetry states, this experiment could also be reinterpreted as a search for a violation of the symmetrization postulate for photons. The connection is not straightforward though and was not pursued in the paper. In [19], the experimental upper limit on the two-gamma decay of Z-boson, $Z \to \gamma\gamma$, was used to establish an upper bound to a possible small violation of the exchange-symmetry for a system of two photons. The same idea, based on what is called Landau-Yang theorem, was exploited in [20] to improve the limit by searching for the forbidden $J = 0 \to J' = 1$ transitions in atoms excited by two photons of the same energy. A limit of $\beta^2/2 \leq 10^{-7}$ was set on the probability that two photons are in an exchange-antisymmetric state. A different approach was followed in [22]. A very tight bound to a violation of statistics for photons was inferred considering photons and electrons as coupled "quons" and relating the bound for photons to that obtained in [15] for electrons. Although this argument is indirect and model-dependent, it is very interesting and it could also be extended to other particles.

Experiments on Nuclei in Molecules

Wigner [23] and Ehrenfest and Oppenheimer [24] showed that a composite system of fermions is a boson or a fermion depending on whether it is made of an even or an odd number of fermions (in fact, for this argument to be valid, it is necessary that the interaction between the composite particles is negligible compared to the internal excitation energy so that the internal structure can be neglected and the system can be considered as a single particle). Considering the total angular momentum resulting from the constituents angular momenta, an extension of the spin-statistics connection to composite systems, such as nuclei, is obtained.

The requirement of symmetry of the wave function under exchange of identical particles has a striking demonstration in the spectra of molecules including identical nuclei. Let us consider a molecule containing two identical spin-0 nuclei as, for example, $^{16}O_2$. According to the Born-Oppenheimer approximation and neglecting the coupling of the nuclear spin with the rest of the molecule (which is not important for these experiments since the spin of the nuclei is zero), the total wave function ψ_t can be written in the form

$$\psi_t = \psi_e \psi_v \psi_r \psi_n \qquad (4)$$

where ψ_e, ψ_v, ψ_r are the electronic, vibrational and rotational functions, respectively, and ψ_n is the nuclear spin function. For integer-spin nuclei, the total wavefunction ψ_t must be symmetric in the exchange of two nuclei. The $^{16}O_2$ molecule represents a particularly simple case because the nuclear spin of ^{16}O is zero and ψ_n is therefore obviously symmetric. The vibrational wave function ψ_v is also unaltered in the exchange of the nuclei because it depends only on the magnitude of the internuclear distance. Since the total wavefunction ψ_t must be symmetric, only the states corresponding to even (odd) rotational quantum numbers are allowed if ψ_e is symmetric (antisymmetric) [25]. In Fig. 1, a simplified scheme is shown of the levels of the $^{16}O_2$ molecule relevant for the experiments described in the following. The ground state is a $^3\Sigma_g^-$ state, which is antisymmetric under the exchange of the two nuclei. The rotational states corresponding to even values of the rotational number K are therefore forbidden. Indeed, since the early work on $^{16}O_2$ spectra [26], it was observed that alternate lines are missing.

The other molecule investigated in this context, the $^{12}C^{16}O_2$ molecule, has the same symmetry properties. Both the ground electronic and vibrational wave functions are in this case symmetric in the exchange of the two ^{16}O nuclei. Therefore only rotational states corresponding to even values of the rotational quantum number are allowed.

The basic idea of the spectroscopic tests described in the following is to search with extremely high sensitivity for (weak) molecular lines involving the forbidden states. Indeed, the effect of a (small) violation of the expected symmetry would be that some molecules could be found in antisymmetric states corresponding to wrong-parity rotational numbers. The possibility of such a test was first suggested in [27] considering the CO_2 molecule. A specific experiment on O_2 was later proposed in [28]. The idea of these experiments is analogous to the one underlying the experiment on electrons in the helium atom [16]. This represents indeed a rare case in which a violation of the spin-statistics connection for integer-spin particles would be detected on a virtually zero background with a sensitivity comparable to the one achieved in experiments on fermions.

Experiments on $^{16}O_2$

In a first series of experiments [29,30,31], the spectrum of the $^{16}O_2$ molecule was investigated searching for transitions between states which are antisymmetric under the exchange of the two nuclei. The purpose of these experiments was to measure or to bound the relative abundance of molecules in "wrong-symmetry" states. I will first describe in some detail the experimental approach and results reported in [30] and then compare it with other experiments performed or planned on O_2.

The choice of the oxygen molecule was motivated by the simplicity of this system which makes the interpretation of the results easier. The 0-0 band of the $X\,^3\Sigma_g^- \rightarrow b\,^1\Sigma_g^+$ system of $^{16}O_2$ spectra around 762 nm was investigated. The observed transitions in this region are weak magnetic dipole transitions with an absorption coefficient of 10^{-6} cm^{-1}. However, the lines are narrow and well isolated, and high sensitivity laser spectroscopy detection methods can be used.

In [30], the laser source was a distributed-feedback (DFB) diode laser emitting 5 mW cw in a single mode. The emission wavelength could be varied between 760 nm

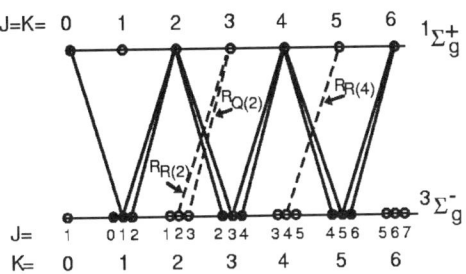

FIGURE 1. Partial scheme of the energy levels and the magnetic-dipole transitions of $^{16}O_2$. There is an ^RR branch ($\Delta J = +1$, $\Delta K = +1$), a ^PP branch ($\Delta J = -1$, $\Delta K = -1$), an ^RQ branch ($\Delta J = 0$, $\Delta K = +1$) and a ^PQ branch ($\Delta J = 0$, $\Delta K = -1$). The observed transitions (full line in the figure) start from the odd-K rotational levels of the electronic ground state, which are symmetric with respect to an exchange of the nuclei. Transitions starting from even numbered levels are missing. The ones shown in the figure, indicated by broken lines, were searched for in [30].

and 762 nm by changing the temperature of the laser. A part of the light was sent to a 7-digits wavelength-meter and to a temperature-stabilized Fabry-Perot interferometer which provided a stable frequency marker. The change of the laser frequency was measured to be less than 20 MHz in a time of 100 s. The absorption cell was a White-type multipass cell. The absorption path was 100 m. A 10-cm focal-length lens focused the laser light emerging from the cell into a Si photodiode-preamplifier. In this experiment, a low-frequency wavelength-modulation detection technique was used. The laser frequency was modulated at a frequency $f = 40$ kHz by adding a modulation to the bias injection current of the diode. The output of the photodetector was sent to a lock-in amplifier and demodulated at a frequency of $2f$ in order to increase the detection sensitivity and to reduce a background signal due to the change of the laser intensity during the frequency scan. The output signal of the lock-in was sent to a digital oscilloscope and to a personal computer for the data acquisition. In order to improve the signal-to-noise ratio, the signal was averaged over several scans of the laser frequency.

The sensitivity achieved with the apparatus described above was high enough to observe, in addition to the lines of $^{16}O_2$, also those of the $^{16}O^{18}O$ and $^{16}O^{17}O$ molecules, with a total pressure in the cell as low as a few mbar. Fig. 2 shows, for example, the signal recorded for the $^RR(2)$ and $^RQ(1)$ line of $^{16}O^{17}O$ and $^{16}O^{18}O$, respectively, in a natural abundance sample with a total pressure in the cell of 20 Torr. Historically, the detection of these weak lines in the atmospheric oxygen spectra led to the identification of the ^{18}O and ^{17}O isotopes of oxygen [26,32]. The spectra shown in Fig. 2 are interesting for two reasons. First, they provide a striking example of the effect of the instinguishability of nuclei: the $^RR(2)$ line can only be observed because the two nuclei in the $^{16}O^{17}O$ molecule differ by one neutron. Second, because of the small natural abundance of these isotopes ($^{18}O = 0.2\%$, $^{17}O = 0.04\%$), their weak lines provided an accurate test of the sensitivity of the apparatus, which is particularly important in an experiment leading to a null result.

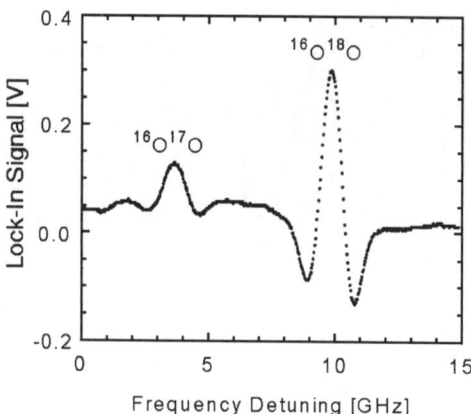

FIGURE 2. Detection of rare isotopes of oxygen. The signals shown correspond to the ^{R}R(2) line of ^{16}O^{17}O and to the ^{R}Q(1) line of ^{16}O^{18}O as observed in a natural abundance sample with a total pressure of 20 Torr in the cell. (From Ref. [30]).

As discussed above, the purpose of this experiment was to set an upper limit to the intensity of possible absorption lines starting from exchange-antisymmetric states, with respect to the intensity of the normal transitions of ^{16}O$_2$ involving symmetric states. The expected position of the missing transitions was calculated, with an uncertainty of less than 300 MHz, assuming the same Hamiltonian as for molecules in allowed states and using the data available in the literature for the relevant molecular parameters [33]. We searched for the following transitions: ^{R}R(2) at 761.6740 nm, ^{R}Q(2) at 761.5626 nm, ^{R}R(4) at 761.3891 nm (vacuum wavelengths). Fig. 3(a) shows the signal recorded as the laser frequency was scanned across the expected position (indicated by the arrow) of the missing ^{R}R(4) line. The total O$_2$ pressure in the cell was 200 Torr. In Fig. 3(b), the absorption signal for the observed ^{R}R(3) line of ^{16}O$_2$ is shown for comparison. In this case, the pressure in the cell was reduced to 5 Torr in order to avoid a broadening of the line due to the optical depth of the sample. The laser modulation width was adjusted in order to keep the same value of modulation index. In Fig. 3(a) the vertical scale was expanded by a factor 10^4 with respect to Fig. 3(b). The noise observed in Fig. 3(a) is due to the laser intensity noise and to small interference fringes caused by the residual reflectivity of the AR-coated optical components.

The spectra recorded in this work showed no evidence of anomalous lines. By comparing the noise level with the intensity of the observed lines, an upper limit can then be deduced for the intensity of forbidden lines with respect to the intensity of observed lines. After scaling the recorded signals for the different values of the pressure in the cell, and taking into account the correction due to the pressure broadening of the lines, an upper bound of $\beta^2/2 \leq (5\pm2)\times10^{-7}$ was set to a possible violation of the expected symmetry for ^{16}O nuclei.

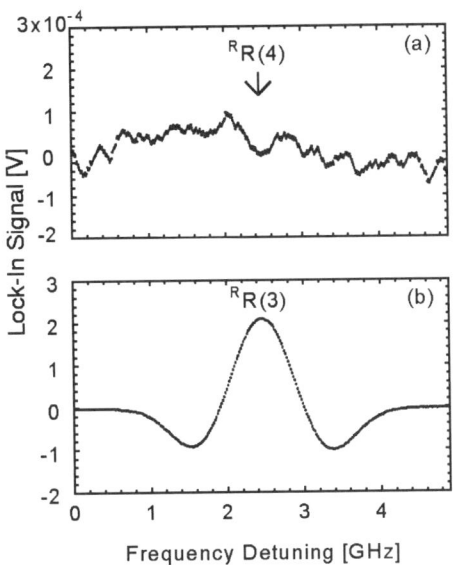

FIGURE 3. (a) The signal recorded as the laser frequency was scanned over the region (indicated by the arrow) where the $^RR(4)$ missing line would be observed (O_2 pressure P = 200 Torr). (b) Signal corresponding to the $^RR(3)$ transition of $^{16}O_2$ (P = 5 Torr). The vertical scale in (a) was expanded by a factor 10^4 with respect to (b). (From Ref. [30]).

A similar result was obtained in [31]. Transitions in the same $X\,^3\Sigma_g^- \rightarrow b\,^1\Sigma_g^+$ band of $^{16}O_2$ were investigated. The apparatus was based on a tunable diode laser source and a 4-m long absorption path. A low-frequency wavelength modulation and a $2f$ demodulation detection scheme was used. The missing line searched for was the $^PQ(22)$ [34]. A limit $\beta^2/2 \leq 10^{-6}$ was obtained.

Experiments on $^{16}O_2$ were later reported in [35,36]. In [35], a cavity-ring-down spectroscopy method was used to perform a broad investigation of O_2 spectra including less abundant isotopes and detecting also electric quadrupole lines. Although the detection sensitivity was not high enough to improve the limit set in [30], this work provided a check of previous results. A significant improvement in sensitivity was obtained in [36] using a high-finesse optical cavity as absorption cell, with an equivalent absorption length of ~1 km, and a radio frequency detection scheme. This allowed to reach a very high detection sensitivity, comparable to that achieved in [37] in the investigation of electric quadrupole lines of O_2 in the sunset spectrum of the Earth's atmosphere. A limit $\beta^2/2 \leq 5 \times 10^{-8}$ was set, which is the best value obtained so far with experiments on $^{16}O_2$. In this experiment, the sensitivity was mainly limited by technical factors and further improvements can be obtained with this method. A different method that would allow, in principle, to reach extremely high sensitivity in the detection of molecules in "wrong-symmetry" states is based on resonant ionization spectroscopy. An experiment is in progress [38]. The possibility of extending the experiments to a search of missing lines in the spectra of $^{18}O_2$ was discussed in [39].

Experiments on $^{12}C^{16}O_2$

The most accurate test on ^{16}O nuclei was performed by investigating the vibrational spectrum of the $^{12}C^{16}O_2$ molecule [40]. The CO_2 molecule has the same symmetry properties of O_2 but, since it is triatomic, it has strong active vibrational bands in the infrared which are lacking in O_2 spectra. In particular, the intensity of the $12^01\text{-}00^00$ combination band around 2 μm, investigated in [40], is more than two orders of magnitude larger than the electronic transitions of oxygen previously investigated; the absorbance of a given population is correspondingly larger and this results in an increased detection sensitivity. Since the ground electronic wavefunction of the CO_2 molecule is symmetric in the exchange of the two ^{16}O nuclei, the rotational wavefunction in the ground vibrational state must be symmetric and only even values for the rotational quantum number J are allowed. The R-branch, for example, should then be composed only of R(2J) transitions.

The position of the missing transitions was calculated from the molecular parameters [41] with an uncertainty smaller than 50 MHz. Only two lines were investigated, the others being too close to transitions belonging to hot bands or rare isotopes. The chosen transitions were the R(25) at 2.001790 μm and the R(33) at 2.000015 μm.

An InGaAsP distributed-feedback (DFB) diode laser was used as laser source, emitting on a single mode around 2 μm, with an output power of 7 mW. The intensity of the observed lines was preliminarily measured by a direct absorption scheme, on a relatively short pathlength (20-80 cm) in a single-pass cell. For each transition, the line profile was recorded for a set of CO_2 pressures and integrated, in order to extract the line intensity S. The following step consisted in the measurement of the intensity of selected weak transitions in the region of the missing lines; this provided an accurate scaling factor to compare the intensity of allowed and missing transitions. This intermediate measurement was performed with a 20-m Herriot-type multipass cell, by means of direct absorption and two-tone frequency modulation techniques.

The central part of the experiment was a high-sensitivity search for the missing lines, using a 130 m pathlength (Fig. 4). A balanced two-tone frequency modulation technique was used. The modulation frequencies were (1300 ± 3.5) MHZ, and the detection was performed at 7 MHz. The effective detection bandwidth could be reduced by averaging a large number of scans. The maximum acquisition time was 10 minutes, corresponding to approximately 500 scans, limited by a slow drift of the laser emission frequency with respect to the absorption profiles. The CO_2 pressure in the cell was set to 30 Torr, corresponding to the maximum sensitivity for this apparatus. The two-tone frequency modulation signal consisted of various replicas of the absorption profile, spaced by 1300 MHz, with intensities scaling as a function of the modulation index. Such replicas were used to further scale the intensity measurement,

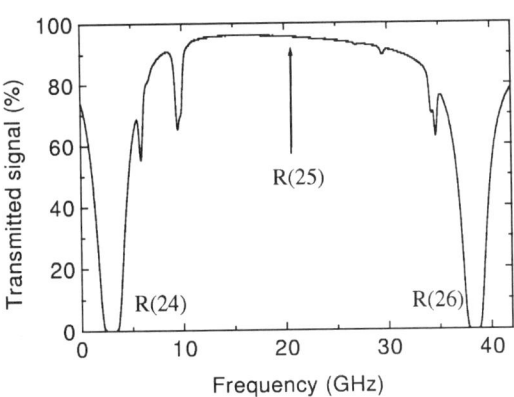

FIGURE 4. Absorption spectrum in the region where the R(25) missing line of the $12^01 - 00^00$ combination band of $^{12}C^{16}O_2$ would be observed. The weak lines belong to hot bands or other isotopes of CO_2. The calculated position of the missing line is indicated by the arrow. The spectrum was recorded with an absorption path of 130 m and 2.5 Torr of gas in the cell. (From Ref. [40]).

down to the noise level. A bound of $\beta^2/2 \leq (2.1 \pm 0.7) \times 10^{-9}$ to the relative population of the forbidden states was deduced in this work. This experiment gives at present the most stringent test of the spin-statistics connection for ^{16}O nuclei (for very recent results, see contribution by G. Modugno et al. in this volume).

It is worth mentioning that in this experiment, several transitions of medium intensity ($S=10^{-28} - 10^{-25}$) were observed which are not assigned in the literature. Most of them could be grouped in three different bands, and all the experimental observations about the Doppler width, pressure broadening coefficient and the spacing between the lines, seemed to assign them to a symmetric isotope of CO_2. A systematic investigation of the dependence of the intensity of these lines on the gas temperature allowed to exclude, however, any connection with symmetry-violating states.

Experiments on Polyatomic Molecules and Test of the Symmetrization Postulate

An interesting prospect is the investigation of spectra of molecules containing more than two identical nuclei. Most of the experiments performed so far involve only two identical particles, therefore providing a test of the spin-statistics connection. In order to search for possible violations of the symmetrization postulate of quantum mechanics, systems including more than two identical particles should be considered. As discussed above, in this case possible symmetries do not reduce to the completely symmetric and antisymmetric cases. Molecules offer indeed the possibility of investigating more and more complex structures and, as for the experiments performed on O_2 and CO_2, high detection sensitivity can be achieved by laser spectroscopy methods. The interest of experiments on molecules including more than two identical nuclei as tests of the symmetrization postulate was first mentioned in [40] and discussed in [42]. A good candidate for this experiment is OsO$_4$, a highly symmetric molecule with up to four identical spin-0 nuclei. In spite of the higher complexity of the spectrum with respect to the simpler molecules investigated previously, high resolution and high sensitivity spectroscopy schemes have been developed, especially in the region around 10 μm which is of metrological interest. The good knowledge of molecular parameters makes it simpler to find the position of the relevant transitions and to separate them from spurious signals. In particular, transition frequencies can be singled out that would represent a signature of a violation of the symmetrization postulate. The expected detection sensitivity is expected to be comparable to that achieved in previous tests on molecules. Another possibility, based on the investigation of molecules with three identical nuclei such as SO$_3$ and NH$_3$, was studied in detail by G. Modugno et al. (see their contribution in this volume). An experiment on OsO$_4$ is in progress and preliminary results are presented in the contribution by Ch. Bordé and Ch. Chardonnet in this volume.

CONCLUSIONS

Several experiments confirm the validity of the spin-statistics connection for various types of particles to a high level of accuracy. In particular, spectroscopic tests on molecules provide an upper bound to a possible violation of the expected symmetry for ^{16}O nuclei. The best results published so far are $\beta^2/2 \leq 5\times10^{-8}$ from spectroscopy of O$_2$ and $\beta^2/2 \leq 2\times10^{-9}$ from spectroscopy of CO$_2$. This provides a proof of the general formalism of quantum mechanics and can be extended also to different particles. In a recent paper [43], the results obtained in [40] for nuclei were used to set a bound on possible violations of the Pauli exclusion principle for nucleons and for quarks.

Spectroscopic tests can be further improved and extended to new systems: the detection sensitivity can be increased, for example, by reducing the technical noise, by increasing the absorption pathlength, or by selecting stronger transitions. The fundamental vibrational band of CO$_2$ around 4.3 μm, which is about two thousand times stronger than the one at 2 μm investigated so far, can provide a significative increase of sensitivity. An increase in the detection sensitivity could also be obtained by a resonant-ionization-spectroscopy detection scheme. Similar tests can be performed on other nuclei as, for example, ^{18}O.

On the other hand, not many experiments have really tested the validity of the symmetrization postulate with high precision. Experiments on molecules including more than two identical nuclei should allow to test the validity of this fundamental postulate of quantum mechanics with an accuracy comparable to that achieved in the spectroscopic tests of the spin-statistics connection reviewed in this paper.

REFERENCES

1. Pauli, W., *General Principles of Quantum Mechanics*, Springer-Verlag, Berlin, 1980.
2. Feynman, R.P., Leighton, R.B., and Sands, M., *The Feynman Lectures on Physics, vol.III: Quantum Mechanics*, Addison-Wesley, 1965.
3. Pauli, W., *Phys. Rev.* **58**, 716-722 (1940).
4. Duck I., and Sudarshan, E.C.G., *Pauli and the Spin-Statistics Theorem*, World Scientific, Singapore, 1997; *Am. J. Phys.* **66**, 284-303 (1998).
5. Messiah, A.M.L., and Greenberg, O.W., *Phys. Rev.* B **136**, 248-267 (1964).
6. Amado, R.D., and Primakoff, H., *Phys. Rev.* C **22**, 1338-1340 (1980).
7. Greenberg, O.W., and Greenberger, D.M., in *Quantum Coherence and Reality*, edited by J.S. Anandan and J.L. Safko, World Scientific, Singapore, 1994.
8. Gentile, G., *Il Nuovo Cimento* **17**, 493-497 (1940).
9. Green, H.S., *Phys. Rev.* **90**, 270-273 (1953).
10. Greenberg, O.W., *Phys. Rev.* D **43**, 4111-4120 (1991).
11. Greenberg, O.W., *Physica* A **180**, 419-427 (1992).
12. Leinaas, J.M., and Myrheim, J., *Il Nuovo Cimento* B **37**, 1-23 (1977).
13. Wilczek, F., *Fractional Statistics and Anyon Superconductivity*, World, Singapore, 1990.
14. Greenberg, O.W., and Mohapatra, R.N., *Phys. Rev.* D **39**, 2032-2038 (1989).
15. Ramberg, E., and Snow, G.A., *Phys. Lett.* B **238**, 438-441 (1990).
16. Deilamian, K., Gillaspy, J.D., and Kelleher, D.E., *Phys. Rev. Lett.* **74**, 4787-4790 (1995).
17. Fivel, D. I., *Phys. Rev.* A **43**, 4913-4922 (1991). For a critique of this work, see: O. W. Greenberg in *Workshop on Harmonic Oscillators*, NASA Conference Pub. 3197, edited by D. Han, Y. S. Kim and W. W. Zachary, NASA, Greenbelt, 1993.
18. Man'ko, V.I., and Tino, G.M., *Phys. Lett.* A **202**, 24-27 (1995).
19. Ignatiev, A.Yu., Joshi, G.C., and Matsuda, M., *Mod. Phys. Lett.* **11**, 871-876 (1996).
20. DeMille, D., Budker, D., Derr, N., and Deveney, E., *Phys. Rev. Lett.* **83**, 3978-3981 (1999).
21. Gerry, C.C., and Hilborn, R.C., *Phys. Rev.* A **55**, 4126-4130 (1997).
22. Greenberg, O.W., and Hilborn, R.C., *Found. Phys.* **29**, 397-407 (1999).
23. Wigner, E.P., *Math. Naturwiss. Anz. Ung. Ak. Wiss.* **46**, 576 (1929).
24. Ehrenfest, P., and Oppenheimer, J.R., *Phys. Rev.* **37**, 333-338 (1931).
25. Herzberg, G., *Spectra of Diatomic Molecules*, D. Van Nostrand Company, Princeton, 1950.
26. Dieke, G.H., and Babcock, H.D., *Proc. Natl. Acad. Sci. U.S.* **13**, 670-8 (1927).
27. Hilborn, R.C., *Bull. Am. Phys. Soc.* **35**, 982 (1990).
28. Tino, G.M., *Il Nuovo Cimento* D **16**, 523-530 (1994).
29. Tino, G.M., de Angelis, M., and Gianfrani, L., in *Laser Spectroscopy XII International Conference*, edited by M. Inguscio, M. Allegrini, A. Sasso, World Scientific, Singapore, 1996, pp. 105-108.
30. de Angelis, M., Gagliardi, G., Gianfrani, L., and Tino, G.M., *Phys. Rev. Lett.* **76**, 2840-2843 (1996).
31. Hilborn, R.C., and Yuca, C.L., *Phys. Rev. Lett.* **76**, 2844-2847 (1996).
32. Giauque, W.F., and Johnston, H.L., *J. Am. Chem. Soc.* **51**, 3528-3534 (1929).
33. Babcock, H.D., and Herzberg, L., *Astrophys. J.* **108**, 167-190 (1948).
34. See note in [35].
35. Naus, H., de Lange, A., and Ubachs, W., *Phys. Rev.* A **56**, 4755-4763 (1997).
36. Gianfrani, L., Fox, R.W., and Hollberg, L., *J. Opt. Soc. Am.* B **16**, 2247-2254 (1999).
37. Brault, J.W., *J. Mol. Spectrosc.* **80**, 384-387 (1980).
38. Eyler, E., private communication.
39. Gagliardi, G., Gianfrani, L., and Tino, G.M., *Phys. Rev.* A **55**, 4597-4600 (1997).
40. Modugno, G., Inguscio, M., and Tino, G.M., *Phys. Rev. Lett.* **81**, 4790-4793 (1998).
41. Rothman, L.S., *et al.*, *J. Quant. Spectrosc. Rad. Tr.* **48**, 537 (1992).
42. Tino, G.M., *Fortschr. Phys.* **48**, 537-543 (2000).
43. Greenberg, O.W., and Hilborn, R.C., *Phys. Rev. Lett.* **83**, 4460-4463 (1999).

The Pauli Principle and Ultrahigh Resolution Spectroscopy of Polyatomic Molecules

Christian J. Bordé and Christian Chardonnet

Laboratoire de Physique des Lasers, CNRS UMR 7538, Université Paris-Nord, Avenue J.-B. Clément, 93430 Villetaneuse, France

Abstract. We review some observed consequences of the Pauli principle in ultrahigh resolution spectroscopy of polyatomic molecules with several identical bosonic or fermionic nuclei. An important feature is the link with parity. Several possible tests of the symmetrization postulate come out of this analysis and we discuss in more detail an experiment which is presently conducted with OsO_4 in our laboratory.

INTRODUCTION

Tests of the Pauli symmetrization postulate have, this far, been performed at a very high sensitivity level, only on simple spectra of diatomic or triatomic molecules with two identical spinless nuclei (specifically ^{16}O) [1–7] (see also the contributions of G. Tino and R.C. Hilborn in this book and references therein). Thanks to modern techniques of ultra-high resolution spectroscopy (saturation spectroscopy, two-photon spectroscopy) it is now possible to resolve individual hyperfine components of vibration-rotation lines of molecules having three or more identical nuclei with a high degree of symmetry: spherical tops such as SF_6 [8–14], OsO_4 [9,12,15–17], SiF_4 [18], symmetric tops such as NH_3 [12,19], PH_3 [20], PF_5 [12]...

We have built several generations of saturation spectrometers with very high resolving power, accuracy and sensitivity in the 9 -12 μm spectral range [9,21]. Recently, a linewidth of 80 Hz (HWHM) has been achieved for OsO_4, thanks to the use of slow molecules [22] and the absolute frequency accuracy is now typically better than 100 Hz. The long absorption path (108 m) of our large cell gives access to very high sensitivity. For high contrast, we have also used recently external Fabry-Perot resonators for a spectroscopic test of parity violation in chiral molecules [23,24]. Also, the spectra of some of these molecules are now very well-known e.g. the ν_3 band of SF_6, which is known to an accuracy better than 40 kHz for J values ≤ 100 [25].

The Pauli principle has important consequences on the hyperfine spectra of these molecules, which we have observed in the course of many studies. We have also emphasized on several examples that there was a strong link between these consequences and parity [26] and we have shown two spectacular examples where this link leads to a parity doubling in PH_3 [20] and SiF_4 [18].

CONSEQUENCES OF THE PAULI PRINCIPLE ON HYPERFINE SPECTRA

In the ground electronic state of molecules the total internal wavefunction is the tensor product of a vibration-rotation wave function and of a nuclear spin wavefunction. The total nuclear spin wave function is built from products of spin wavefunctions for each nucleus, which are symmetrized in the point group of the molecule [27,28]. Using a double tensor notation in $^{(L)}O(3) \times G \subset ^{(L)}O(3) \times ^{(M)}O(3)$ [13,28] and the notations of Berger and Landau [29,30], we shall write this wave function $\Psi_{NS}^{(I_+, C_{NS})}$, where I is a value of the total nuclear spin corresponding to the point group symmetry species C_{NS} and where the subscript + corresponds to even parity[1]. As an example, for SiF_4, $I = 2$ for $C_{NS} = A_1$, $I = 1$ for $C_{NS} = F_2$ and $I = 0$ for $C_{NS} = E$. For SF_6, I can take the two values 1 and 3 for $C_{NS} = A_{1g}$, 1 and 2 for E_g, 0 and 2 for F_{1u}, $I = 1$ for F_{2g} and for F_{2u} and finally $I = 0$ for $C_{NS} = A_{2g}$ or A_{2u}. Similarly, we write the vibration-rotation wave functions $\Psi_{VR}^{(J_\tau, R_\lambda C_{VR})} = \left[\Psi_R^{(J_\tau, J_\tau)} \otimes \Psi_V^{(0_+, l_\rho)} \right]^{(J_\tau, R_\lambda C_{VR})}$, where R is the pure rotation angular momentum, which comes from the coupling of the total orbital momentum J with the vibrational angular momentum l and where $\tau, \lambda, \rho = \pm$ are parity indices. The rovibrational states are also characterized by their symmetry species C_{VR} which is obtained from the reduction of the representation $D^{(R_\lambda)}$ in G: $D^{(R_\lambda)} \downarrow G = \sum_j C_j^{(n_j)}$. The rotational wave functions are degenerate with respect to parity.

The total internal wave function in the ground electronic state is obtained by coupling the vibration-rotation and the nuclear spin parts as

$$\Psi_{tot}^{(F_\tau, C)} = \left[\Psi_{VR}^{(J_\tau, R_\lambda C_{VR})} \otimes \Psi_{NS}^{(I_+, C_{NS})} \right]^{(F_\tau, C)}.$$

For any molecular system containing identical nuclei, the only allowed states are either totally symmetric with respect to permutations (case of bosons) or totally antisymmetric (case of fermions). The Pauli principle implies therefore that the only populated vibration-rotation states are such that $C = A_1$ for bosons and $C = A_2$ for fermions (A_{2u} for SF_6). This introduces a correspondence between the vibration-rotation symmetry species and the total spin value I and hence the number of allowed hyperfine components $2I + 1$. One can check in Fig.1 that, for SiF_4, we have indeed five hyperfine components for A_2 lines and three for F_1 lines. This correspondence between the values of I allowed by the Pauli principle and the

---

[1]) This parity, which depends on the intrinsic parity of the nuclei, is fixed and is chosen to be + for simplicity.

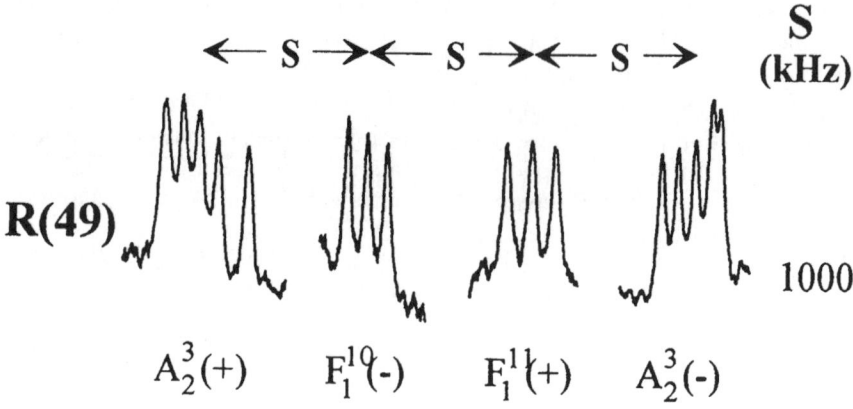

FIGURE 1. Superfine and hyperfine structures of a trigonal ($O_3 \uparrow O$) cluster in the ν_3 spectrum of SiF_4. The hyperfine structure of each component spreads over a few kilohertz, whereas the superfine splitting S is one MHz in this case.

symmetry species of the vibration-rotation lines is well verified on the hyperfine structure of all the spectra observed this far, in symmetric and spherical tops. A first clear consequence for hyperfine spectra, if other permutation symmetries were allowed, is that one would have extra-components violating this correspondence and one could look for these missing lines, whose positions can be easily calculated from the theory. For example, we have seen that for SF_6, A_1 lines require $I = 0$ for both the components A_{1u} and A_{1g} and, as a matter of fact, only doublets are observed in this case [12,13], but one could certainly look for additional weak components corresponding to other values of I, which would result from a violation of the Pauli principle. This test will require high resolution, which implies a limited signal-to-noise. A simpler experiment is presently underway which can be performed with very high signal-to-noise and will be presented in the next section. Let us mention another important consequence of the Pauli principle in such spectra and discussed in detail in references [13,18,20,26]. Some lines are parity doublets while others are parity singlets, but states, which differ by their overall parity, differ also by some other part of their wave function, either through their rotational or through their nuclear spin wave functions. Even if inversion is negligible, two states of opposite parity cannot be strictly degenerate owing to rovibrational or hyperfine interactions.

A HIGH SENSIVITY TEST ON A DARK BACKGROUND

In the case of spherical tops, an important consequence of the internal dynamics is that the vibration-rotation lines tend to cluster together in multiplets (doublets, triplets or quadruplets) corresponding to rotation about one of the symmetry axes with a degeneracy equal to the number of equivalent axes [31]. Furthermore these multiplets are split by tunnel effect in a tumbling motion between these equivalent axes giving rise to regular spacings between their components [9,13]. An example is shown in Fig.1 where a typical quadruplet is displayed corresponding to rotation around a three-fold axis in SiF_4.

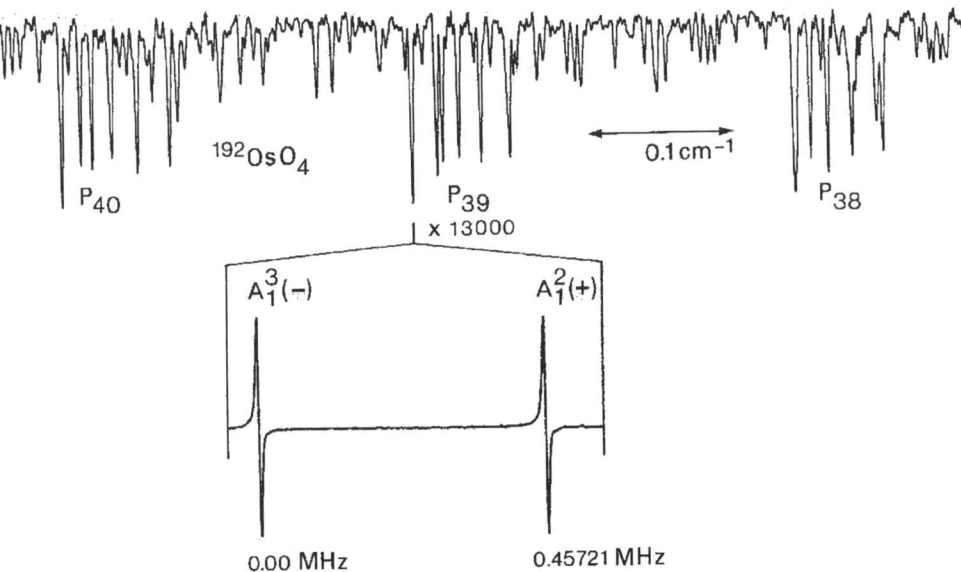

FIGURE 2. Top: Fourier transform spectrum of $^{192}OsO_4$ showing the fine structure of three J manifolds of the ν_3 band. Bottom: high-resolution derivative saturation spectrum of a superfine doublet. The F lines present in Fig.1 are missing here because of the Pauli principle.

For SiF_4, the Pauli principle allows the existence of superfine quadruplets $A_2(+)F_1(-)F_1(+)A_2(-)^2$, whereas for $Os^{16}O_4$ it forbids the states with F vibration-rotation symmetry species since $C_{NS} = A_1$ and, as a matter of fact, only $A_1(+)A_1(-)$ doublets are observed [12]. In this case, the principle of our test consists in looking for the missing lines of the quadruplet (at one third and two thirds of the splitting) with the ultimate sensitivity that we can reach with our

---

[2] These representations are those of the point group T_d and differ from those of the permutation-inversion group e.g. $A_1(-)$ corresponds to A_2, $A_1(+)$ to A_1....

long absorption cell (108 m) or with our Fabry-Perot resonators. In the past, we have indeed been able to record very weak lines e.g. those corresponding to the CO_2 laser lines themselves [32] or those corresponding to weak crossover resonances [33]. Here, we look for transitions from unpopulated states with a transition moment similar to the strong observed lines. We have thus to optimize the S/N ratio of the observed lines and look for the missing lines in the same conditions. Already, from Fig.2, published in reference [12], we can estimate that these lines are at least a factor 1000 smaller than the allowed ones, but we can certainly improve this result by many orders of magnitude. The intensities of the lines in saturation spectroscopy can be calculated [34–36,22] and the sensitivity can be easily estimated [37]. The linear absorption coefficient is given by [35,21]:

$$k_\nu = \frac{4\pi^2}{3}\alpha\nu\frac{1}{\sqrt{\pi}u/\lambda}\frac{\mu^2}{e^2}\frac{N}{Z}\exp(-E/kT)$$

where α is the fine structure constant, ν the transition frequency, u/λ the Doppler width, μ the transition moment, e the electron charge, N the number density, Z the partition function and $\exp(-E/kT)$ the Boltzmann factor. In the case of the OsO_4 lines under consideration, we find 0.5 cm^{-1}/Torr. In order to optimize the contrast of the saturation resonances, we have to use a gas pressure such that the linear absorption satisfies $k_\nu L \sim 1$; with a path length $L = 108$ m, this leads to a typical pressure of a few 10^{-4} torr. The pressure broadening is still reasonable (a few kilohertz) and quite compatible with typical $A_1(+)A_1(-)$ splittings.

The maximum laser power that we may then apply to the molecules, is such that the saturation parameter should not exceed a few units, to optimize the size of the saturation peaks. In the transit-time regime, it is given by the π pulse power [21] 1.36 x 10^{-5} W, but in the collisional regime [37], this power will rise typically up to several milliwatts. Since the photon noise is in the picowatt range, a typical sensitivity of $10^{-8}-10^{-9}$ is expected.

ACKNOWLEDGEMENTS

The authors are grateful to G. Tino for his encouragements to use $Os^{16}O_4$ for a test of the Pauli principle.

REFERENCES

1. Hilborn R.C., *Bull. Am. Phys. Soc.* **35**, 982 (1990).
2. Tino G.M., "Proposed search for small violations of Bose statistics in molecular spectra", *Nuovo Cimento* D **16**, 523-530 (1994).
3. de Angelis M., Gagliardi G., Gianfrani L. and Tino G.M., "Test of the symmetrization postulate for spin-0 particles", *Physical Review Letters* **76**, 2840-2843 (1996).
4. Modugno G., Inguscio M. and Tino G.M., "Search for small violations of the symmetrization postulate for spin-0 particles", *Phys. Rev. Lett.* **81**, 4790-4793 (1998).

5. Hilborn R.C. and Yuca C.L., "Spectroscopic test of the symmetrization postulate for spin-0 nuclei", *Physical Review Letters* **76**, 2844-2847 (1996).
6. Gianfrani L. Fox R.W. and Hollberg L., "Cavity-enhanced absorption spectroscopy of molecular oxygen", *J.O.S.A.* **B16**, 2247-2254 (1999) .
7. Tino G.M., "Testing the symmetrization postulate of quantum mechanics and the spin-statistics connection", *Fortschr. Phys.* **48**, 537-543 (2000).
8. Bordé Ch.J., Ouhayoun M. and Bordé J., "Observation of magnetic hyperfine structure in the infrared saturation spectrum of $^{32}SF_6$", *J. Mol. Spectr.* **73**, 344 (1978).
9. Bordé Ch.J., Ouhayoun M., Van Lerberghe A., Salomon C., Avrillier S., Cantrell C.D. and Bordé J., "High resolution saturation spectroscopy with CO_2 lasers. Application to the ν_3 bands of SF_6 and OsO_4", in *Laser Spectroscopy IV*, eds H. Walther and K.W. Rothe, Springer-Verlag (1979).
10. Bordé J., Bordé Ch.J., Salomon C., Van Lerberghe A., Ouhayoun M. and Cantrell C.D., "Breakdown of the Point-Group Symmetry of vibration-rotation states and optical observation of ground-state octahedral splittings of $^{32}SF_6$ using saturation spectroscopy", *Phys. Rev. Lett.* **45**, 14-17 (1980).
11. Salomon Ch., Bréant Ch., Van Lerberghe A., Camy G. and Bordé Ch.J., "A phase-locked waveguide CO_2 laser for broad-band saturation spectroscopy with kilohertz resolution and absolute frequency accuracy. First observation of superhyperfine structures in the ν_3 band of SF_6", *Applied Physics* **B29**, 153 (1982).
12. Bordé Ch.J., Bordé J., Bréant Ch., Chardonnet Ch., Van Lerberghe A. and Salomon Ch., "Internal dynamics of simple molecules revealed by the superfine and hyperfine structures of their infrared spectra", in *Laser Spectroscopy VII*, eds T.W.Hänsch and Y.R. Shen, Springer Verlag (1985) pp.108-114.
13. Bordé J.and Bordé Ch.J., "Superfine and hyperfine structures in the ν_3 band of SF_6", *Chemical Physics* **71**, 417-441 (1982).
14. Bordé J. and Bordé Ch.J., "Addendum to Superfine and hyperfine structures in the ν_3 band of SF_6: evidence for a tensor spin-vibration interaction in spherical tops", *Chemical Physics* **84**, 159-165 (1984).
15. Chardonnet Ch. and Bordé Ch.J., "Saturated absorption spectroscopy: update of $Os^{16}O_4$ measurements", in *Handbook of Infrared Standards II*, G.Guelachvili and K.Narahari Rao eds, Academic Press (1993) pp. 611-616.
16. Chardonnet Ch. and Bordé Ch.J., "Hyperfine interactions in the ν_3 band of Osmium Tetroxide. Part I: Precise determination of the spin-rotation constant by crossover resonance spectroscopy", *Journal of Molecular Spectroscopy* **167**, 71-98 (1994).
17. Chardonnet Ch., Palma M.L. and Bordé Ch.J., "Hyperfine interactions in the ν_3 band of Osmium Tetroxide. The electric quadrupole interaction in $^{189}OsO_4$",*Journal of Molecular Spectroscopy*, **170**, 542-566 (1995).
18. Pfister O., Chardonnet Ch. and Bordé Ch.J., "Hyperfine-induced lifting of parity degeneracy in semi-rigid tetrahedral molecules", *Phys. Rev. Lett.* **76**, 4516-4519 (1996).
19. Salomon Ch., Chardonnet Ch., Van Lerberghe A., Bréant Ch. and Bordé Ch.J., "Première observation de la structure hyperfine magnétique dans le spectre infrarouge de l'ammoniac", *J. Physique Lettres*, **45**, L-1125-L-1129 (1984).
20. Butcher R.J., Chardonnet Ch. and Bordé Ch.J., "Hyperfine lifting of parity degeneracy and the question of inversion in a rigid molecule", *Physical Review Letters* **70**,

2698-2701 (1993).
21. Bordé Ch.J., "Développements récents en spectroscopie infrarouge à ultra-haute résolution", in *Revue du Cethedec, Ondes et Signal* **NS83-1**, 1-118 (1983).
22. Chardonnet Ch., Guernet F., Charton G. and Bordé Ch.J., "Ultra-high resolution saturation spectroscopy of slow molecules in an external cell, *Applied Physics* **B59**, 333-343 (1994).
23. Chardonnet Ch., Daussy Ch., Marrel T., Amy-Klein A., Nguyen C.T. and Bordé Ch. J., "Parity violation test in chiral molecules by laser spectroscopy", in *"Parity Violation in Atomic Physics and Electron Scattering"*, World Scientific, eds B. Frois and M.A. Bouchiat (1999) pp. 325-355.
24. Daussy Ch., Marrel T., Amy-Klein A., Nguyen C.T., Bordé Ch.J. and Chardonnet Ch., "Limit on the parity nonconserving energy difference between the enantiomers of a chiral molecule by laser spectroscopy", *Phys. Rev. Lett.* **83**, 1554-1557 (1999).
25. Acef O., Bordé Ch.J., Clairon A., Pierre G. and Sartakov B., "New accurate fit of an extended set of saturation data for the ν_3 band of SF_6. Comparison of Hamiltonians in the spherical and in the cubic tensor formalisms", *J. Mol. Spectr.* **199**, 188 (2000).
26. Chardonnet Ch., Amy-Klein A., Bernard V., Durand P.-E., George T., Guernet F., Nicolaisen H., Pfister O. and Bordé Ch.J., "Ultra-high resolution spectroscopy at 10 μm: applications and new trends", in *Laser Spectroscopy XII*, eds M. Inguscio and A. Sasso, World Scientific, 208-211 (1996).
27. Bordé J., "Determination of the total nuclear spin in a rovibronic state. Application to the molecules SF_6 and PF_5", *J. Phys.(Paris)* **39**, L-175 (1978).
28. Michelot F., Bobin B. and Moret-Bailly J., "Nuclear hyperfine interactions in spherical tops in their ground electronic state", *J. Mol. Spectr.* **76**, 374-411 (1979).
29. Berger H., "Classification of energy levels for polyatomic molecules", *Journal de Physique (Paris)* **38**, 1371-1375 (1977).
30. Landau L.D. and Lifshitz E.M., *Quantum Mechanics* (Pergamon, New York, 1976).
31. Harter W.G. and Patterson C.W., "Orbital level splitting in octahedral symmetry and SF_6 rotational spectra. I. Qualitative features of high J levels", *J. Chem. Phys.* **66** 4872-4885 (1977).
32. Chardonnet Ch., Van Lerberghe A. and Bordé Ch.J., "Absolute frequency determination of super-narrow CO_2 saturation peaks observed in an external absorption cell", *Optics Comm.* **58**, 333-337 (1986).
33. Chardonnet Ch. and Bordé Ch.J., "Strong-Field Saturation Spectroscopy of Weak Hyperfine Crossover Resonances", *Europhysics Letters*, **9**, 527-532 (1989).
34. Bordé J. and Bordé Ch.J., "Theory of relative intensities of hyperfine components in saturation spectroscopy", *J. of Molecular Spectroscopy*, **78**, 353-378 (1979).
35. Bobin B., Bordé Ch.J., Bordé J. and Bréant Ch., "Vibration-rotation molecular constants for the ground and $(v_3 = 1)$ states of $^{32}SF_6$ from saturated absorption spectroscopy", *J. Mol. Spec.* **121**, 91-127 (1987).
36. Bordé Ch.J., "The Physics of Optical Frequency Standards Using Saturation Methods", in: *Frequency Standards and Metrology"*, A. de Marchi (Ed.) Springer-Verlag (1989) 196-205.
37. Bordé Ch.J., "Comments on photoacoustic and photothermal spectroscopy of gases compared to optical methods", *Journal de Physique*, **44**, C6-593-C6-601 (1983).

Towards an Improved Test of Bose-Einstein Statistics for Photons

Damon Brown[†], Dmitry Budker[†,*], and David P. DeMille[‡]

[†]*Department of Physics, University of California at Berkeley,
Berkeley, California 94720-7300, USA*
debrown@physics.berkeley.edu

[*]*Nuclear Science Division, Lawrence Berkeley National Laboratory
Berkeley, California 94720, USA*
budker@socrates.berkeley.edu

[‡]*Department of Physics, Yale University, New Haven, CT 06520-8120, USA*
david.demille@yale.edu

Abstract. In this contribution we describe our current efforts at Berkeley to improve upon a limit on Bose-Einstein statistics violation for photons obtained earlier (D. P. DeMille, D. Budker, N. Derr, and E. Deveney, *Phys. Rev. Lett.* **83**(20), 1378 (1999)). The principle differences between the new and the previous experiments are the use of a narrow-band cw laser, an atomic beam, a different transition in atomic Ba, and possibly a different detection technique. We also discuss the ultimate limits that one can hope to obtain with this degenerate two-photon transition technique. Some potential conceptual difficulties with the interpretation of this experiment and the ways to avoid them are discussed as well.

I EXPERIMENTAL TECHNIQUE AND CHOICE OF THE TRANSITION

The principle of the technique and details of the earlier experiment [1] that used pulsed lasers and a barium vapor cell are discussed in D. P. DeMille's invited contribution in the current proceedings [2].

Our present choice of the two-photon transition is shown in Fig. 1. This scheme differs from that of Ref. [1] in its particular choice of levels, but is otherwise very similar to it. It offers a convenient excitation wavelength accessible to a dye laser, and a strong fluorescence decay channel at a sufficiently shorter wavelength, which is convenient for fluorescence detection. A slightly smaller energy defect Δ (93 cm^{-1} vs -163 cm^{-1} in Ref. [1]) enhances the two-photon rate. Note, however, that in addition to this blessing, a smaller Δ may also carry a curse of relatively high

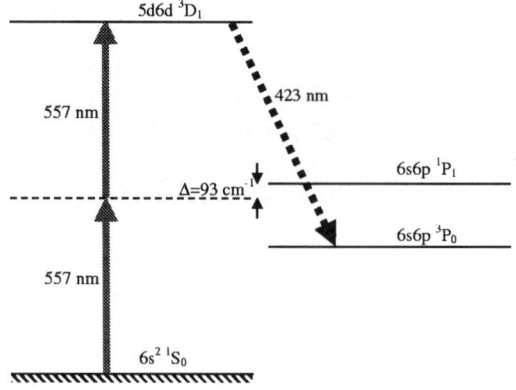

FIGURE 1. Selected levels and transitons in Ba I.

backgrounds, as discussed below. We emphasize that a variety of schemes involving $J = 0 \rightarrow J = 1$ two-photon transitions are available in Ba, Yb, and other atoms.

The schematic of the new experiment is shown in Fig. 2. Ba atoms are produced in a stainless steel source fitted with a multi-channel effusor constructed out of crinkled stainless steel foil. The atomic beam is further collimated with an external collimator constructed as a set of parallel, appropriately spaced thin metal foils. The atoms in the collimated beam interact with an intense light beam in a power build-up cavity. An unusual feature of the cavity is that there are intracavity quarter-wave plates in front of each mirror. The purpose of the waveplates is to ensure that light waves propagating in opposite directions have opposite polarizations.

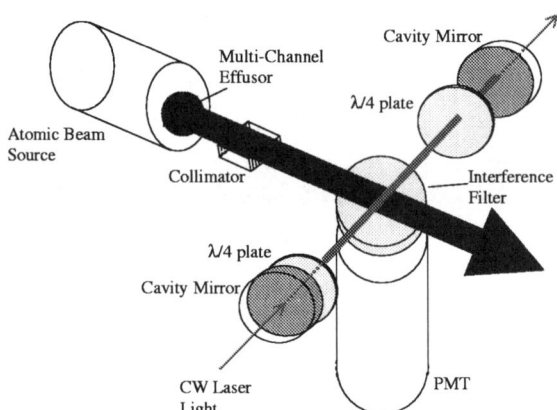

FIGURE 2. Experimental arrangement.

We introduce this feature in the experiment since in the absence of Bose-Einstein (BE) statistics violation and for photons of different energy, $J = 0 \to J = 1$ two-photon transitions are still forbidden if the photons are of the same polarization. (This point is further discussed below.) Fluorescence from the excitation region in appropriate decay channels is used to measure the excitation rate. In principle, more sophisticated detection techniques can be considered for future work (photon bursts, selective laser photoionization, etc.), that will provide higher detection efficiency and selectivity. However even with a straightforward detection system consisting of a cooled phototube installed near the interaction region and a light collection mirror (not shown in Fig. 2), the beam apparatus will provide considerable improvement in detection efficiency compared to that of Ref. [1]. In addition, collisional backgrounds present in the cell experiment [1] will be eliminated. Most importantly, small transverse Doppler width and the narrow bandwidth of the light reduce, by many orders of magnitude, the main limitations of the pulsed laser experiment [1] related to the two-photon transitions arising from the finite laser bandwidth ($\Delta \nu \approx 3$ GHz) and Doppler width (≈ 300 MHz).

It is a well known feature of two-photon transitions, in which two equal energy photons are picked up by an atom from counter-propagating light beams, that the first-order Doppler effect is absent [3]. However, the energies of the two photons are different as seen by the atom with a non-zero velocity component along the axis of the cavity. Consequently, the atomic beam has to be collimated to reduce the background due to the non-degenerate two-photon transitions. We are planning to collimate the beam so that $\Gamma_D \equiv 2k \cdot \Delta v_t \sim \Gamma_0$. Here k is the light wave number, Δv_t is the spread of the transverse velocities in the atomic beam, and $\Gamma_0 \approx 2\pi \cdot (6.9 \text{ MHz})$ is the natural line width of the two-photon transition. Note that, as can be seen from the structure of the two-photon amplitudes (with expressions given in Refs. [1,2] amended in a straightforward manner by replacing $E_k \to E_k - i\Gamma_k/2$, where E_k and Γ_k are the energy and width of the state k), the finite level widths do not lift the ban on degenerate two-photon transitions. Thus, the non-degenerate two-photon background continues to decrease if the atomic beam is collimated even tighter, so that the Doppler width Γ_D becomes smaller than Γ_0. However, collimation to the level $\Gamma_D \sim \Gamma_0$ appears adequate for the projected level of sensitivity for the current version of the experiment (see estimates below). Future work may benefit from transverse laser cooling of the atomic beam or the use cold atomic samples where the Doppler width is negligible in place of the atomic beam.

II PARAMETER OPTIMIZATION

The number of two-photon transitions in the interaction region scales as

$$\frac{(\text{Light Power}, P)^2}{\text{Beam Area}, A}.$$

This scaling comes about because the transition rate per atom goes as the square of the light intensity, and the number of atoms in the interaction region is proportional to A. Thus, at first sight, in order to maximize the counting rate, one would need to increase light power as much as possible, and tightly focus the light beam. However, there exists a "parasitic" effect, the AC Stark shift, that limits possible gains to be achieved by following this route. For the scheme shown in Fig. 1, while both the upper and lower levels of the two-photon transition shift in the presence of the strong light field, the shift of the lower (ground) state is expected to be dominant because the energy defect is very small and the $6s^2\ ^1S_0 \to 6s6p\ ^1P_1$ transition is fully allowed (the $6s6p\ ^1P_1 \to 5d6d\ ^3D_1$ amplitude is suppressed by an order of magnitude). The ground state AC Stark shift can be estimated as:

$$\Delta_{AC} = -\frac{d_{gi}^2 E^2}{4\Delta} \approx -0.1 MHz \cdot \frac{2P(W)}{A(mm^2)}.$$

Here $d_{gi} \approx 3.2\ e \cdot a$ is the dipole matrix element of the $^1S_0 \to\ ^1P_1$ transition (known from the 8.4 ns lifetime of the $^1P_1$ state), E is the light electric field amplitude, and P is the power in each of the counter-propagating light waves. If all atoms were exposed to a uniform electric field, the AC Stark shift would not present much of a problem – it would simply shift the position of the two-photon resonance. However, since atoms passing through different regions of the light beam "see" different intensities this leads to a non-uniform AC Stark shift, and an effective inhomogeneous line broadening. Limiting this broadening at the level of the homogeneous width of the transition, sets a limit on the maximum ratio of P and A:

$$|\Delta_{AC}| = \Gamma_0 \Rightarrow \frac{P(W)}{A(mm^2)} \approx 30.$$

With this, we can now discuss the parameters of the power build-up cavity. Since the P/A ratio is fixed and the two-photon signal goes as P^2/A, it actually turns out beneficial to work with a large beam area. We are planning to employ a confocal cavity and use the degeneracy between transverse modes specific for this cavity configuration. This degeneracy allows one to excite multiple transverse modes simultaneously to obtain a large spot size and achieve transverse beam dimensions $\sim$1 mm ($A = 1$ mm^2; see e.g. [4] for a discussion of excitation of multiple transverse modes in a cavity). With $A = 1$ mm^2 and $\frac{P(W)}{A(mm^2)} = 30$, we need $P = 30$ W. The light power available from a dye laser system (Coherent CR-699 pumped by a Spectra Physics 2040 Ar-ion laser) is on the order of $P_{in} = 0.3$ W, thus the cavity should provide a build-up of $\frac{2P}{P_{in}} = 200$, corresponding to cavity finesse of $200 \cdot \pi/2 \approx 300$.

We now turn to estimating the two-photon rates. It will be convenient to introduce an allowed two-photon transition rate on resonance given by

$$W_{gf} \sim \frac{E_0^4}{\Gamma_0} \cdot \frac{d_{gi}^2 d_{if}^2}{\Delta^2} \sim 3 \cdot 10^2 \left(\frac{P(W)}{A(mm^2)}\right)(s^{-1}).$$

Here we used an estimated value $d_{if} = 0.1\ e \cdot a$. With the density of atoms in the interaction region $N \sim 10^{10}$ cm^{-3} and the length of the interaction region $l = 1$ cm, with $P = 30$ W and A=1 mm^2, we get $\sim 10^{13}$ allowed transition per second.

This estimate shows that if one is able to maintain high detection efficiency with no background counts for a time t, it will be possible to limit the relative probability of the forbidden transition at the level of $\nu \lesssim (10^{13} \cdot t(s))^{-1}$. Here ν corresponds to the relative fraction of the anti-symmetric photon pairs in our power build-up cavity out of all possible photon pairs (compare to $\nu \lesssim 10^{-7}$ obtained in Ref. [1]). This estimate shows a potential for high statistical sensitivity, and also indicates the necessity to ensure background-free operation. Note that if background counts are present, the limit on ν would only improve as $t^{-1/2}$ rather than as t^{-1} in the case of background-free operation.

III BACKGROUNDS AND CALIBRATION

Possible sources of background can be classified into *resonant* and *non-resonant* with respect to the detuning of the light frequency from the two-photon resonance. The dominant resonant background comes from the already mentioned two-photon transitions originating from the finite Doppler width. Inspection of the expressions for the BE-allowed transition rate (see e.g. Refs. [1,2]) shows that the *relative* background rate with respect to the allowed rate scales as $(\Gamma_D/\Delta)^2$ and comprises $\sim 10^{-11}$ with the chosen value $\Gamma_D \sim \Gamma_0$. We note that once the experiment using the current transition scheme becomes limited by this background, it may be beneficial to switch to a scheme with larger Δ. (Of course, the allowed two-photon rate itself scales $\propto \Delta^{-2}$, and will generally also be reduced for larger Δ. Another way to reduce the relative contribution of the background is to use $\Gamma_D < \Gamma_0$, see Section I.)

An example of non-resonant background is spontaneous hyper-Raman scattering which arises due to the population of virtual even parity states with $J = 0, 2$ followed by spontaneous emission at the detection wavelength. The relative rate for such process scales as

$$\frac{\Gamma_0 \cdot \Gamma(J \neq 1) \cdot B.R.}{\Delta_2^2} \cdot \left(\frac{d_{iJ \neq 1}}{d_{if}}\right)^2.$$

Here $\Gamma(J \neq 1)$ is the width of the corresponding even parity state, $B.R.$ is the branching ratio for the decay of this state into the final state of the detection transition, Δ_2 is the energy difference between the $J \neq 1$ state and the upper $J = 1$ state of the two-photon transition under investigation, and $d_{iJ \neq 1}$ is the appropriate matrix element. For the scheme used in Ref. [1], the relative rate for this process is $\sim 10^{-13}$, however, for the present scheme there is a further suppression $\sim \alpha^4$, where α is the fine structure constant, due to the fact that a $J = 2 \rightarrow J = 0$ decay can only go as a magnetic quadrupole transition (M2).

Another background process, hyperfine-interaction-induced two-photon transitions, is present only for Ba isotopes with non-zero nuclear spin (^{135}Ba, 6.6% natural abundance and ^{137}Ba, 11.2%, both with nuclear spin $I = 3/2$), and is negligible if one works near the resonance corresponding to the most abundant spin-less isotope (^{138}Ba, 71.7%), which is expected to be well separated in frequency due to isotope shift. However, the hyperfine-induced transition provide a convenient way of calibrating the apparatus. The relative rate for such a process contains a factor characterizing the admixture of the $J \neq 1$ states to the $J = 1$ upper state due to off-diagonal hyperfine interaction $\sim (\Delta_{\mathrm{hfs}}/\Delta_2)^2 \sim 10^{-8}$. Another way to calibrate the apparatus and to carry out initial measurements of the spectroscopic parameters of the two-photon transition (isotope shift, hyperfine structure, etc.) is to tilt the axis of the power build-up cavity, so it forms a large angle with the atomic beam. Such a geometry enhances the Doppler effect-induced transitions.

IV CURRENT STATUS

The laser system, the vacuum chamber, and the Ba beam source are all in place, and the first fluorescence signals (with one-photon excitation) have been observed. The external atomic beam collimator design has been successfully tested in another experiment carried out in our group which uses an atomic beam of Yb with similar parameters [5]. The optical cavity design is currently in progress.

V LIGHT POLARIZATION AND BOSE-EINSTEIN STATISTICS

The present experiment (as well as the earlier work [1]) is explicitly designed to ensure that we use a configuration where the two-photon transition is allowed for non-degenerate photons. To this effect, the polarizations of the two counter-propagating light waves are chosen orthogonal to each other. However, it is not obvious that this is necessary for the observation of BE-violating transitions. In fact, thinking naively in terms of classical fields, the two counter-propagating light waves of the same frequency form a standing wave, for which at each spatial location, there is *some* steady polarization. Since we are studying E1-E1 two-photon transition, the propagation direction would appear irrelevant, and the transitions would seem forbidden simply by a polarization argument, and not by BE-statistics. This is particularly clear for the case of two opposite circular polarizations. The resulting standing wave is locally linearly polarized at each point. The two-photon transition appears forbidden by the property of the Clebsch-Gordan coefficients (Fig. 3).

In our opinion, this example clearly illustrates the connection between BE statistics violation and light polarization. If BE statistics is violated, the naive polarization addition picture used in the above example must fail also. This highlights

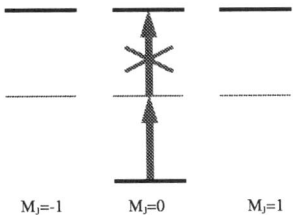

$M_J=-1 \quad M_J=0 \quad M_J=1$

FIGURE 3. Why $J = 0 \to J = 1$ transitions are forbidden for linear light polarization (in the absence of BE-statistics violation).

a need for a theory that would incorporate possible small statistics violations for photons, and would generalize the usual electrodynamics and optics.

VI CONCLUSION

With the straightforward spectroscopic experiment now underway, it appears feasible to push the limit on Bose-Einstein statistics violation for photons to the level of $\nu \lesssim 10^{-11}$, with possible further improvements, perhaps by several orders of magnitude, down the road. However, certain basic questions concerning the nature, meaning and implications of statistics violation for photons remain open.

We are indebted to D. F. Kimball, C.-H. Li, A.-T. Nguyen, and J. E. Stalnaker for many useful discussions. This research is supported by NSF CAREER grant PHY-9733479.

REFERENCES

1. DeMille, D. P., Budker, D., Derr N., and Deveney, E., *Phys. Rev. Lett.* **83**, 3978-3981 (1999).
2. DeMille, D. P., Budker, D., Derr N., and Deveney, E.,"Experimental tests of the symmetrization postulate for photons," in the current proceedings.
3. Vasilenko, L. S., Chebotayev, V. P., and Shishayev, A. V., *JETP Lett.* **12**, 113-116 (1970).
4. Budker, D. Rochester, S. M., and Yashchuk, V. V., *Rev. Sci. Instr.* **71**, 341-346 (2000).
5. A detailed description of this work and references are available at our group web site: http://phylabs.berkeley.edu/budker/.

Testing the Pauli Exclusion Principle With Accelerator Mass Spectrometry

D. Javorsek II[1], M. Bourgeois[2], D. Elmore[1], E. Fischbach[1],
D. Hillegonds[2], J. Marder[3,4], T. Miller[1], H. Rohrs[5,2],
M. Stohler[1], and S. Vogt[6,2]

[1] *Department of Physics, Purdue University, West Lafayette, IN 47907*
[2] *Department of Chemistry, Purdue University, West Lafayette, IN 47907*
[3] *Isotech LLC, Euclid, OH 44132*
[4] *Brush Wellman Inc., Cleveland, OH 44110*
[5] *Department of Chemistry, Washington University, St. Louis, MO 63130*
[6] *International Atomic Energy Agency, Vienna, 1400 Austria*

Abstract. We report the results of a new experimental search for the Pauli-forbidden $1s^4$ state of Be, denoted by Be'. Using the Accelerator Mass Spectrometer facility at Purdue, we set limits on the abundance of Be' in metallic Be, Be ore, natural gas, and air (10^{-14}). Our results improve on those obtained in a previous search for Be' by a factor of approximately 300.

INTRODUCTION

The Pauli Exclusion Principle (PEP) has played a central role in quantum mechanics since its formulation in 1925 [1]. Notwithstanding its many successful predictions, the PEP remains somewhat enigmatic, particularly with respect to the question of whether small deviations from it are possible. Recently this question has become the subject of renewed investigation both theoretically [2-10] and experimentally [11-20]. Greenberg and Mohapatra [5], extended the analysis of a single oscillator performed by Ignat'ev and Kuz'min [4] into a quantum field theory which mainly obeys the usual spin-statistics connection but which permits small violations with amplitude β. One of the most interesting conclusions to emerge from the theoretical efforts is what Greenberg and Mohapatra (GM) [6] term the "surprising rigidity" of the PEP. By this GM refer to the apparent impossibility of formulating a consistent local field theory of small PEP violations within the framework of a positive-metric Hilbert space.

In another attempt to formulate a consistent theory permitting PEP violations Mohapatra [7] and Greenberg [8] both introduce the quon algebra which can be described as an interpolation between the usual Fermi-Dirac and Bose-Einstein statistics. This interpolation is achieved by first multiplying the anticommutation relation governing fermions, $\{a_k, a_l^\dagger\} = \delta_{kl}$, by $(1-q)/2$ and the commutation relation governing bosons, $[a_k, a_l^\dagger] = \delta_{kl}$ by $(1+q)/2$, and summing the two. The resulting quon algebra for creation and annihilation operators is

$$a_k a_l^\dagger - q a_l^\dagger a_k = \delta_{kl} \tag{1}$$

where for $q = 1$ and $q = -1$ we recover Bose and Fermi statistics respectively. For $-1 < q < 1$ this formalism solves the problem of negative norms encountered earlier but unfortunately is shown to be non-local [8,21].

Nonetheless, since the PEP makes clear predictions which have direct experimental implications, tests of the PEP are possible even without a fully consistent theoretical framework. However, in the absence of such a framework comparisons among different experiments are not straightforward, even though the interpretation of individual experiments may be clear, as has been emphasized recently by Baron, Mohapatra, and Teplitz [10]. We will return to this point below when we discuss our experimental results.

Generally speaking, tests of the PEP fall into one of four classes: Searches for PEP-forbidden electronic [12,15,19,20] or nuclear [16] states, or for PEP-forbidden electronic [11,13,17] or nuclear [14,18] transitions. Evidently, a search for a PEP-forbidden electronic configuration can also be viewed as a test of the shell structure of atoms which underlies the periodic table. In an early experiment of this genre Fischbach, Kirsten, and Schaeffer [12] set limits on the abundance of Be' in air, where Be' is beryllium in which the PEP-forbidden electronic configuration $1s^4$ replaces the usual configuration $1s^2 2s^2$. Although originally conceived as a test for electrons with an additional quantum number, this experiment can also be viewed as a test for PEP violation [5,7,10,15]. We note that the underlying assumption in both Ref. [12] and the present paper is that the chemistry of Be' is similar to that of He.

In contrast to previous experiments in which the PEP is violated by a single additional fermion in a forbidden state, a search for Be' probes for a PEP violation by two electrons. Potentially relevant discussions regarding the conservation of statistics theorem have also been made in the context of the quon formalism introduced above [8,9,22,23]. Initially developed by Wigner [24] and Ehrenfest and Oppenheimer [25] the conservation of statistics theorem is an application of the spin-statistics theorem to composite systems and states that composites of an even number of fermions and bosons are bosons while an odd number are fermions. The application of this theorem to a more complicated system, like a Be atom, is more convoluted and depends directly on the model used for the interaction Hamiltonian. However, in the absence of a consistent theoretical framework, we are unable to determine with certainty how such particles would violate the PEP and must include the possibility that fermions may favor violations in pairs. To recapitulate, although such a violation by two fermions may appear at first sight to be suppressed relative to that for a single fermion, in the absence of a consistent fundamental theory of PEP violation one cannot exclude the possibility that the reverse may be true.

SIGNIFICANCE

The object of the present paper is to improve the limits obtained in Ref. [12] by modifying that experiment in two significant ways: 1) Rather than using conventional mass spectrometry techniques as in Ref. [12], the present experiment utilizes the Purdue Accelerator Mass Spectrometer (AMS) which has far greater sensitivity. 2) In addition, we have started from samples where the probability of Be' retention is highest. In contrast to Ref. [12], which set a limit on Be' in air, the present search utilizes pure Be metal, Be ore, and a sample of natural gas containing He, in addition to a sample of laboratory air.

Searching for violations to the Pauli principle using Be is also attractive because the valence shell is directly involved in the possible violation of the PEP by the electrons. Because the anomalous electrons are found in the 1s shell we avoid potential difficulties in explaining how shielding of the nucleus by lower levels may effect the way the atom behaves chemically. We also note that another unique aspect of using Be is that all four electrons have crowded into the ground state of the atom exerting behavior consistent with bosonic particles. In this sense searching in Be provided the opportunity to test for even numbers of fermions violating the PEP while also providing a case with all electrons in the atom exhibiting bosonic behavior. Of course a similar search for anomalous lithium could be performed looking for violations involving only a single particle or odd number of fermions.

THE SAMPLES

Standards and Blank Samples

To set a limit on the presence of Be' in each of our unknown samples, it was necessary to determine the efficiency of the AMS in detecting Be in a known gaseous sample. The needed calibrations were carried out by preparing gaseous samples of a volatile Be compound, $Be(C_5HF_6O_2)_2$, containing a known concentration of Be. Seven standard samples were prepared by mixing $Be(C_5HF_6O_2)_2$ with He to produce mixtures with known Be/He concentrations, as shown in Table 1. (In this paper all concentrations are quoted as ratios of the number of Be or Be' atoms to the number of He atoms in the sample.) In addition a blank sample was prepared with pure He.

Unknown Samples

Due to the lack of a complete theory providing a mechanism for the creation of anomalous electrons and hence anomalous Be we may greatly increase our chances of finding Be' by looking in a variety of Be samples. As mentioned above, one of the novelties of this particular experiment is the diversity of samples in which the presence of Be' was tested. We tested for the presence of Be' in a pure metal sample produced in 1967. We also looked for Be' in beryl which has a nominal chemical formula $3BeO \cdot Al_2O_3 \cdot 6SiO_2$. Both of these samples were obtained by liberating and collecting any Be' gas which may have been trapped inside. The gas field sample containing a significant amount of He (and possibly Be') was obtained from a natural gas field in Texas. The assumption behind the use of the gas field sample is that a geological region capable of trapping He might also trap Be', which may arise from outgassing of the Earth's crust or through direct cosmic ray production. In addition to the gas field sample, we also ran a sample of laboratory air in order to compare to previous experiments [12]. Because of the lack of a consistent theoretical framework any predictions about how this anomalous form of Be is created are merely speculation.

Since the samples obtained from the Be metal, the ore, and the gas field sample were prepared in the same way as the standards, with similar He partial pressures, a comparison of the Be counting rates in the samples and the standards determined the Be/He ratio in each sample (a more detailed discussion of the samples is provided in Ref. [26]).

PRINCIPLE OF THE EXPERIMENT

Each gas sample is admitted into the ion source through at tube attached to the back of a specially designed cathode sample holder containing a SmO surface. When the gas exits the cathode surface it is exposed to a 5 keV Cs^+ ion beam. Any Be in the sample is then combined with the oxygen produced by the SmO surface and creates BeO^- ions which eventually enter the accelerator.

Since the potential which produces the 5 keV Cs^+ beam is much greater than the typical binding energies of atomic electrons, it can be presumed that nearly all the Be' atoms initially present in the unknown samples described above are first converted to Be through the highly-energetic collisions with the Cs^+ ions. It should be noted that although we are assuming that the probability P for the Cs^+ beam to convert Be' to Be is ≈ 1, it could be possible that $P < 1$. Hence the limits quoted in Table 2 should be understood as corresponding to $P = 1$.

Thus for the unknown samples any Be atoms detected above background (see below) can then be identified as having come from the gas samples, and hence are candidates for Be'.

MEASUREMENTS AND RESULTS

Before running the unknown samples it was necessary to calibrate the sensitivity of the AMS to Be concentration and determine the background inherent in the system. Since the Be concentration of each standard sample was known, we created a calibration plot after running each standard on the AMS and carrying out a least-squares fit to the resulting data provided in Table 1 (see Fig. 1a). As expected the

TABLE 1. Summary of Be counting rates in counts-per-minute (cpm) for standards and blanks. For the blank and each of the seven standards S1–S7, we exhibit the Be concentration, [Be]/[He], and the corresponding 1σ errors.

Type	Be Concentration [$\times 10^{-6}$]	Counting Rate [cpm]
Blank	0.00 ± 0.05	0.18 ± 0.08
Standard, S1	0.24 ± 0.05	1.77 ± 0.14
Standard, S2	0.54 ± 0.05	14.56 ± 0.47
Standard, S3	0.97 ± 0.10	26.14 ± 0.32
Standard, S4	2.49 ± 0.10	65.02 ± 0.47
Standard, S5	4.84 ± 0.10	124.91 ± 0.82
Standard, S6	8.47 ± 0.20	212.81 ± 0.56
Standard, S7	16.94 ± 0.20	426.84 ± 0.56

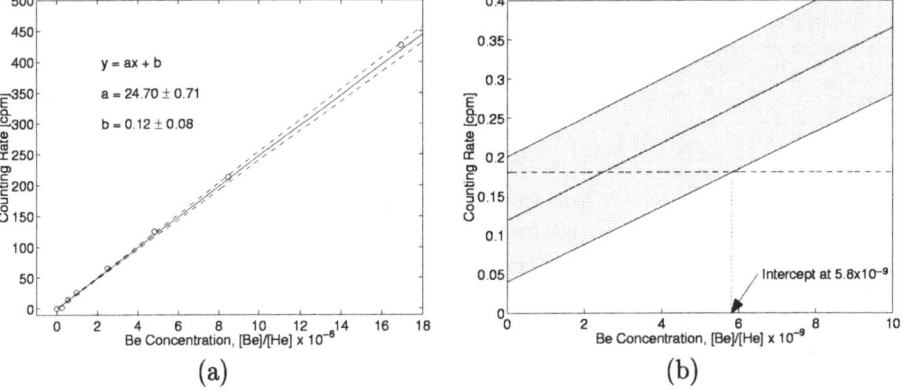

FIGURE 1. (a) Calibration of the AMS for gaseous Be. The open circles represent the experimental data for the blank and the seven Be Standards exhibited in Table 1. The solid line is the result of the least-squares fit shown in the figure and the 1σ errors are represented by the dashed lines. (b) The shaded region is an enlarged version of the lower end of (a) with the 1σ errors included. The horizontal dashed line is the measured counting rate for Be, which is presumed to arise from the presence of Be' in the samples. The limits on [Be']/[He] for the beryl and gas field samples are identical, as shown in Table 2, and are obtained from the intersection of the horizontal line and the 1σ band as shown. The limits from the other samples are obtained in a similar manner.

Be counting rate in the detector varied linearly with the concentration.

The blank sample, made purely of He, was then run to determine the background Be concentration. Finally, the unknown samples were introduced to the AMS. The intersection of the horizontal band corresponding to the Be counting rate in the unknown sample with the calibration line then determined the minimum detectable Be concentration, as illustrated in Fig. 1b. No signal for Be above background was detected in any of the unknown samples (Be metal, ore, gas field, and laboratory air).

CONCLUSIONS

Our results are shown in Table 2. We see from Fig. 1b that since no signal for Be' was detected above background, this translates into a 1σ limit on the concentration of Be' relative to He in each of the samples, as shown in Table 2. For air this gives a limit on the Be' concentration, ρ', in air of $\rho' < 3 \times 10^{-14}$, which improves on the limit in Ref. [12] by a factor of almost 300. However, the more significant results are those from the Be metal, beryl, and gas field sample, since these were obtained from sources where Be' was more likely to be found than in air. In the absence of a rigorous fundamental theory of PEP violation, it is unclear how to convert the results from these specific samples into meaningful limits on the PEP-violating parameter $\beta^2/2$ often quoted in the literature [6] or into limits on $(1+q)/2$ [23], the square of which represents the probability of finding an electron in a symmetric state within the constructs of the quon formalism. For this reason we list only the quantities directly measured by our experiment in Tables 1 and 2, and defer the interpretation of these results to a future complete theory. That theory could, of course, indicate that PEP violation would not be expected to arise in some or all of the systems studied to date, in which case a new class of experiments may be called for.

TABLE 2. Summary of unknown sample concentrations. The first column gives the Be' concentration and the second column expresses the Be' concentration relative to Be or air. All results are at the 1σ level and assume that the probability of converting Be' to Be in the ion source is unity.

Sample	Be' Concentration, $\frac{[Be']}{[He]}$	Final Be' Concentration
Be Metal	$< 2 \times 10^{-8}$	$\frac{[Be']}{[Be]} < 9 \times 10^{-12}$
Beryl	$< 6 \times 10^{-9}$	$\frac{[Be']}{[Be]} < 1 \times 10^{-11}$
Gas Field	$< 6 \times 10^{-9}$	—
Air	$< 5 \times 10^{-9}$	$\rho' = \frac{[Be']}{[air]} < 3 \times 10^{-14}$

ACKNOWLEDGEMENTS

The authors are deeply indebted to Professor Graham Cooks for his help in the initial stages of this experiment, and for many useful suggestions. We also wish to thank Professor Don Gaines and Dr. Dovas Saulys of the Chemistry Department, University of Wisconsin for providing us with the sample of $Be(C_5HF_6O_2)_2$. We wish to express our appreciation to the Brush Wellman Corporation for the samples of Be metal and Be ore, to Bill Moore of the U.S. Bureau of Mines for providing us with the Bush Dome gas field sample and for determining its chemical composition, and to Adam Carr, Otto Furuta, and James Klaunig for helpful discussions. This work was supported in part by the U.S. Department of Energy under Contract No. DE-AC02-76ER01428, and National Science Foundation Grant No. 9809983-EAR.

REFERENCES

1. Pauli, W., *Z. Phys.* **31**, 765–783 (1925).
2. Green, H.S., *Phys. Rev.* **90**, 270–273 (1953).
3. Messiah, A.M., and Greenberg, O.W., *Phys. Rev.* **136**, B248–267 (1964).
4. Ignat'ev, A.Yu., and Kuz'min, V.A., *Sov. J. Nucl. Phys.* **46**, 444–446 (1987).
5. Greenberg, O.W., and Mohapatra, R.N., *Phys. Rev. Lett.* **59**, 2507–2510 (1987); *Phys. Rev.* **D39**, 2032–2038 (1989).
6. Greenberg, O.W., and Mohapatra, R.N., *Phys. Rev. Lett.* **62**, 712–714 (1989).
7. Mohapatra, R.N., *Phys. Lett.* **B242**, 407–411 (1990).
8. Greenberg, O.W., *Phys. Rev.* **D43**, 4111–4120 (1991).
9. Greenberg, O.W., *Phys. Lett.* **A209**, 137–142 (1995).
10. Baron, E., Mohapatra, R.N., and Teplitz, V.L., *Phys. Rev.* **D59**, 036003 (1999).
11. Goldhaber, M., and Goldhaber, G.S., *Phys. Rev.* **73**, 1472–1473 (1948).
12. Fischbach, E., Kirsten, T., and Schaeffer, O.A., *Phys. Rev. Lett.* **20**, 1012–1014 (1968).
13. Reines, F., and Sobel, H.W., *Phys. Rev. Lett.* **32**, 954 (1974).
14. Logan, B.A., and Ljubičić, A., *Phys. Rev.* **C20**, 1957–1958 (1979).
15. Novikov, V.M., et al., *Phys. Lett.* **B240**, 227–231 (1990).
16. Nolte, E., et al., *Nuc. Inst. and Meth. in Phys. Res.* **B52**, 563–567 (1990).
17. Ramberg, E., and Snow, G.A., *Phys. Lett.* **B238**, 438–441 (1990).
18. Kishimoto, T., et al., *J. Phys.* **G18** 443–448 (1992).
19. Deilamian, K., Gillaspy, J.D., and Kelleher, D.E., *Phys. Rev. Lett.* **74**, 4787–4790 (1995).
20. Barabash, A.S., et al., *JETP Lett.* **68**, 112–116 (1998).
21. Fredenhagen, K., *Commun. Math. Phys.* **79**, 141–151 (1981).
22. Greenberg, O.W., and Hilborn, R.C., *Foundations of Phys.* **29**, 397–407 (1999).
23. Greenberg, O.W., and Hilborn, R.C., *Phys. Rev. Lett.* **83**, 4460–4463 (1999).
24. Wigner, E.P., *Math. Naturwiss. Anz. Ung. Ak. Wiss.* **46**, 576 (1929).
25. Ehrenfest, P. and Oppenheimer, J.R., *Phys. Rev.* **37**, 333–338 (1931).
26. Javorsek II, D., et al., *Phys. Rev. Lett.* (in Press).

Spectroscopic Tests of the Symmetrization Postulate and of the Statistics for Nuclei in Molecules

Giovanni Modugno*, Davide Mazzotti*, Michele Modugno*,
Nathalie Picqué*, Giovanni Giusfredi[†], Pablo Cancio Pastor[†], Paolo
De Natale[†], and Massimo Inguscio*

* *INFM, LENS and Dipartimento di Fisica, Universitá di Firenze, Largo Enrico Fermi 2, I-50125 Firenze, Italy.*
[†] *Istituto Nazionale di Ottica Applicata, Largo Enrico Fermi 6, I-50125 Firenze, Italy.*

Abstract. We discuss the implications of the spin-statistics connection and of the symmetrization postulate on the quantum states of three identical nuclei arranged in rigid symmetrical molecules. We show that in the case of three identical nuclei a whole class of rotational states are forbidden also by the symmetrization postulate, and not only by the spin-statistics connection. We then present the experiments we are performing on NH_3, to search for the possible existence of these states forbidden by the symmetrization postulate, and on CO_2, to improve the sensitivity for detection of exchange-antisymmetric states of spin-0 nuclei. Preliminary results of the latter experiment indicate that violations of the statistics for oxygen nuclei, if present, are smaller than 10^{-11}, with an improvement of more than two orders of magnitude with respect to previous experiments.

INTRODUCTION

The principle of the latest experiments on systems of identical particles have already been discussed by G. Tino in his contribution to this volume (see also [1]). Here we want to present in more detail the principle of an experiment to be performed on systems composed by three particles, which show new features with respect to the two-particles systems mainly considered up to now. Since we are interested in experiments to look for possible violations of the basic postulates in the quantum mechanical description of identical particles, namely the symmetrization postulate (SP) and the spin-statistics connection (SSC), we consider a prototypical system that can also be found in nature, which is composed of three identical nuclei in a plane molecule with D_3 symmetry. Due to the increased dimension of the permutation group with respect to the two-particles case, some classes of rotational states of the three nuclei molecule can be built without a defined symmetry under permutation of particles, thereby forbidden by the SP. Although none of such states have been observed so far in real molecules, a high-sensitivity spectroscopic investigation of simple molecules would allow to look for possible tiny violations of the SP or, in other words, to look for particles which are not bosons nor fermions. We note that in the frame of ordinary quantum mechanics this issue is even more general than the search for half-integer spin particles which behave like bosons, or vice-versa. An analogous experimental test, to be performed on molecules with an even larger number of identical nuclei, has been proposed in [1]. In this contribution we also report the present state of the experiments running in Firenze to look for violations of both the SSC and the SP in molecules with two and three identical nuclei.

UNSYMMETRICAL STATES IN THREE-NUCLEI MOLECULES

The concept of indistinguishability of identical particles is formally expressed by requiring no change to any quantum observables under permutation of the particles [2]. In general, given a N-particles system, $N!$ permutation operators $\hat{P}$ forming the permutation group S_N can be defined. The operators $\hat{A}$ corresponding to observables are therefore permutation-invariant

$$[\hat{P}, \hat{A}] = 0 \tag{1}$$

and since the evolution operator $\hat{U}(t)$ is related to the Hamiltonian of the system $\hat{H}$, which is a physical observable, the above condition is fulfilled at any instant of time. An important consequence of these relations is the *super-selection rule*: the Hilbert space $\mathcal{H}$ can be written as a direct sum of orthogonal subspaces, each one invariant under the permutation group, the orthogonality being preserved along the evolution of the system.

For $N = 2$ the Hilbert space can be decomposed exactly as $\mathcal{H} = \mathcal{H}_+ \oplus \mathcal{H}_-$, which are respectively the symmetric and the antisymmetric subspaces. For $N > 2$ this is no longer true, and we have

$$\mathcal{H} = \mathcal{H}_+ \oplus \mathcal{H}_- \oplus \mathcal{H}'_1 \oplus \mathcal{H}'_2 \oplus \cdots, \tag{2}$$

where $\mathcal{H}'_j$ are permutation-invariant subspaces. Contrarily to $\mathcal{H}_+$ and $\mathcal{H}_-$, which have a definite symmetry and are one-dimensional representations of the permutation group, the subspaces $\mathcal{H}'_j$ do not posses a definite symmetry and have dimension greater that one. The SP, requiring physical states to be either symmetric or antisymmetric, seems therefore to be a *natural* assumption, due to the particular properties of $\mathcal{H}_+$ and $\mathcal{H}_-$. Nevertheless, there is no stringent reason to exclude *a priori* the possibility of physical states not obeying SP, since their presence do not violate any basic principle of quantum mechanics. The only difference to take into account is the lack of a correspondence between physical states and vectors, since the subspaces $\mathcal{H}'_j$ have dimension greater that one and do not admit a complete set of mutually commutating physical observables.

In the following we work out in more detail the permutation properties for the special case $N=3$ [3]. The symmetrization group S_3 is formed by the 3! cyclic permutations of the labels $(1, 2, 3)$, belonging to three distinct classes

$$P_{(1)(2)(3)} = I, \; P_{(2,3)}, P_{(1,3)}, \; P_{(1,2)}, \; P_{(1,2,3)}, \; P_{(3,2,1)}. \tag{3}$$

This group is isomorphic to the *dihedral group* D_3, generated by the symmetry transformation shown in Figure 1. In fact, reflection about each axis (a, b, c) is equivalent to a two label permutation $P_{(i,j)}$, while rotations around the center by angles $2\pi/3$ and $-2\pi/3$, lead to cyclic permutation of all the three labels. The visualization of the group S_3 by its association to the geometrical symmetries of D_3 is very helpful since we are interested just in this realization on a physical system. The matrix representation of the group S_3 can be reduced, since it must contain a number of irreducible representations equal to the number of distinct classes. From general results [3] we know that these representations are the symmetric one, the antisymmetric one, and a 2-dimensional representation occurring two times. It is possible to find two pairs of vectors which generate such 2-dimensional irreducible representations of $\mathcal{H}'$, by diagonalizing the permutation operators $P_{(1,2,3)}$ and $P_{(3,2,1)}$ [4]. We should note that the eigenvalues of these permutations on the generators of $\mathcal{H}'$ are of the form

$$\lambda_{\pm} = \exp(\pm i \frac{2\pi}{3}). \tag{4}$$

Moreover, the eigenvalues for the pair of generators of each of the two irreducible subspaces $\mathcal{H}'$ are always different (λ_+ and λ_-) and therefore the sign of the argument of the exponential cannot be an observable of the system.

We want now to consider the prototypical molecule, composed by three identical nuclei at the vertices of an equilater triangle. To simplify the discussion, we consider

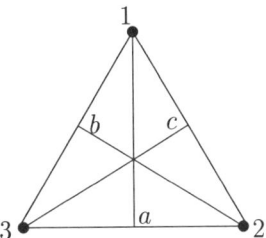

FIGURE 1. The nontrivial symmetry operations on this triangular configuration are: (i) reflection around the axes a, b, c and (ii) rotation around the center by angles $2\pi/3$ and $-2\pi/3$.

only the ground electronic and vibrational states of the molecule, which we suppose to be of species Σ, and therefore completely symmetric under any permutation of the nuclei. In the frame of the Born-Oppenheimer approximation the overall symmetry under permutation is therefore dictated by the nuclear rotational and spin states. The generic state belongs to the direct product of the Hilbert space for the nuclear coordinates and of that for the nuclear spin

$$\mathcal{H} = \mathcal{H}_S \otimes \mathcal{H}_R. \tag{5}$$

Each Hilbert space is 6-dimensional, with the general properties discussed above, the quantum labels for the states being respectively the mean coordinates and the spin components for the three nuclei

$$|\mathbf{x}_1, \mathbf{x}_2, \mathbf{x}_3\rangle, \; |S_{z1}, S_{z2}, S_{z3}\rangle. \tag{6}$$

The first case we want to consider is that of spin-0 nuclei: only one spin state can be defined, which clearly belongs to the symmetric subspace, and therefore the rotational states alone determine the overall symmetry of the molecular states. As is well known, the rotational states can be described in terms of eigenfunctions of angular momenta; for a symmetric molecule like the one we are considering they are a somewhat complicated superposition of eigenfunctions of the total angular momentum $\mathbf{J}$ and of its projection $\mathbf{K}$ along the threefold symmetry axis perpendicular to the plane of the molecule. The eigenvalues of the energy have the form

$$E(J, K) = BJ(J+1) - (B-C)K^2, \tag{7}$$

where B and C are the rotational energy constants, and K can assume the values $-J, -J+1, \cdots, J$. To determine the symmetry of the rotational states under permutation of particles, we want to compare such permutations to particular rotations of the molecule. Indeed, as we already noted, the permutation group is isomorphic to the D_3 group, to which the symmetry operations of our molecule belong. Moreover a cyclic permutation of the nuclear coordinates corresponds to a physical rotation of the molecule. In particular, the cyclic permutations of the three labels are equivalent to rotations in the plane of the molecules by angles $\theta = \epsilon 2\pi/3$, with $\epsilon = \pm 1$, while the exchanges are equivalent to rotations about the twofold axes of symmetry in the plane by $\phi = \epsilon \pi$. The general transformation rule of the rotational states under a rotation by an angle $\alpha \mathbf{u}_\alpha$ is

$$\Psi_R \to \Psi_R e^{i\alpha \mathbf{u}_\alpha \cdot \mathbf{J}}, \tag{8}$$

which must be compared with the transformation rule under the corresponding permutation, i. e. to the eigenvalues of the permutation. For the first kind of rotations, only the component $\mathbf{K}$ of the angular momentum must be considered, and by comparing Eqs.(4) and (8) it appears that the rotational states with $K=3q$, with q integer, belong to $\mathcal{H}_\pm$, while the ones with $K=3q\pm 1$ necessarily belong to $\mathcal{H}'$. In the case of rotations about the symmetry axes in the plane, only for the case of $K=0$ the angular momentum along such axes is a constant of the motion, and we obtain that odd- and even-J states belong to $\mathcal{H}_-$ and $\mathcal{H}_+$ respectively. In conclusion, in this kind of molecules it is possible to define not only exchange-antisymmetric states, like in diatomic molecules, but also completely unsymmetrized states, which would be allowed only in presence of a violation of the SP. We note that for the states in $\mathcal{H}'$ it is not possible to define the sign of the angular momentum **K**, due the multidimensionality of the Hilbert subspace (Eq.(4)), and in fact such a sign cannot be measured as long as the molecule is symmetric, since the pairs of states with **K** pointing in opposite direction are degenerate (Eq.(7)).

The discussion can be extended to spin-1/2 nuclei: two doublet states and one quartet can be defined, with total spin $I=1/2$ and $I=3/2$ respectively. The doublets are easily assigned to just one of the unsymmetric 2-dimensional Hilbert subspaces $\mathcal{H}'_s$, while the quartet is symmetric. Clearly it is not possible to build up completely antisymmetric states with three spin components, each having only two values; for such purpose a larger individual spin would be needed ($S \geq 1$). The overall symmetry of rotational and spin states can be evaluated by using the general rules for decomposition of the direct product into irreducible representations [3]. As a result, it is always possible to find the proper combination of unsymmetrical rotational and spin wavefunctions to build up antisymmetrical total states, as required by the SP and the SSC. Therefore, the states with $K=3q\pm 1$ are no longer forbidden, at least if the hyperfine structure of the states, due to the coupling of I and J, is not resolved. On the contrary, the states with $K=0$, J even, are a superposition of symmetric and unsymmetric states, and are therefore forbidden by both the SSC and the SP. These results are summarized in Table 1.

It is possible to extend further these arguments to molecules containing additional nonidentical nuclei in or out of the plane (point groups D_{3h} and C_{3v}), whose rotational states can be assigned in an analogous way to the various Hilbert sub-

TABLE 1. Appropriate Hilbert subspaces for the rotational and spin states of a D_{3h} molecule containing three spin-0 or spin-1/2 identical nuclei.

Rotational state		Spin-0 nuclei		Nuclear spin	Spin-1/2 nuclei	
K	J	Subspace	Forbidden by	I	Subspace	Forbidden by
$3q, q\neq 0$	any	$\mathcal{H}_+, \mathcal{H}_-$	-	1/2	$\mathcal{H}'$	SP
				3/2	$\mathcal{H}_+, \mathcal{H}_-$	-
$3q\pm 1$	any	$\mathcal{H}'$	SP	1/2	$\mathcal{H}_+, \mathcal{H}_-, \mathcal{H}'$	-
				3/2	$\mathcal{H}'$	SP
0	even	$\mathcal{H}_+$	-	1/2	$\mathcal{H}'$	SP
				3/2	$\mathcal{H}_+$	SSC
0	odd	$\mathcal{H}_-$	SSC	1/2	$\mathcal{H}'$	SP
				3/2	$\mathcal{H}_-$	-

spaces. The main difference is found for nonplanar molecules, which in general have twice the number of states of their planar counterparts, when considering $K=0$ rotations. These rotations, by angles $\phi=\epsilon\pi$, are no longer equivalent to permutations of two identical nuclei, as it was for a planar molecule, since also the additional nuclei out of the plane has rotated. On the other hand, the equivalence can be reestablished by considering the combination of such a rotation with an inversion of the coordinates of all the nuclei. For more details see [4,5].

EXPERIMENTAL TESTS OF THE SP AND THE SSC

Following the discussion of the previous section, it is possible to think of experiments searching possible violations of the SSC and the SP. The principle of the experiment would be, as usual, to probe electromagnetic transitions looking for molecules possessing the forbidden rotational states. If high-sensitivity spectroscopic techniques are used, tiny violations not observed so far can possibly be detected. In order to test the validity of the SP, i.e. to look for the existence of unsymmetrical three-particle states, two particular molecules appear interesting. Specifically, in the case of spin-0 nuclei the SO_3 molecule represents the prototypical plane system studied above, with an entire class of rotational states ($K=3q\pm1$) which belong to $\mathcal{H}'$. The nonplanar NH_3 molecule is instead composed of three spin-1/2 nuclei with the proper symmetry, and therefore some of the rotational states are

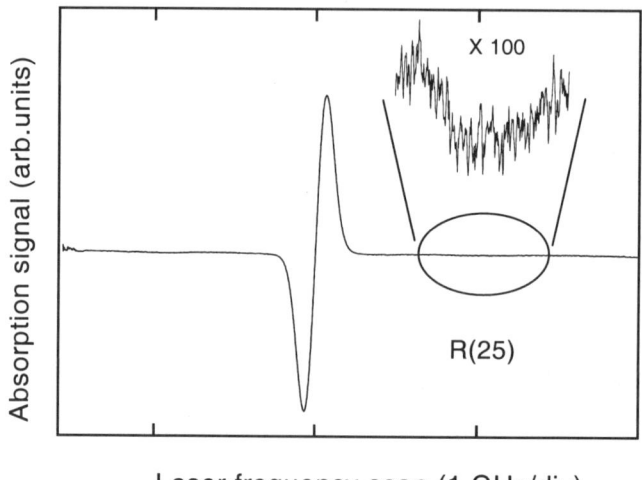

FIGURE 2. Experimental spectrum of CO_2 around the expected frequency of a transition starting from the SSC-forbidden $J=25$ rotational state. No absorption is visible above the noise level. The nearby absorption signal, due to a weak allowed transition, is used for calibration purposes.

forbidden by both the SSC and SP (if the hyperfine structure is not resolved, part of the $K=0$ states can belong only to $\mathcal{H}_+$ or $\mathcal{H}'$). Both these molecules have strong vibrational absorption bands in the infrared (see for example [4]), which can be probed with high sensitivity using low-noise semiconductor laser sources. We are presently devising in Firenze an experiment to look for possible small violations of the SP on both these systems. We plan to start the investigation from NH_3, which is a chemically stable molecule with very strong absorption bands studied extensively in the past, for which a detailed spectral database is present. We note that such an experiment would represent a test of the SP and the SSC for protons, which could be compared with the recent theoretical predictions [6].

As for the experimental tests of the SSC, a linear molecule with only two identical nuclei is certainly preferred, because of the much simpler rotational spectrum, that implies also an increased statistical weight of the single rotational states. To significantly improve the sensitivity achieved in a recent experiment performed in our group, using a diode laser resonant with the 2 μm band of CO_2 [7], we decided to investigate transitions in the fundamental vibrational band of this molecule, which is about 2000 times stronger than the previous one. For this purpose, we developed a new coherent source tunable around the band center at 4.25 μm [8]. Its peculiar features have recently allowed to observe sub-Doppler molecular spectra [9], and a new acquisition routine has been especially developed to co-add spectra over long periods of time. Preliminary results for a transition starting from the state with $J=25$ put an upper limit to the SSC violation at 10^{-11}, which is about 200 times lower than our previous test. In Figure 2 the experimental recording is shown. It was obtained using an absorption pathlength of 112 m, a gas pressure of 2 Torr and an acquisition time of 2 hours. This experiment provides the highest sensitivity ever reached for detecting exchange-antisymmetric states for two spin-0 nuclei and confirms the validity of the SSC.

We thank G. M. Tino for stimulating discussions.

REFERENCES

1. Tino, G. M., *Fortschr. Phys.* **48**, 537-543 (2000).
2. Messiah, A., *Quantum Mechanics*, North-Holland, Amsterdam, 1961.
3. Wu-Ki Tung, *Group Theory in Physics*, World Scientific, Singapore, 1985.
4. Modugno, G., and Modugno, M., *Phys. Rev. A*, accepted; e-print: quant-ph/0003118.
5. Herzberg, G., *Infrared and Raman Spectra of Polyatomic Molecules*, D. Van Nostrand Company, Princeton, 1950.
6. Greenberg, O. W. and Hilborn, R. C., *Phys. Rev. Lett* **83**, 4460-4463 (1999).
7. Modugno, G., Inguscio, M., and Tino, G. M., *Phys. Rev. Lett.* **81**, 4790-4793 (1998).
8. Mazzotti, D., De Natale, P., Giusfredi, G., Fort, C., Mitchell, J., and Hollberg, L., *Appl. Phys.* **B 70**, 747-750 (2000).
9. Mazzotti, D., De Natale, P., Giusfredi, G., Mitchell, J., and Hollberg, L., *Opt. Lett.* **25**, 350-352 (2000).

VI. Philosophical Issues

Putting a New Spin on Particle Identity

Steven French

*Division of History and Philosophy of Science,
School of Philosophy,
University of Leeds,
Leeds LS2 9JT,
UK
Email: s.r.d.french@leeds.ac.uk*

Abstract. Almost immediately after the development of quantum statistics, physicists were concerned with the implications for the individuality of particles. Such concerns have been further explored from a variety of perspectives during the past fifteen years or so. My intention in this paper is to briefly lay out the various positions that have been adopted and examine the philosophical consequences of the more general forms of statistics that have also been developed. The status of the spin-statistics theorem within this context has been considered in recent work by Hilborn and Yuca and I shall discuss their analysis within the philosophical framework concerning particle individuality. In particular I shall argue that a well-known philosophical position on this issue appears to be less accommodating to generalised forms of quantum statistics and I shall conclude by discussing the philosophical consequences of the supposed relationship between quantum statistics and spin.

INTRODUCTION: THE PHILOSOPHY OF INDIVIDUALITY

It is commonly accepted that rocks, chairs, people etc. are all individuals. What it is that forms the basis of this individuality - sometimes referred to as the *Principle of Individuality* - has been debated by philosophers for centuries and one can, rather crudely, delineate the following broad positions (an excellent and much more detailed discussion can be found in [1]):

One might first begin by noting the obvious point that we *distinguish* rocks, chairs, people etc., through certain of their properties. Perhaps, then, we can refer to such properties to ground the individuality of things. Unfortunately properties, by their very nature, are multiply instantiable so, as this positions stands, there is no guarantee that two or more things won't possess the same set of properties and in such a case the possibility of individuation would be lost. One possible guarantor of individuality would be the infamous Principle of the Identity of Indiscernibles (PII) which, put bluntly, states that two things simply *cannot* possess the same set of (relevant) properties. The debate then shifts to the status of this Principle - whether it is necessarily true or only contingently so. Are there properties which only one thing and no other could have? One such would be the property each thing has of being identical with itself, written formally as '$a=a$', where a is the appropriate name or label; this does indeed make the Principle true, but trivially so. A more interesting property is that of spatio-temporal location: if we assume some form of *impenetrability* then no two things can have exactly the same spatio-temporal properties and if the latter are included in our set, then PII appears to hold. Spatio-temporal location then performs a nice unifying role in our metaphysics: it allows us to distinguish even those things which share all their other properties and it acts as a Principle of Individuality.

However, there is an alternative tradition which insists that, at least conceptually, distinguishability and individuality must be kept distinct. So, for example, it has been argued that whereas the former has to do with a thing's relationships with other things, the latter has to do with the thing taken on its own. According to this venerable tradition, what makes a thing an individual cannot be any multiply instantiable property - these relate to distinguishability only - but rather something else, over and above or *transcending* such properties. According to certain, Scholastic, philosophers, this something else is a thing's 'haecceity', reincarnated in modern times as its 'primitive thisness' which has been further explicated in terms of self-identity ('$a=a$' again). According to others, it is substance, famously characterised by Locke as 'something we know not what'. Those of a broadly empiricist persuasion have not taken too kindly to such notions, but they have proven quite hard to do away with.

Now, can these metaphysical accounts of individuality be extended to the particles of modern physics?

QUANTUM MECHANICS AND THE LOSS OF INDIVIDUALITY

There is a well-known way of answering this question which proceeds by examining the aggregate behaviour of such particles. We begin by taking particles of the same 'kind', which can be regarded as 'identical', in the sense of possessing the same 'intrinsic' or state-independent properties*. We then consider the distribution of these particles over states - two particles over two one-particle states, say - and it is assumed that each resulting arrangement is accorded equal probability. In classical statistics, the situation where we have one particle in each state is given a weight of two, corresponding to the two arrangements that may be formed by a permutation of the particles. This gives us the standard Maxwell-Boltzmann statistics, of course. That a permutation of the (identical) particles is included in the count of possible arrangements is taken to imply that the particles are *individuals*, in some sense. Since the particles are identical in the above sense, their intrinsic properties cannot serve to individuate them and hence this individuality must be cashed out in terms of 'primitive thisness', or substance, or, more typically, their spatio-temporal location.

It is not difficult to see how the examination is going to proceed in the quantum case. Here, standardly, we have two forms of statistics, of course - Bose-Einstein and Fermi-Dirac - but in both cases the above situation of one particle in each state is given a weight of one. This is typically taken to reflect the fact that arrangements obtained by particle permutations do not feature in the relevant counting in quantum statistics. The implication, then, is that the particles can no longer be considered to be individuals - that they are, in some sense, *'non-individuals'*.

This argument forms the basis of what I shall call the 'Received View' of particles in Quantum Mechanics (QM). It emerged in tandem with the formal development of

---

* Philosophers tend to prefer the term 'indistinguishable', reserving 'identical' for when two apparently distinct things turn out to be the same thing. For the purposes of this paper, I'm going to stick with the physicists' usage.

quantum statistics. Thus, it can be found already present in the work of Born in 1926 ([2]) and was expressed by Weyl, in the following, typically evocative fashion,

'... the possibility that one of the identical twins Mike and Ike is in the quantum state E_1 and the other in the quantum state E_2 does not include two differentiable cases which are permuted on permuting Mike and Ike; it is impossible for either of these individuals to retain his identity so that one of them will always be able to say "I'm Mike" and the other "I'm Ike." Even in principle one cannot demand an alibi of an electron!' ([3])

It was subsequently promulgated by Schrödinger, in particular, who argued that quantum particles should be considered as having lost their identity. By this he insisted that he did not mean merely the impossibility one has of recognizing a particle, once observed, after it has interacted with another of the same species:

'I beg to emphasize this and I beg you to believe it: It is not a question of being able to ascertain the identity in some instances and not being able to do so in others. It is beyond doubt that the question of 'sameness', of identity, really and truly has no meaning.' ([4], pp. 17-18)

This 'Received View' has been the subject of considerable philosophical interest ([5]; [6]; [7]; [8]) and has motivated a number of attempts to accommodate this loss of identity or 'non-individuality' within non-standard logical or mathematical frameworks ([9]; [10]). However, resistance has come from two directions, and in both cases the existence of a fundamental metaphysical distinction between classical and quantum particles has been denied. Thus it has been suggested that in the case of the former, like that of the latter, particles of the same kind are not only identical but are also 'indistinguishable' in the stronger sense of there being no provision for the particles to carry any *identification marks* ([11]). However, a defender of the Received View would object to this last claim: in the classical case at least, it is precisely such provision that is provided by some principle of individuality which transcends the properties of the particle. To note that classical particles do not carry such empirically superfluous, dynamically irrelevant marks in practice and to infer, on those grounds, that they are indistinguishable, like their quantum counterparts, is to confuse individuality with distinguishability.

Others have adopted a more positivistic approach by insisting that the meaning of 'indistinguishability', in this stronger sense, is determined experimentally ([12]). Such 'indistinguishability' is then seen as not restricted to quantum particles, since it features in the resolution of Gibbs' Paradox: if like gases at the same pressure and temperature are mixed, then there is no change in the experimental entropy, which disagrees with the result obtained for the statistical entropy, based on the above sorts of permutation considerations. The standard resolution, of course, is to exclude the relevant permutations from the calculation of the statistical entropy. The question now is, on what grounds can one justify this exclusion? If it is argued that the grounds are experimental, rather than metaphysical (*ibid.*), then one could simply turn the force of the argument around and insist that what this shows is that the world is actually quantum in nature, as we would expect[†].

Proceeding from the other direction, one can also argue that classical and quantum particles are metaphysically similar in that both may be treated as *individuals*. In this case, one must account for the difference in the ways in which particle permutations are treated and this difference can best be approached by considering the more general

---

[†] Further discussion of the Gibbs Paradox in this context can be found in [13], pp. 271-280.

framework offered by non-standard statistics. It is this alternative to the Received View that I shall focus on here.

NON-STANDARD STATISTICS AND QUANTUM INDIVIDUALITY

The Received View assumes the standard account of quantum statistics based on the so-called Symmetrisation Postulate (SP), which effectively dictates that the appropriate wave-functions for an assembly of identical particles be either symmetric or antisymmetric with regard to permutations of the particle indices, or labels, giving Bose-Einstein and Fermi-Dirac statistics respectively. However, it was realised early on, by Dirac and Weyl for example, that other forms of statistics are also mathematically possible. The history of these 'non-standard statistics' is largely unwritten (further details can be found in [13]) but what I want to emphasise here is that they offer a much broader perspective on this issue of particle individuality. The work of Greenberg and Messiah in particular, important as it is for physics, also has a certain philosophical significance [17].

Focusing on the relationship between standard quantum statistics and particle indistinguishability, they noted that SP is '... an extremely strong condition, very much stronger than what is implied by the indistinguishability of identical particles' (*ibid.*, p. 248) and, furthermore, that the usual ways of introducing SP into the quantum mechanical formalism are suspiciously ad hoc in nature. They therefore proposed replacing SP by the weaker 'Indistinguishability Postulate' (IP), which states that if a particle permutation is applied, there is no way of distinguishing the resultant permuted wavefunction from the original unpermuted one by any means of observation at any time. The former is sufficient but not necessary for the latter and SP can be regarded as a restriction on the states for all observables, whereas IP is a restriction on the observables for all states. This offers a more comprehensive framework for particle statistics and permits the introduction of multi-dimensional sub-spaces of Hilbert space, corresponding to para-statistics as well as the more standard forms. Regarding the apparent evidence for symmetrisation, Greenberg and Messiah argued that many experimental results which have been taken to be tests of SP are actually just tests of IP (*ibid.*, pp. 259-267).

The implication from quantum statistics to non-individuality can now be resisted by giving another account of the reduction in statistical weight attaching to the arrangement of one particle in each state in the argument above. This alternative account is beautifully illustrated using parastatistics (see [13], pp. 210-218): as is well known, IP acts as a super-selection rule which partitions Hilbert space into a number of irreducible sub-spaces. Given IP, Schur's Lemma then implies that all matrix elements of an observable connecting different representations are zero. Hence, if there exist states corresponding to subspaces of different representations then transitions between such states must be forbidden. Hilborn and Yuca call this the 'Fundamental Theorem' ([18], p. 30) because of its importance in the context of possible violation of the spin-statistics theorem. We shall return to consider it in the context of their account below*.

---

* States corresponding to subspaces of the same representation are not separated in this way and for certain kinds of three-body interactions, transitions may occur between them. This has implications for the relevant weighting assignments in parastatistics which turn out to be different from those for Gentile's intermediate statistics [19].

Hence, what we have is a situation in which IP imposes a restriction on the states such that once a particle is in a given set of states, the other sets are rendered inaccessible to the it. If we consider the time evolution of the system as effected by some Hamiltonian, then IP implies that if the system starts in a subspace corresponding to a particular irreducible representation, then it will always remain in that subspace: bosons will always be bosons, fermions will always be fermions etc. From this perspective, the explanation for the reduction in statistical weight in my simple example above lies with the inaccessibility of certain states: if, in this case, the restriction is imposed that the state of the system be either symmetric or anti-symmetric then only one of the two possible states formed by a permutation of the particles is ever available to the system and so the statistical weight corresponding to the distribution of one particle in each (one-particle) state is half the classical value. On this view, the reduction in weight has nothing to do with the supposed non-individuality of the particles and so one can continue to regard them as *individuals* which are simply prevented from occupying certain states [20]. The two components of this view - the accessibility constraints and the individuality of the particles - can then be understood *philosophically* in ways that emphasise the commonality with the classical situation.

Thus IP can be thought of as a kind of extra postulate of the theory, or a form of initial condition in the specification of the situation. And of course, the notion of state accessibility restrictions can also be found in classical statistics, albeit in a somewhat restrictive form. The most important and most obvious has to do with the available energy but other uniform integrals of the equations of motion may exist for a particular assembly which have the effect of restricting its representative point to a certain region of phase space. With regard to thermodynamic consequences, only the energy integral is significant and imposes constraints on accessible regions of the phase space. In quantum statistics, we have these further constraints and the symmetry type of any suitably specified set of states is an absolute constant of motion equivalent to an exact uniform integral in classical terms ([21], p. 213). This goes some way towards explicating the greater role played by symmetry in the quantum case as compared to the classical*.

With regard to the individuality of the particles, how one understands this *philosophically* may still be problematic. Given that the particles are still regarded as possessing the same set of intrinsic properties, they cannot be individuated in terms of these nor in terms of their state-dependent properties†, unless one were to accept some form of 'empirically superfluous' or dynamically irrelevant property again [23]. One could of course continue to maintain that this individuality should be understood in terms of some kind of 'primitive thisness' or underlying substance and live with the metaphysical discomfort that accompanies such a position. Finally, articulating this sense of individuality in terms of spatio-temporal location runs into the well known problems to do with trajectories in quantum mechanics. Nevertheless, one can detect elements of this last possibility in the configuration approach to indistinguishable

---

* When it comes to ordinary statistics, the question as to what states are accessible is straightforward. In the paraparticle case it is slightly more complicated: for particles obeying Gentile's paragas statistics, the number of accessible states will simply be intermediate between the number accessible to fermions and bosons. This number will increase as the statistics becomes less fermionic and more bosonic. In the case of parastatistics, where there is a possibility of transitions occurring between states carrying the same representation (i.e. between the triangular sub-spaces above) it is a linear combination of states that now becomes accessible to the particles.

† Thus it has been argued that the Principle of Identity of Indiscernibles (PII) is violated in quantum mechanics [22].

particles and a connection can be made between this approach and perhaps the most well known attempt to retain distinct trajectories in QM, namely Bohmian mechanics.

INDIVIDUALITY AND THE CONFIGURATION SPACE APPROACH

In order to accommodate particle permutations, points corresponding to such a permutation are identified within the full configuration space formed by the N-fold Cartesian product of three-dimensional Euclidean space, yielding the reduced quotient space [24]. Here, '... the mutual indistinguishability of the particles is coded into the topology of configuration space itself, and the different statistical types then correspond to different choices of boundary conditions on the wave function ...' ([25], p. 687). This description covers two inter-related issues: first, how do the above remarks concerning indistinguishability and (non-) individuality look from this perspective? And secondly, do we still recover non-standard statistics?

These issues can be seen to be related by reflecting on the problem of collisions: the reduced configuration space is not in general a smooth manifold since it possesses singular points where two or more particles coincide. This leads to two technical difficulties: first, it is not clear how one might define the relevant Hamiltonian at such singularities ([25], p. 687) and secondly, it would appear that their presence implies Bose-Einstein statistics only. The obvious and now standard, solution is to simply remove from the configuration space the subcomplex consisting of all such coincidence points, yielding a smooth manifold. The relevant group for n particles is then the n-string braid group and the irreducible unitary representations of this group can be used to label the different statistics that are possible. Imbo, Shah Imbo and Sudarshan have provided a definition of the 'statistical equivalence' of two such representations in terms of which they obtain not merely ordinary statistics, parastatistics and fractional or anyon statistics but more exotic forms which they call 'ambistatistics' and 'fractional ambistatistics' [26]. Interestingly, at the conclusion of their paper they note a possible statistics problem for three-dimensional geons and suggest that it can be solved through the application of fractional ambistatistics ([26] p. 107). There is a nice, if obvious, comparison here with Greenberg's application of parastatistics to the problem of quark statistics [27].

However, it has been claimed that one of the advantages of the configuration space approach is that it actually *excludes* the possibility of non-standard statistics [28]. This claim is based on the work of Leinaas and Myrheim [24] which apparently demonstrates that for a space of dimension 3 or greater only the standard statistics are possible. The conclusion drawn is that '... the (anti-) symmetrisation condition on the wave function is now seen to be related to the dimensionality of space, in contrast to the Messiah and Greenberg analysis wherein the (anti-)symmetrisation condition receives the status of a postulate' ([28], p. 3). Nevertheless, the claim is problematic. It involves two crucial steps: the first takes into account the topology of the reduced configuration space and the basic point is that in the three dimensional case, the reduced space (minus the singularity) is doubly connected, whereas in the case of two-dimensions it is infinitely connected. The next all-important step is to apply an appropriate quantisation and here Leinaas and Myrheim follow what they take to be the simplest way, namely the 'Schrödinger quantisation scheme' ([24], p. 10). Crucially, this associates a one-dimensional complex Hilbert space with each point in the configuration space which means that the familiar permutation operator P yields the well known phase factor $\exp[i\xi]$, where ξ is real and independent of the point **x**. For bosons and fermions ξ = o and π, respectively but in the cases of one-dimensional and two-dimensional configuration spaces a continuum of intermediate statistics is permitted. In the three dimensional case, however, the fact that the space is doubly connected means that the further condition must be imposed on P that $P^2 = 1$ so that P = ±1 and, of course, only Bose-Einstein and Fermi-Dirac statistics are allowed. In other words, non-standard statistics are eliminated not just because the relevant configuration space is doubly connected but because a standard quantisation procedure is assumed which allows one-dimensional sub-spaces only and so it should come as no surprise that para- and ambi-statistics cannot arise. Effectively what Leinaas and Myrheim have done is to ignore the 'kinematical ambiguity' inherent in the quantisation procedure which derives from the (mathematical) fact that the set of irreducible representations of the permutation group contains not just the trivial representation manifested above but also others corresponding to exotic statistics ([26], pp. 103-104).

What about the more philosophical issue concerning particle individuality? Let us consider, first of all, the justification for the removal of the coincidence points. An obvious basis for this would be to refer to the *impenetrability* of the particles. This in turn can be understood as due to certain repulsive forces holding between them. Such a conjecture has been made in the case of anyons [29] and more generally this has been taken to confer a further advantage on the configuration space approach, in the sense that

'... it allows particle statistics to be understood as a kind of 'force' in essence similar to other interactions with a topological character, like the interaction between an electric and magnetic charge in three spatial dimensions, or the type of interaction in two dimensions which is responsible for the Bohm-Aharonov effect and fractional statistics ([25], p. 687).

Brown et. al. reject this understanding of impenetrability as suspiciously ad hoc in the general case of indistinguishable particles ([28], p. 5). Their claim is that this ad hocness is removed in the framework of de Broglie-Bohm pilot wave theory. Here, as is well known, there is a dual ontology of point particles plus pilot wave, where the role of the latter is to determine the instantaneous velocities of the former through the so-called 'guidance equations'. Since these equations are first-order, the trajectories of two particles which are non-coincident to begin with will never coincide. In effect the impenetrability of the particles is built into the guidance equations and the singularity points remain inaccessible. The conclusion drawn is that '... within the topological approach to identical particles the removal of the set ... of coincidence points from the reduced configuration space ... thus follows naturally from de Broglie-Bohm dynamics as it is defined in the full space ...' ([28], p. 6)*.

It is part of the attraction of the Bohm theory that it retains, or appears to retain, aspects of classical ontology, with particles traversing well-defined spatio-temporal trajectories guided by the pilot wave. This obviously meshes well with the above metaphysical package in which particles are regarded as individuals and Brown *et. al.* explicitly note this point. However, they ground the relevant sense of individuality in

---

* There is the worry that this might exclude the possibility of Bose-Einstein statistics, since it has been claimed that the de Broglie-Bohm trajectories cannot in fact cross. Brown et. al. argue that this latter claim is simply not correct since symmetry considerations demonstrate that if the particles coincide at all, they coincide forever. If the bosons are initially separated, then the relative velocity vanishes at the coincident point which together with the first order nature of the guidance equations means that they can never cross ([28], pp. 6-7).

the *intrinsic properties* of the particles ([28], p. 7) and this is problematic. First of all, as they note, certain interference experiments - interpreted within this framework - seem to imply that these properties actually belong to the pilot wave rather than the particles themselves, thus undermining the classical nature of the ontology. Brown *et. al.* advocate a 'principle of generosity' which assigns these properties to both the particle *and* the pilot wave ([28], pp. 7-8) and which retains some element of classicality*. Secondly, however, setting aside the concern whether this is generous enough, Brown et. al. are looking in the wrong place for individuality: particles of the same kind will possess the same intrinsic properties and are classically *identical* in that sense. The choice is either to go for something like 'primitive thisness' or, more appropriately in the Bohmian context, insist that the particles are both distinguished and individuated by their distinct spatio-temporal trajectories, as in the case of classical particles†.

My point is not to advocate the de Broglie-Bohm view but merely to emphasise that the configuration space approach can also provide a suitable formal framework for this package of individual particles subject to accessibility constraints (we recall Bourdeau and Sorkin's remark above about different choices of boundary condition on the wave-function). However, it is also important to recognise that this approach offers nothing new, philosophically speaking, since one could also maintain the alternative view that the particles are non-individuals. Indeed, one of the motivations given by Leinaas and Myrheim is that within this approach one can dispense with the whole business of introducing particle labels and then effectively emasculating their ontological force by imposing appropriate symmetry constraints via SP ([24]., p. 2; cf. also Bourdeau and Sorkin [25], p. 687)‡. Of course, there is still the issue of the singular points, whose removal on the grounds of impenetrability considerations meshes so nicely with the alternative metaphysics. One possibility is to tackle the problem of collisions directly. Bourdeau and Sorkin, for example [25], focus on the Hamiltonian, which they require to be self-adjoint and in the two-dimensional case they show that for fermions, the self-adjoint extension of the Hamiltonian to cover the singularities is unique, so that collisions are strictly forbidden, whereas in the case of both Bose-Einstein and

---

* Further criticisms of the view that de Broglie-Bohm theory is philosophically classical can be found in [30].
† There is still the problem of bosons which start out on coincident trajectories of course.
‡ There is an interesting disagreement between the proponents of this approach which relates to my earlier discussion of the import of permutation considerations. Leinaas and Myrheim [24] agree with Mirman that the word 'permutation' has no physical meaning, in the sense that particles are not actually regarded as physically interchanged. Not surprisingly, they adopt a broadly positivistic approach in which the introduction of particle indices is seen as bringing unobservable elements into the theory and is thus to be avoided. As with Hestenes, and others, whose analysis of Gibbs Paradox they follow (as do Brown *et. al.*) they consider 'indistinguishability', in the strong sense, or what I've called non-individuality, to be characteristic of classical particles also. Bourdeau and Sorkin, on the other hand, insist that 'the physical meaning of statistics resides in processes of *exchange* of identical partners ...' ([25], p. 687) and hence, since such exchange involves paths in the configuration space away from the singular points, the statistics is defined independently of what happens when the particle coincide and the existence of the singular points cannot imply the impossibility of non-trivial statistics. Interestingly, the history of the exchange integral is a history of a shift from thinking of exchange in physical terms to conceptualising it in terms of indistinguishability [31]. On the latter understanding of 'exchange', giving indistinguishability an essentially operationalist meaning might be seen as collapsing an important distinction between the classical and quantum situations. Furthermore, as I have indicated, once impenetrability is introduced into the configuration space analysis, an element of individuality has been introduced within it.

fractional statistics there are a range of alternative extensions, some of which allow collisions but some of which do not*. By requiring that the wave-function remains finite at the coincident point they argue that a unique choice of Hamiltonian can then be made and it turns out that collisions are allowed only in the case of Bose-Einstein statistics. Thus, whereas for fermions it doesn't really matter whether the singular points are retained or not, for bosons and anyons, *on this account*, it does, since these points are either the locations of collisions in the boson case or the locations of vanishing wave-function for anyons. There is clearly more to say about these and related issues - such as the generalisation of this line to higher dimensions and parastatistics, for example ([25], p. 694) - but I want to move on to concerns that are more directly philosophical.

WHICH PACKAGE SHOULD WE BUY?

The upshot is that the relevant physics appears to support two quite different metaphysical packages: one in which the particles are non-individuals and another in which they are individuals. The physics itself cannot tell us which one to go with (see [20]; [23] and [32]). Whether you find this conclusion bothersome or not depends on your philosophical predilections. If you think that physics should tell us how the world *is*, where such a tale will include saying whether the particles are individuals or not, then this 'underdetermination' of the metaphysics by the physics presents a problem. One could, of course, abandon this realist aim for that of a selection of stories as to how the world *could be*, with our two packages as just two possible endings [23]. Or one could try to argue for the relative superiority of one package over another.

One could argue, for example, that the non-individuals package meshes better with the framework of quantum field theory and is thus to be preferred on these grounds alone [7]. There are two aspects to this 'meshing' which need to be separated. The first emphasises that on standard accounts of quantum field theory particle indices are not assigned from the word go and thus particle permutations are undefined ([33]; [34]; [8]). However, state spaces for QFT can be constructed which do involve such indices ([35]; [23]), although the existence of states that are superpositions of particle number presents a problem for this attempt ([36]; [37]).

The second aspect is more methodological in nature. If the particles are regarded as individuals and particle indices introduced, then as well as the standard symmetrised and anti-symmetrised wave functions, one can obtain, of course, not only mixed-symmetric combinations but also non-symmetric ones. Yet these non-standard forms are not realised in nature - they are just so much 'surplus structure'. And a package that does not result in such surplus structure is to be preferred over one that does ([33]; [34]; [8]). Now this methodological principle of 'avoid surplus structure' is an interesting one. 'Surplus structure' arises when a theory of physics is embedded in some mathematical framework and we get more mathematics than we need to express that particular theory [38]. Giving a physical interpretation to certain elements of this surplus mathematical structure can then lead to fruitful extensions of the theory, as in the case of the positive and negative energy solutions of the Dirac equation, for example. From this perspective, it comes as no surprise that when we embed quantum mechanics in group theory, we get 'surplus' representations that we don't apparently need to represent the particles that we know of [39]. One could then take the attitude that the surplus structure is innocuous and can be simply ignored, since it is not

---

* Thus they argue that simply cutting out the singular points results in a loss of information.

physically realised (we can find this attitude in Weyl and, more recently, Huggett [37]).

But the point might be pressed that this leaves a mystery as to *why* this structure is not realised and that one of the jobs of a theory is to explain away such mysteries. One can draw an analogy with statistical mechanics: this theory can explain why we never observe a cold cappuccino starting to boil, whereas the quantum-particles-as-individuals view cannot explain why we never observe states of non-standard symmetry [40]. Here we need to tread carefully, however. It would be a bad move, I think, to adopt a methodological principle that ruled out all non-standard wave functions, including those relevant to parastatistics, fractional statistics, ambistatistics and so on (cf. [18], pp. 39-40). If violations of the spin-statistics theorem were to be discovered experimentally, then something like Greenberg's q-mutator formalism might well be the appropriate one to adopt [15]. If we understand surplus structure as it was originally intended, then such structure in these cases might prove rather fruitful. Of course, the non-symmetric functions are unlikely to prove at all fruitful in this way, but it is not clear whether, as a methodological principle, 'avoid surplus structure' is honed enough to cut these out whilst leaving the more interesting combinations.

Furthermore, the analogy itself might be viewed as problematic. Let us reflect again on our cold cup of cappuccino. Statistical mechanics does not tell us that it will never boil but only that the probability of this occurring is extremely low. And this in turn is explicated in terms of the very low number of states which correspond to the coffee boiling compared to the vast number of states for which it remains cold. If we were to press on and demand to know why there should be this disparity in the number of states accessible to the cappuccino - or, equivalently, why we should find ourselves in a region of entropy increase - then an answer might be given in terms of the initial conditions pertaining to our region of the universe. Why do we never observe entropy decreases in our cups of coffee? Because that's the way our region of the universe is, as expressed in the initial conditions. Why do we never observe non-symmetric states, given this metaphysics of individual particles? Because that's the way the universe is. At some point the explanatory buck has to stop and the theory itself can't do all the work. Of course, you might prefer the alternative explanation in which non-symmetric states are ruled out from the word go, but then you might legitimately be urged to flesh out the metaphysics and logic of 'non-individuality'. No explanation of this sort comes without a price attached.

SYMMETRY AND THE EXPLANATION OF STATISTICS

This issue of accommodating non-standard statistics and possible violations of the spin-statistics theorem has been further explored by Hilborn and Yuca [18]. In particular they discuss the question as to how the statistics, or more fundamentally, the permutation symmetry, of the particles should be regarded. Should permutation symmetry be understood as a fundamental, 'primary' property or as a 'secondary' one, reducible to the former? The promise of reducibility is held out by the spin-statistics theorem but non-standard statistics and the possibility of violations render this problematic. If, on the other hand, permutation symmetry is understood as an intrinsic property and intrinsic properties are understood as delineating natural kinds, then electrons, say, obeying different statistics would constitute different such kinds. This also leads to problems, as Hilborn and Yuca spell out.

First of all, the only thing that would distinguish these electrons of different kinds would be their permutation symmetry; in all other respects they are identical. Suppose then, that two or more such electrons of different kinds were to come together in an atomic system - which statistics would dominate ([18], p. 41)? More fundamentally, as Hilborn and Yuca emphasise, we don't measure the permutation symmetry of separate particles; more than one particle is needed. Thus the permutation symmetry of the individual particle is unobservable and as a property in this intrinsic sense it can be regarded as 'empirically superfluous', *until a collective is formed*, whereupon it suddenly springs into action as it were.

For these sorts of reasons, Hilborn and Yuca prefer a 'holistic' perspective which can accommodate this 'emergent' quality. On this view, permutation symmetry emerges at the collective level as a property, not of the particles themselves, but of the quantum state. Here they draw an analogy with energy, noting that we allow for electrons to exist in different energy states without thinking of them as different *kinds* of particle. Pursuing this analogy amounts to regarding permutation symmetry as an *extrinsic* property of the particles, with the caveat that whereas we can observe the energy of a particular electron, we cannot observe its permutation symmetry.

As they point out, the possibilities of non-standard statistics and violations of the spin-statistics theorem in general can be accommodated quite naturally within this framework:

'This holistic point of view is both more faithful to the possibilities of physics (including possible violations of the spin-statistics connection) and a stronger philosophical stance. It also has the merits of simplicity and efficiency. On this account, permutation symmetry is a property of the collective state of the identical particles, not an intrinsic property to be associated with each particle.' ([18], p. 44).

However, the notion of 'emergent' properties is philosophically murky[*]. Furthermore, Hilborn and Yuca understand Redhead and Teller to be arguing in favour of the holistic stance, but Redhead and Teller themselves understand the metaphysics of this holism in terms of non-individual quanta and deploy the above 'surplus structure' argument to support such a view. What I want to urge here is the point that an understanding of permutation symmetry can also be accommodated within the alternative metaphysics of particles as individuals. In this case, as I have indicated, the state space breaks up into sub-spaces of different symmetry, with transitions between such sub-spaces suitably prohibited. As Hilborn and Yuca emphasise, this prohibition also holds not only for para-particles but also for Greenberg's 'quons' ([18], pp. 48-50). In this case, as I have indicated and they spell out, the particles have a certain probability of being found in an antisymmetric state and a certain probability of being found in a symmetric state, but once in such a state they cannot get out of it. Hence, the conceptual analysis I have already given for parastatistics seems equally applicable to the q-mutator case. Indeed, the latter can be taken as an even more striking exemplification of the individuals + state accessibility restrictions package than the regular and para-statistics cases through the explicit disassociation of the statistical type of the particle from its intrinsic properties: electrons, delineated as such through their charge, rest mass etc., can be metaphysically regarded as individuals for which there is a certain probability that they end up in symmetric states and a certain probability that they end up in anti-symmetric ones. Which state a particular electron ends up in

---

[*] Humphreys has attempted to develop a framework for representing emergence - albeit in a somewhat different context - which introduces the notion of 'fusion' according to which the causal effects of the fused whole cannot be reduced to or represented by the causal effects of the component parts [41]. Interestingly, he offers the example of entangled states in QM but it is not immediately apparent whether this framework can be extended to cover statistics.

depends not on its own intrinsic nature but on other conditions which obtain. Electrons obeying different statistics would still belong to different subspaces*.

The point, then, is that, as in the case of the configuration space approach, considerations such as Hilborn and Yuca's cannot unequivocally support one of the above metaphysical packages over another; or, in other words, the physics of non-standard statistics and violations of the spin-statistics theorem does not force us to adopt a particular metaphysics of particle individuality.

There is a further way of approaching this issue which focuses on the role of *symmetry*, already noted above. Huggett has argued that permutation invariance should be regarded straightforwardly as a symmetry on a par with rotational symmetry, for example, rather than as reflecting some kind of mysterious 'brute fact' about the universe or as a manifestation of the supposed non-individuality or loss of identity of quantum particles [42]. The central idea here is that both the rotation group and the permutation group are 'elementary state covariant' (ESC), where a symmetry group is said to be ESC if and only if the particle state vectors transform according to the unitary representation of the group ([42], p. 338). The (philosophical) point then is that this gives an account of the relationship between quantum statistics and permutations which, Huggett claims, is identical to that which is given for spin and rotations in non-relativistic quantum physics: if permutations are included in the full group of symmetries and it is postulated that this group is elementary state covariant, then only those many-particle states are allowed which are appropriately symmetrised or para-symmetric. The advantages of this are that it appears to avoid the above metaphysical packages altogether and, furthermore, provides a unified treatment of quantum statistics and spin in terms of a fundamental symmetry principle ([42], pp. 339-340).

Now, of course, as Huggett acknowledges, permutation symmetry is very different from rotational symmetry, in that permutations are not just indistinguishable to similarly permuted observers but to *all* observers. Nevertheless, he argues, both permutation symmetry and space-time symmetries obey what he calls 'global Hamiltonian symmetry' which implies that the relevant symmetry operator commutes with the relevant Hamiltonian†. With regard to the permutation group, of course, permutations of a sub-system are permutations of the whole system and this 'global Hamiltonian symmetry' very straightforwardly implies permutation invariance, without any additional assumptions concerning the structure of state space ([42], pp. 344-345). Hence, Huggett concludes,

'... we should view permutations in a similar light to rotations: we should not take [permutation invariance] as a fundamental symmetry principle in order to explain quantum statistics. Instead we should recognize that it is a particular consequence of global Hamiltonian symmetry given the group structure of the permutations. Further, if we accept the similarity of permutation and rotation

---

* Of course, if transitions between such statistics are prohibited, then one might be tempted to regard such particles as belonging to different natural kinds. Indeed, I would suggest that this is the crucial issue, rather than that of some intrinsic-extrinsic distinction: a property counts as a natural kind property if transitions between states corresponding to different values of this property are prohibited. Energy is not such a property, charge, spin and symmetry are.

† What we take the relevant Hamiltonian to cover is crucial here because, again as Huggett acknowledges, the principle would appear to be violated in the case where, for example, we have a noncentral potential term in the Hamiltonian of an atomic system, but, he insists, the symmetry is restored if we consider the 'full' Hamiltonian of system plus field, which does commute with the operators of the rotation group.

symmetry, it becomes natural to see quantum statistics as a natural result of the role symmetries play in nature, *via* [elementary state covariance]' ([42], p. 346).

However, as Huggett notes, permutation symmetry *is* different from space-time symmetries and permutation invariance only follows from his general symmetry principle given the particular structure of the permutation group. This generates the obvious question: why should the group structure be this way and not like that of the rotation group? Or, better perhaps, since the question could be answered by simply insisting 'that's the way the maths is', why should *this particular piece of maths* be applicable? One obvious answer is to say that it reflects the nature of the objects themselves, as non-individual quanta, and so we fall back to that particular metaphysical package. Alternatively, we might insist that it has nothing to do with the *objects*, which can still be regarded as individuals but is a reflection of certain kinds of initial conditions, represented by IP. Huggett wants to avoid both sets of metaphysics but pointing to the symmetry alone is not enough for explanatory purposes, unless the symmetry itself is elevated to the status of a piece of fundamental metaphysics[*]. Just such a metaphysics has begun to be explored in a form of 'structural realism' which regards the world as ultimately and metaphysically *structural* in nature ([44]; [39]). Permutation symmetry - and hence quantum statistics, whether standard or non-standard - would then be simply one aspect of this structure.

Non-standard quantum statistics enlarge the horizons of both physical and philosophical possibility. What I have tried to do here is outline some of the alternative metaphysical frameworks within which these new horizons can be explored.

ACKNOWLEDGEMENTS

I would like to thank Candice Yuca and Robert Hilborn, in particular, for letting me have a pre-print of their forthcoming paper, to which this work is in large part a response. I've also benefited enormously from discussions with Katherine Brading, Harvey Brown, Jeremy Butterfield, Nick Huggett, Simon Saunders and Paul Teller. Any blame for any confusions or lack of clarity, in the physics or philosophy, rests entirely on my shoulders of course.

REFERENCES

1. Gracia, J.J.E., *Individuality*, SUNY Press, Ithaca, 1998.
2. Born, M., *Zeitschrift für Physik* **38**, 803-827 (1926).
3. Weyl, H., *The Theory of Groups and Quantum Mechanics*, Methuen and Co., London, 1931 (English trans., 2nd ed., 1928).
4. Schrödinger, E., *Science and Humanism*, Cambridge University Press, Cambridge, 1952.
5. Cassirer, E., *Determinism and Indeterminism in Modern Physics*, Yale University Press, New Haven, 1956 (English trans., 1935).
6. Hesse, M., *Models and Analogies in Science*, Oxford University Press, Oxford, 1963.
7. Post, H., *The Listener* **70**, 534-537 (1963).
8. Teller, P., *An interpretative introduction to quantum field theory*, Princeton University Press, Princeton, 1995.

---

[*] For further discussion of the notion of symmetry and the role it plays in modern physics, see Castellani [43].

9. Dalla Chiara, M. L. and Toraldo di Francia, G., "Individuals, kinds and names in physics," in *Bridging the gap: philosophy, mathematics, physics*, edited by G. Corsi et al., Kluwer Academic Publishers, Dordrecht, 1993, pp. 261-283.
10. Krause, D., *Notre Dame Journal of Formal Logic* **33**, 402-411 (1992).
11. Sudarshan, E.C.G. and Mehra, J., *International Journal of Theoretical Physics* **3**, 245-253 (1970).
12. Hestenes, D., *American Journal of Physics* **38**, 840-845 (1970).
13. French, S., *Identity and Individuality in Classical and Quantum Physics*, PhD Thesis, University of London, London, 1985.
14. Gentile, G., *Nuovo Cimento* **17**, 493-497 (1940).
15. Greenberg, O. W., *Physica* **180**, 419-427 (1992).
16. Green, H. S., *Physical Review* **90**, 270-273 (1953).
17. Greenberg, O. W. and Messiah, A. M. L., *Physical Review* **136B**, 248-267 (1964).
18. Hilborn, R. C. and Yuca, C. L., (forthcoming).
19. French, S., *International Journal of Theoretical Physics* **26** 1141-1163 (1987).
20. French, S., *Australasian Journal of Philosophy* **67**, 432-446 (1989).
21. Dirac, P. A. M., *The Principles of Quantum Mechanics*, Cambridge University Press, Cambridge, 1978 (1930).
22. French, S. and Redhead, M., *British Journal for the Philosophy of Science* **39**, 233-246 (1988).
23. van Fraassen, B., *Quantum mechanics: an empiricist view*, Oxford University Press, Oxford, 1991.
24. Leinaas, J. M. and Myrheim, J., *Nuovo Cimento* **37B**, 1-23 (1977).
25. Bourdeau, M. and Sorkin, R. D., *Physical Review* **D45**, 687-696 (1992).
26. Imbo, T. D., Shah Imbo, C. and Sudarshan, E.C.G., *Physics Letters* **B234**, 103-107 (1990).
27. Greenberg, O. W., *Physical Review Letters* **15**, 598-602 (1964).
28. Brown, H., Sjöqvist, E. and Bacciagaluppi, G., *Physics Letters* **A 251**, 229-235 (1999).
29. Aitchison, I. J. R. and Mavromatos, N. E., *Contemporary Physics* **32**, 219-233 (1991).
30. Bedard, K., *Philosophy of Science* **66**, 221-242 (1999).
31. Carson, C., *Studies in History and Philosophy of Modern Physics* **27**, 23-45 (1996).
32. Huggett, N., *The Monist* **80**, 118-130 (1997).
33. Redhead, M. and Teller, P., *Foundations of Physics* **21**, 43-62 (1991).
34. Redhead, M. and Teller, P., *British Journal for the Philosophy of Science* **43**, 201-218 (1992).
35. de Muynck, W., *International Journal of Theoretical Physics* **14**, 327-346 (1975).
36. Butterfield, J., *Studies in History and Philosophy of Science* **24**, 443-476 (1993).
37. Huggett, N., "What are Quanta, and Why Does it Matter?" in *Proceedings of the Philosophy of Science Association: PSA 1994*, edited by D. Hull et. al., Philosophy of Science Association, Vol. 2, East Lansing, 1995, pp. 69-76.
38. Redhead, M. L. G., *Synthese* **32**, 77-112 (1975).
39. French, S., "Models and Mathematics: The Application of Group Theory to Physics," in *From Physics to Philosophy*, edited by J. Butterfield and C. Pagonis, Cambridge University Press, Cambridge, 1999, pp. 187-207.
40. Teller, P., "Quantum Mechanics and Haecceities," in *Interpreting Bodies: Classical and Quantum Objects in Modern Physics*, edited by E. Castellani, Princeton University Press, Princeton, 1998, pp. 114-141.
41. Humphreys, P., *Philosophy of Science* **64**, 1-17 (1997).
42. Huggett, N., *British Journal for the Philosophy of Science* **50**, 325-347 (1999).
43. Castellani, E., *Simmetria e Natura: Dalle armonie delle figure alle invarianze delle leggi*, Roma-Bari, Laterza, 1999.
44. Ladyman, J., *Studies in History and Philosophy of Science* **29**, 409-424 (1998).

List of Participants

Paolo ANIELLO, Università di Napoli

Michael V. BERRY, University of Bristol

Christian BORDÉ, Université Paris-Nord

Damon English BROWN, University of California, Berkeley

Dmitry BUDKER, University of California, Berkeley

Enrico CELEGHINI, Università di Firenze

Masud CHAICHIAN, University of Helsinki and Helsinki Institute of Physics

David DEMILLE, Yale University

Fay DOWKER, Queen Mary and Westfield College, University of London

Steven R. D. FRENCH, University of Leeds

Gianluca GAGLIARDI, Seconda Università di Napoli

Hendrik GEYER, University of Stellenbosch

GianCarlo GHIRARDI, Università di Trieste

Livio GIANFRANI, Seconda Università di Napoli

John GILLASPY, NIST, Gaithersburg

O. W. GREENBERG, University of Maryland

Jonathan HARRISON, University of Bristol

Robert C. HILBORN, Amherst College

Francesco IACHELLO, Yale University

Dan JAVORSEK, Purdue University

Ilya KAPLAN, Universidad Nacional Autonomo de Mexico

Jacob KATRIEL, Technion - Israel Institute of Technology

Stefan KIRCHNER, Universität Karlsruhe

You-Quan LI, Augsburg University / Zhejiang University

Vladimir MAN'KO, P.N. Lebedev Physical Institute of Russian Academy of Sciences

Italo MANNELLI, Scuola Normale Superiore, Pisa

Giuseppe MARMO, Università di Napoli

Ashok MISHRA, Institute of Mathematical Sciences, CIT Campus

Giovanni MODUGNO, European Laboratory for Nonlinear Spectroscopy

Alexander OTTE, University of Stuttgart

Mikhail PLYUSHCHAY, Universidad de Santiago de Chile / IHEP, Protvino

Bruno PREZIOSI, Università di Napoli

Mario RASETTI, Politecnico di Torino

Jonathan ROBBINS, BRIMS, Hewlett-Packard Laboratories and University of Bristol

Salvatore SOLIMENO, Università di Napoli

Allan I. SOLOMON, The Open University

E. C. George SUDARSHAN, University of Texas, Austin

Guglielmo M. TINO, Università di Napoli

Michael YORK, Greenbrae, CA

Candice YUCA, Cambridge University

Francesco ZACCARIA, Università di Napoli

Anthony ZEE, Institute for Theoretical Physics, University of California, Santa Barbara

AUTHOR INDEX

B

Berry, M., 3
Bordé, C. J., 274
Bourgeois, M., 288
Brown, D., 281
Budker, D., 227, 281

C

Celeghini, E., 29, 59
Chaichian, M., 219
Chardonnet, C., 274

D

Demichev, A., 219
DeMille, D. P., 227, 281
De Natale, P., 295
Derr, N., 227
Deveney, E., 227
Dobaczewski, J., 190
Dowker, H. F., 205

E

Elmore, D., 288

F

Fischbach, E., 288
French, S., 305

G

Geyer, H. B., 190
Ghirardi, G. C., 16
Gillaspy, J. D., 241
Giusfredi, G., 295
Greenberg, O. W., 113

H

Harrison, J. M., 67
Hilborn, R. C., 128
Hillegonds, D., 288

I

Iachello, F., 179
Inguscio, M., 295
Inomata, A., 155

J

Javorsek, D., 288

K

Kaplan, I. G., 72
Katriel, J., 79
Kirchner, S., 155

L

Li, Y.-Q., 85

M

Mahler, G., 98
Man'ko, V. I., 92
Mannelli, I., 253
Marder, J., 288
Marmo, G., 92
Mazzotti, D., 295
Miller, T., 288
Mishra, A. K., 162, 169
Modugno, G., 295
Modugno, M., 295

N

Navrátil, P., 190

O

Otte, A., 98

P

Pastor, P. C., 295
Picqué, N., 295
Plyushchay, M. S., 197
Prešnajder, P., 219

R

Rajasekaran, G., 162, 169
Rasetti, M., 29, 59
Robbins, J. M., 3, 67
Rohrs, H., 288

S

Solomon, A. I., 142
Sorkin, R. D., 205
Stohler, M., 288
Sudarshan, E. C. G., 40, 92

T

Tino, G. M., 260

V

Vogt, S., 288

Y

York, M., 104

Z

Zaccaria, F., 92
Zee, A., 55